CAMBRIDGE MONOGRAPHS ON PHYSICS

GENERAL EDITORS

M. M. Woolfson, D.Sc.

Professor of Theoretical Physics, University of York

J. M. Ziman, D.Phil., F.R.S.

Henry Overton Wills Professor of Physics, University of Bristol

SYMMETRY PRINCIPLES IN ELEMENTARY PARTICLE PHYSICS

SYMMETRY PRINCIPLES IN ELEMENTARY PARTICLE PHYSICS

W. M. GIBSON
Reader in Physics, University of Bristol

B. R. POLLARD
Lecturer in Theoretical Physics, University of Bristol

CAMBRIDGE UNIVERSITY PRESS

CAMBRIDGE
LONDON NEW YORK NEW ROCHELLE
MELBOURNE SYDNEY

CAMBRIDGE UNIVERSITY PRESS
Cambridge, New York, Melbourne, Madrid, Cape Town, Singapore,
São Paulo, Delhi, Dubai, Tokyo, Mexico City

Cambridge University Press
The Edinburgh Building, Cambridge CB2 8RU, UK

Published in the United States of America by Cambridge University Press, New York

www.cambridge.org
Information on this title: www.cambridge.org/9780521299640

First published 1976
First paperback edition 1980
Re-issued 2010

A catalogue record for this publication is available from the British Library

ISBN 978-0-521-20787-4 Hardback
ISBN 978-0-521-29964-0 Paperback

CONTENTS

CHAPTER 4

Lorentz invariance

CHAPTER 5

Parity

PREFACE

The study of elementary particles and their interactions has brought to light symmetries and relationships which are nowadays objects of study in themselves. In this book we have attempted to present these symmetry principles and their associated conservation laws as a set of interrelated physical principles, explained in terms of the simplest appropriate mathematics. Mathematical excursions into the more abstract aspects of the subject have been avoided; thus in particular, we have not made explicit use of the formal apparatus of group theory. Similarly we have omitted all descriptions of the experimental methods by which quoted results have been obtained.

The level is thus intended to meet the needs of a graduate student working in particle physics, who wants an accurate but not too abstract explanation of the principles commonly quoted in the literature of his subject. It is to be hoped that many such readers would afterwards progress further with the aid of more advanced literature.

We have drawn heavily on material used by both of us for postgraduate lectures in Bristol, and we acknowledge the contribution which these postgraduate classes have made to our own powers of understanding and explanation.

Our thanks are due also to the many colleagues and friends who have over the years shed light on difficult topics through discussion, and to Miss Alma Dawes, Miss Margaret James, Miss Anna Love and Mrs Nancy Thorp who have typed our outpourings. We should also like to thank Dr J.W. Alcock for assistance with proof-reading and the editorial staff of the Cambridge University Press for their assistance at all stages.

W.M. Gibson
B.R. Pollard

Bristol, August 1974

Note on 1980 reprint. Opportunity has been taken to correct a number of errors and to add a postscript on Recent Developments.

CHAPTER 1

INTRODUCTION TO ELEMENTARY PARTICLES

1.1 Perspective

The general field to which this book contributes is known to some as the Physics of Elementary Particles, to others as High Energy Physics, and has recently been given the additional name of Sub-nuclear Physics. These titles are almost synonymous, emphasising respectively the search for basic components of which matter is made up, the need for high energies to probe the inner structure of matter, and the fact that the search leads us deeper than the atomic nucleus.

Many textbooks in this field have presented the elementary particles and their properties, leading from the regularity of these properties to the gradually uncovered theory of the fundamental interactions. We, however, take as our subject not the particles themselves but the symmetry principles and conservation laws by which their properties are governed. The understanding of these laws and principles has of course grown from the experimental study of the actual particles, and this fact must continually bring us back to the basis of observed fact, as we work through the essentially mathematical framework of the symmetry principles and conservation laws.

1.2 The particles

A purely empirical classification of the elementary particles may be made according to mass, with baryons having mass of the order of that of the proton, leptons having small or zero rest mass, and mesons intermediate mass.

It soon becomes clear, however, that properties other than mass can lead us to classification schemes of a more fundamental nature. First we have the question of spin and statistics: the important distinction here is between particles of half-integral spin which obey Fermi–Dirac statistics and particles of zero or integral spin which obey Bose–Einstein statistics. The former, known as fermions, can be created only as pairs with corresponding antiparticles, so that the total number of a given type of fermion is conserved (see §1.4). It is an observed fact, so far

Table 1.1. *Baryons and antibaryons*

		I_3	I	J	P	S	Y	B
N	p	$+\frac{1}{2}$	$\frac{1}{2}$	$\frac{1}{2}$	+	0	+1	+1
	n	$-\frac{1}{2}$						
	Λ^0	0	0	$\frac{1}{2}$	+	−1	0	+1
Σ	Σ^+	+1	1	$\frac{1}{2}$	+	−1	0	+1
	Σ^0	0						
	Σ^-	−1						
Ξ	Ξ^0	$+\frac{1}{2}$	$\frac{1}{2}$	$\frac{1}{2}$	+	−2	−1	+1
	Ξ^-	$-\frac{1}{2}$						
$\overline{\text{N}}$	$\bar{\text{p}}$	$-\frac{1}{2}$	$\frac{1}{2}$	$\frac{1}{2}$	−	0	−1	−1
	$\bar{\text{n}}$	$+\frac{1}{2}$						
	$\overline{\Lambda}^0$	0	0	$\frac{1}{2}$	−	+1	0	−1
$\overline{\Sigma}$	$\overline{\Sigma}^+$	+1	1	$\frac{1}{2}$	−	+1	0	−1
	$\overline{\Sigma}^0$	0						
	$\overline{\Sigma}^-$	−1						
$\overline{\Xi}$	$\overline{\Xi}^+$	$+\frac{1}{2}$	$\frac{1}{2}$	$\frac{1}{2}$	−	+2	+1	−1
	$\overline{\Xi}^0$	$-\frac{1}{2}$						

Table 1.2. *Mesons* ($B = 0$)

	I_3	I	G	J	P	C_n	$S = Y$
π^+	+1	1	−	0	−	+	0
π^0	0						
π^-	−1						
K^+	$+\frac{1}{2}$	$\frac{1}{2}$		0	−		1
K^0	$-\frac{1}{2}$						
$\overline{K}^0$	$+\frac{1}{2}$	$\frac{1}{2}$		0	−		−1
K^-	$-\frac{1}{2}$						
η	0	0	+	0	−	+	0
ρ^+	+1	1	+	1	−	−	0
ρ^0	0						
ρ^-	−1						
ω	0	0	−	1	−	−	0
ϕ	0	0	−	1	−	−	0

Table 1.3. *Leptons and antileptons* ($J = \frac{1}{2}$)

	L	Helicity
e^-	+1	$\pm\frac{1}{2}$
ν_e	+1	$-\frac{1}{2}$
e^+	−1	$\pm\frac{1}{2}$
$\bar{\nu}_e$	−1	$+\frac{1}{2}$
μ^-	+1	$\pm\frac{1}{2}$
ν_μ	+1	$-\frac{1}{2}$
μ^+	−1	$\pm\frac{1}{2}$
$\bar{\nu}_\mu$	−1	$+\frac{1}{2}$

unexplained, that all fermions have a non-zero baryon or lepton number, and are distinguishable from their antiparticles by the opposite values of these numbers. Bosons, on the other hand, can be created in numbers which are limited only indirectly by other conservation laws. For example high energy neutron–proton scattering can be accompanied by the creation of one, two, three or more pions. The baryons and the leptons are fermions, while all the strongly interacting mesons are bosons. The muon, originally classed as a meson on account of its mass, is classed as a lepton by virtue of its spin $\frac{1}{2}$ and (see later) its weakly interacting nature.

As has been hinted above, a further useful classification of particles may be made according to the nature of their interactions. This leads us to group the baryons and mesons together as hadrons, strongly interacting particles (see §1.3), while the leptons remain apart as weakly interacting as well as light in mass.

The actual known particles are listed, according to the above principles, in tables 1.1 to 1.3. Quoted in these tables are the values of the quantum numbers which are discussed in §1.4.

By reason of its relation to the electron and the muon via the weak interaction, the neutrino (having zero rest-mass) is classed as a lepton, while the other object of zero rest-mass, the photon, has to be treated in a class of its own, as the quantum of the electromagnetic field.

1.3 Types of interaction

Different types of interaction between particles may be distinguished by their values of the coupling constant, a dimensionless number related to the strength of the interaction, and also to the typical value of cross-section for processes proceeding via this interaction.

There is an element of convention in the specification of the coupling constant for different types of interaction, but the general aim is to express the interaction energy as a fraction of the mass which is equivalent to the range of the interaction.

Strong interaction. For the strong interaction, the mutual energy of two particles separated by a distance r may be expressed as

$$E = \frac{g^2}{r} e^{-r/a}$$

where g is analogous to electric charge, and a is the range of the interaction, expressible as the Compton wavelength of a particle (actually the pion) of mass m given by

$$a = \frac{\hbar}{mc}$$

Thus the interaction energy when $r = a$ may be put as

$$E = \frac{g^2 mc}{\hbar}$$

whence

$$\frac{E}{mc^2} = \frac{g^2}{\hbar c}$$

This is the quantity generally used as the coupling constant for the strong interaction; it has value

$$\frac{g^2}{\hbar c} \sim 15$$

Electromagnetic interaction. The electromagnetic interaction has a strength characterised by the quantity $e^2/\hbar c$, which is known as the fine structure constant and has value 1/137. One may describe this quantity by an argument similar to that used above for the strong interaction; but since there is no unique range for a force obeying an inverse square law, one must say that $e^2/\hbar c$ is the interaction energy of two electronic charges separated by a general distance r, expressed as a fraction of the rest-energy of an object which would have Compton wavelength r.

Weak interaction. For the weak interaction we have to use the fact that decay rates lead us to a dimensional measure of interaction strength

$$G = 1.4 \times 10^{-49} \text{ erg cm}^3$$

The range is unknown, so to get a dimensionless number it is necessary to introduce a standard of length, such as the Compton wavelength of the pion or of the proton ($\hbar/m_p c$). This gives a coupling constant

$$\frac{G}{\hbar c}\left(\frac{m_p c}{\hbar}\right)^2$$

of order 10^{-5} (or 2×10^{-7} if one uses $\hbar/m_\pi c$). In fact the true range may be much smaller than $\hbar/m_p c$, in which case the interaction energy would be greater than the number 10^{-5} suggests.

Gravitational interaction. It is of interest to compare the three interactions which are important in elementary particle physics with a fourth, the gravitational interaction which is far too weak to have any significance in this field. If we use G' as the gravitational constant, and consider two electrons, we get a gravitational coupling constant

$$\frac{G'm^2}{\hbar c} \sim 10^{-45}$$

a number which amply demonstrates the difference in scale between gravitational effects on the one hand and electromagnetic or nuclear effects on the other.

1.4. Conservation laws

Many of the regularities observed in physics may be expressed as conservation laws, each of which states that the magnitude of some quantity is constant. The most familiar such laws are the laws of conservation of energy and momentum, which are universally valid, in quantum mechanics as in classical mechanics. Equally rigid are the laws of conservation of angular momentum and of electric charge. Conservation laws of this type are to be distinguished from those which apply in idealised systems to which real situations may or may not approximate. Laws of this latter type arise in the quantum-mechanical description of the interactions between elementary particles, and the principal ones will form a basis for our consideration of the symmetry principles.

To set up a few signposts to the topics under review, we may draw attention to the quantum numbers quoted for the individual particles in tables 1.1 to 1.3. The baryons, which can undergo transitions into each other, are given a baryon number $B = +1$, while their antiparticles are given a value $B = -1$. The fact that baryons, being fermions, can be created and annihilated only in particle–antiparticle pairs is then expressed by saying that the total baryon number B is conserved. This law appears to be in the universally valid class.

The other class of fermions, the leptons, appears to obey a similar but independent law of conservation of total number. For this purpose we assign lepton numbers $L = +1$ for e^+, μ^+, ν and $L = -1$ for $e^-, \mu^-, \bar{\nu}$. The conservation of lepton number expressed in this way appears to be as valid as the corresponding conservation of baryon number. It appears, further, that we may divide the leptons and antileptons into electronic ($e^+, \nu, e^-, \bar{\nu}$) and muonic ($\mu^+, \nu_\mu, \mu^-, \bar{\nu}_\mu$), the numbers of which are conserved separately. We could thus allocate electronic and muonic lepton numbers separately, and say that their totals were conserved separately in all known processes.

The intrinsic parity P of the particles may be linked with the parity $P = (-)^l$ associated with orbital angular momentum l in the relative motion of the particles, to calculate the total parity of a system. The law of conservation of parity, stating that total parity is conserved, is valid for processes which occur through the strong nuclear interaction, or through the electromagnetic interaction, but is violated in the weak interaction.

Also giving rise to conditionally obeyed conservation laws, to be discussed in later chapters of this book, are the quantum numbers listed as I (isospin), C (charge conjugation symmetry) and G (G-parity), together with S (strangeness) or Y (hypercharge).

CHAPTER 2

QUANTUM MECHANICS AND INVARIANCE PRINCIPLES

In this chapter we begin by reviewing the principles of quantum mechanics, with the basis of which the reader is assumed to be familiar.

We start with the non-relativistic Schrödinger quantum mechanics and indicate later what aspects are more generally valid in the relativistic situation. The concept of the $\mathcal{S}$-matrix is introduced.

The connection between symmetry or invariance principles and conservation laws in quantum mechanics is then described.

To each transformation of the space–time coordinates used to describe a system, there corresponds a unitary transformation of the wavefunctions of the system. If the coordinate transformation depends continuously on a parameter as in the important cases of translations and rotations in space, the associated unitary operator can be expressed in terms of a Hermitian observable quantity called a generator. The important physical observables such as linear and angular momentum can be interpreted in this way as generators. Finally, if the coordinate transformation leaves the system invariant then the corresponding generator commutes with the Hamiltonian of the system; that is, it is a constant of motion.

2.1 Principles of quantum mechanics

2.1.1 *States and observables*

Let us start by considering a system consisting of a fixed number of particles. In quantum mechanics, the state of such a system is completely specified at each instant of time by a normalized wavefunction $\Psi(x, y, z)$ which is a function of the coordinates of all the particles. For simplicity only the coordinates of one particle are indicated. The normalization condition is

$$\iiint \mathrm{d}\tau |\Psi(x, y, z)|^2 = 1 \tag{2.1}$$

where $\mathrm{d}\tau$ denotes the product $\mathrm{d}x\mathrm{d}y\mathrm{d}z$ of the coordinate differentials for all the particles of the system.

Experiments such as those on electron interference lead to the postulate that wavefunctions satisfy the *principle of superposition*, namely that if $\Psi(x,y,z)$ and $\Phi(x,y,z)$ describe possible states of the system, then the linear combination

$$a\Psi(x,y,z) + b\Phi(x,y,z)$$

where a and b are complex numbers, also describes a possible state of the system. This is expressed mathematically by saying that the states of the system form a complex linear vector space of the type called Hilbert space. For this reason a wavefunction is also referred to as a state vector.

It is possible to choose a *complete orthonormal set of state vectors* denoted by $\psi_1, \psi_2, \ldots$ in the Hilbert space of states, with the following properties

(*a*) Each ψ_n is *normalised*

$$\int d\tau |\psi_n(x,y,z)|^2 = 1 \quad n = 1, 2, \ldots$$

(*b*) The state vectors are orthogonal to one another, that is

$$\int d\tau \psi_n^*(x,y,z)\psi_m(x,y,z) = 0 \quad n \neq m$$

(*c*) The set is *complete*, that is every possible state vector of the system can be expressed as a linear superposition of the basic states ψ_n with suitably chosen coefficients c_n. Thus

$$\Psi(x,y,z) = \sum_{n=1}^{\infty} c_n \psi_n(x,y,z)$$

where c_n is given by

$$c_n = \int d\tau\ \psi_n^*(x,y,z)\Psi(x,y,z)$$

and * denotes the complex conjugate.

Such a complete orthonormal set of functions will also be called a *basis*.

The expression for c_n is called the *overlap integral* of, or inner product of ψ_n and Ψ and the right-hand side is conveniently denoted by (ψ_n, Ψ).

In quantum mechanics, an observable quantity is represented by a Hermitian operator. One speaks interchangeably of the observable and the operator representing it. A Hermitian operator is one for which the equality

$$\int d\tau \Psi^* A \Phi = \int d\tau (A\Psi)^* \Phi \tag{2.2}$$

i.e.

$$(\Psi, A\Phi) = (A\Psi, \Phi)$$

holds true for any pair of state vectors Φ and Ψ.

More generally to any operator A corresponds its Hermitian conjugate denoted by $A^\dagger$, which is defined so that the equality

$$(\psi, A^\dagger \phi) = (A\psi, \phi)$$

is true for any two wavefunctions ϕ and ψ. Thus a Hermitian operator is an operator which is equal to its Hermitian conjugate,

$$A^\dagger = A$$

An *eigenvalue* of A is a value of a for which the equation

$$A\phi = a\phi$$

has a solution and the corresponding solution ϕ is the *eigenfunction*. There are in general many solutions which are written $\phi_n, n = 1, 2, \ldots$, and

$$A\phi_n = a_n\phi_n, \quad n = 1, 2, \ldots \tag{2.3}$$

The Hermitian property (2.2) is sufficient to show that all the a_n are real. It further follows from (2.2) that eigenfunctions belonging to distinct eigenvalues are orthogonal.

It can happen that more than one eigenfunction belongs to the same eigenvalue, i.e. some of the a_n may be equal. If r distinct eigenfunctions belong to the same eigenvalue, the eigenvalue is said to be *r-fold degenerate*. A linear superposition of two eigenfunctions corresponding to the same eigenvalue, itself corresponds to that eigenvalue. Hence we may arrange, for example by means of the Schmidt orthogonalisation process, that the distinct eigenfunctions belonging to the same eigenvalue are mutually orthogonal.

It is further assumed that the eigenfunctions of an observable are complete, so that they form a basis, as defined above.

According to the interpretative postulates the only possible outcome of the measurement of an observable A is one of its eigenvalues a_n; furthermore, if the system is in the state Ψ, then the probability of obtainin the result a for the measurement of A is given by

$$P(a) = \sum_n |c_n|^2 \tag{2.4}$$

where c_n is the coefficient of ϕ_n in the expansion of the given normalised Ψ in terms of the eigenfunctions of A

$$\Psi(x, y, z) = \sum_n c_n\phi_n(x, y, z) \tag{2.5}$$

and where the sum in (2.4) goes over all n for which $a_n = a$.

In the non-degenerate case

$$P(a_n) = |c_n|^2$$

and we can say that the probability amplitude (whose squared modulus is $P(a_n)$) for finding the system in the state $\phi_n(x, y, z)$ given that it was in $\Psi(x, y, z)$, is given by the overlap integral

$$(\phi_n, \Psi) = \int \mathrm{d}\tau \phi_n^*(x, y, z)\Psi(x, y, z)$$

For a large class of measurements the following situation holds. Suppose A is measured and the result a_n is obtained. If only one eigenfunction ϕ_n corresponds to the eigenvalue a_n then the system immediately after the measurement will be in the state described by ϕ_n.

More generally, if there are r eigenfunctions $\phi_{n_1}, \ldots, \phi_{n_r}$ corresponding to the eigenvalue a_n then the system will be in a state of the form

$$\sum_{i=1}^{r} c_i \phi_{n_i}$$

and nothing more precise can be asserted without further measurements.

2.1.2 *Simultaneous measurement of two observables*

Operators do not in general commute. For example, the coordinate operator x and the momentum operator

$$P_x = -i\hbar \frac{\partial}{\partial x} \tag{2.6}$$

satisfy

$$[x, P_x] = xP_x - P_x x = i\hbar$$

Here as elsewhere, operator equations are to be interpreted as statements that the two sides give the same result when applied to an arbitrary state vector of the system.

It is often useful to label state vectors by the eigenvalues of the observables of which they are eigenstates. Thus a plane wave

$$\Psi_p(x) = e^{ipx/\hbar} \tag{2.7}$$

can be characterised as an eigenfunction of the momentum operator (2.6) belonging to the eigenvalue p. In this case a single observable characterises the state. In general further labels will be required in the form of eigenvalues of additional operators. We therefore consider the problem of when two observables A and B can simultaneously possess definite values.

We shall say that A and B can be simultaneously measured if we measure A and get the result a, then measure B and get b, then re-measure A and get a again, and so on in principle. As an example, we might measure the momentum of a particle by time of flight and then measure the spin component by a Stern–Gerlach apparatus. Thus the measurement of B must not throw the system out of the state into which it was forced by the measurement of A. This must mean that the state is an eigenvector of both A and B, and if this result is to hold for any state it must be possible to find a complete orthogonal basis $\phi_1, \phi_2, \ldots$ every member of which is a simultaneous eigenstate of A and B

$$A\phi_n = a_n\phi_n, \quad B\phi_n = b_n\phi_n; \quad n = 1, 2, \ldots$$

It may be shown that the condition for this to be possible is that A and B commute

$$[A, B] = 0 \tag{2.8}$$

The proof of this may be found in the standard text-books on quantum mechanics.

We end this section with some remarks on notation. When labelling a state by the eigenvalues of observables of which it is an eigenstate, as with the plane wave $\psi_p(x)$, it is sometimes convenient to suppress the coordinate dependence. In the Dirac notation this is taken a step further and the state with momentum eigenvalue p is written $|p\rangle$ called a *ket* vector. An arbitrary state $\psi(r)$ is written $|\psi\rangle$ or even $|\rangle$. A basis previously denoted by $\phi_1(r), \phi_2(r), \ldots$ would now be written $|1\rangle, |2\rangle, \ldots$

The complex conjugate of a ket $|n\rangle$ is denoted by $\langle n|$ and called a *bra* vector ($\phi_n^*(r)$ in the earlier notation). The overlap integral of $\psi(r)$ with $\phi_n(r)$ previously denoted by (ϕ_n, ψ) and is now written

$$\langle n|\psi\rangle = \int d\tau \phi_n^*(r)\psi(r)$$

and the complex conjugate is

$$\langle n|\psi\rangle^* = \langle\psi|n\rangle$$

The matrix element of an operator A which is the overlap of $A\psi(r)$ with $\phi(r)$ is now written

$$\langle\phi|A|\psi\rangle$$

In general, we shall reserve the Dirac notation for the internal states of particles, in particular for isospin and $SU(3)$.

2.1.3 *Time development of a system and the $\mathcal{S}$-matrix*

The concepts in the two preceding sections refer to any single instant of time. In practice we observe directly or indirectly how the system develops in time.

The time development of the wavefunction is described by the Schrödinger equation

$$i\hbar \frac{\partial \psi}{\partial t} = H\psi \tag{2.9}$$

where H is the Hamiltonian operator for the system. In non-relativistic quantum mechanics, H is the total energy of the system expressed as a function of coordinates and momenta with the substitution

$$p = -i\hbar \boldsymbol{\nabla}$$

for each momentum variable.

The applications of the Schrödinger equation fall into two broad classes: (*a*) bound states, and (*b*) scattering processes. In the study of elementary particles almost all our information comes from scattering processes. The only relativistic bound state problem dealt with successfully is the positronium bound state of electron and positron. Let us therefore consider the quantum-mechanical description of scattering processes.

In a typical scattering or reaction between elementary particles

$$a + b \rightarrow c + d + e + \ldots$$

we can identify three stages. Initially a and b are prepared with definite momenta and possibly with definite spin orientations, they then interact in a region of dimensions of the order of 10^{-13} cm. Finally the reaction products c, d, e, ... emerge and by means of counters etc. are detected when widely separated and freely moving. What we obtain directly from such an experiment is the probability distribution $W(p_c, p_d, p_e, \ldots)$ of the momenta (and spins, if observed) of the final particles relative to the initial particles.

Since the Schrödinger equation (2.9) gives complete information about the development of the system wavefunction in time, it allows in principle a calculation of $W(p_c, p_d, p_e \ldots)$. This programme can be carried through in non-relativistic quantum mechanics for elastic scattering $a + b \rightarrow a + b$. (See for example, chapter 8 of Mandl (1957).) In the relativistic case where, in addition, particles can be annihilated and created, a Schrödinger equation of the form (2.9) still exists as we shall show in §2.2.3, but ψ is no longer a simple function of particle positions, and solutions are possible only in perturbation theory. It has been found useful, particularly in strong interaction studies, to consider as the basic variables of the theory, the scattering or reaction amplitudes such as that of which the modulus squared is the probability W referred to above.

The reaction amplitude is also called the $\mathcal{S}$-matrix element. We now give a direct definition of this quantity. We have already observed that under the conditions of a scattering experiment the initial and final states consist of essentially free particles. Such states can be completely specified by giving for each of the particles present the momentum, spin projection, and other internal quantum numbers such as isospin. We shall lump these together in a single index a which labels the state. For simplicity we shall assume a runs over a discrete set of values.

Let ϕ_a denote the initial state of the system. Long after the interaction has taken place the situation will be described by a definite state vector ψ. ψ will be complicated in general, for it will have components describing various inelastic final states as well as components describing elastic scattering. However the principle of superposition and the fact that the equation (2.9) is linear, require that ψ can be written in the form

$$\psi = \mathcal{S}\phi_a \tag{2.10}$$

where $\mathcal{S}$ is an operator, called the scattering operator. For if the system is prepared in state $\phi_{a'}$ and the corresponding final state is ψ', then the initial state $\alpha\phi_a + \beta\phi_{a'}$ will develop into the final state $\alpha\psi + \beta\psi'$. This requires that ψ and ϕ_a be related by an equation of the form (2.10). Attaching the time dependences to the states ϕ_a and ψ is a delicate matter which we shall not enter into here. For this the reader is referred to textbooks on scattering theory. However $\mathcal{S}$ is in some sense a function of the Hamiltonian H.

Let ϕ_b denote the state which our counters are set up to detect: freely moving particles of definite kinds with well defined momenta. Then the probability amplitude to find the state ϕ_b in the actual state ψ is, by the general principles of quantum mechanics, (ϕ_b, ψ) or by (2.10)

$$(\phi_b, \mathcal{S}\phi_a) = \mathcal{S}_{ba}$$

$\mathcal{S}_{ba}$ is called the $\mathcal{S}$-matrix element.

We assume that as a ranges over all momenta and all internal quantum numbers, the states ϕ_a are orthonormal and form a complete set.

These assumptions are expressed by

$$(\phi_b, \phi_a) = \delta_{ba} \tag{2.11}$$

$$\sum_a \phi_a \phi_a^* = 1 \tag{2.12}$$

where 1 denotes the unit operator.

Any normalised initial state can be expressed in the form

$$\phi = \sum_a c_a \phi_a$$

with suitable c_a, satisfying

$$\sum_a |c_a|^2 = 1 \tag{2.13}$$

According to the discussion above, a system initially in such a state will be found in the final state ϕ_b with probability amplitude, equal to

$$(\phi_b, \mathcal{S}\phi) = \sum_a c_a(\phi_b, \mathcal{S}\phi_a)$$
$$= \sum_a \mathcal{S}_{ba} c_a$$

and hence with probability

$$|\sum_a \mathcal{S}_{ba} c_a|^2 = \sum_a \sum_{a'} \mathcal{S}_{ba} \mathcal{S}^*_{ba'} c_a c^*_{a'}$$

Since the initial state is normalised and the total probability of the system being found in some final state is 1, on summing over b we must have

$$1 = \sum_b \sum_a \sum_{a'} \mathcal{S}_{ba} \mathcal{S}^*_{ba'} c_a c^*_{a'} \tag{2.14}$$

If the conservation of probability is to hold for every initial state ϕ, (2.14) must hold for all possible values of the coefficients c_a subject only to (2.13). It follows that

$$\sum_b \mathcal{S}_{ba} \mathcal{S}^*_{ba'} = \delta_{aa'} \tag{2.15}$$

must hold.

Written in full (2.15) is

$$\sum_b (\phi_{a'}, \mathcal{S}^\dagger \phi_b)(\phi_b, \mathcal{S}\phi_a) = \delta_{aa'}$$

or

$$(\phi_{a'}, \mathcal{S}^\dagger \mathcal{S}\phi_a) = \delta_{aa'} \tag{2.16}$$

where we have used the completeness relation (2.12).

Equation (2.16) is equivalent to the operator relation

$$\mathcal{S}^\dagger \mathcal{S} = 1 \tag{2.17}$$

The related equation

$$\mathcal{S}\mathcal{S}^\dagger = 1 \tag{2.18}$$

may be shown to follow from the less transparent condition that every normalised final state must arise from some normalised initial state. An operator satisfying (2.17) and (2.18) is said to be unitary. The unitarity of $\mathcal{S}$ expresses the conservation of probability.

If we consider a system of non-interacting particles, then the final state is essentially the same as the initial state, i.e. the same kinds of particle are present with the same momenta. It follows that in such a case we have

$$\mathcal{S} = 1$$

It is convenient therefore to introduce the *transition operator* $\mathcal{T}$ defined by

$$\mathcal{S} = 1 + \mathrm{i}\,\mathcal{T}$$

The matrix elements of $\mathcal{T}$ describe the interaction between the constituents of the system.

In the succeeding chapters we shall show how invariance principles lead to conditions on $\mathcal{S}$- and $\mathcal{T}$-matrix elements. Experimental tests of these conditions then serve to test the assumed invariance principles.

In quantum electrodynamics and weak interaction theory it is shown how to derive $\mathcal{S}$-matrix elements from more fundamental quantities: the interaction Hamiltonian and the quantised field operators occurring in it. For these developments, which we shall not make use of, the reader is referred to specialised texts.

Given a system described by a Hamiltonian H, it is of importance to find observables Q which commute with H,

$$[H, Q] = 0 \tag{2.19}$$

for by what was said above, the eigenstates of H are also eigenstates of Q. Thus the stationary states of the system can be labelled by the eigenvalues of Q. This is the typical situation in atomic and nuclear physics. An operator Q satisfying (2.19) is called a constant of the motion.

Such quantities continue to play an important role in connection with the $\mathcal{S}$-matrix. This is expressed precisely by the following theorem.

If Q is an observable which commutes with the Hamiltonian H of a system

$$[Q, H] = 0$$

then Q commutes with the $\mathcal{S}$-matrix for that system

$$[Q, \mathcal{S}] = 0$$

If we accept that $\mathcal{S}$ is in some sense a function of H, then this theorem follows from the mathematical theorem that if Q commutes with H then it commutes with any function of H.

Our theorem has the following useful physical consequence. If the initial state ϕ_a of a system is an eigenstate of Q with eigenvalue q_a so that

$$Q\phi_a = q_a\phi_a$$

then the final state is also an eigenstate of Q for, with $\psi = \mathcal{S}\phi_a$ we have

$$Q\psi = Q\mathcal{S}\phi_a = \mathcal{S}Q\phi_a = q_a\mathcal{S}\phi_a = q_a\psi$$

We may put this in a more useful form as follows.

Consider the matrix element between the states ϕ_b and ϕ_a of the equality $Q\mathcal{S}=\mathcal{S}Q$,

$$(\phi_b, Q\mathcal{S}\phi_a) = (\phi_b, \mathcal{S}Q\phi_a)$$

Applying Q to ϕ_b on the left-hand side and to ϕ_a on the right we have

$$q_b(\phi_b, \mathcal{S}\phi_a) = (\phi_b, \mathcal{S}\phi_a)q_a$$

from which it follows that

$$(\phi_b, \mathcal{S}\phi_a) = 0$$

unless $q_b = q_a$.

Thus if Q is conserved and if the system is initially in an eigenstate of Q with eigenvalue q_a then it can only make transitions to final states belonging to the same eigenvalue of Q.

2.1.4 *Relativistic quantum mechanics*

Now let us consider whether the principles of quantum mechanics discussed above must be modified when we extend them from the domain of atomic phenomena (scale 10^{-8} cm) to that of elementary particles (scale less than 10^{-13} cm). At the velocities attained in elementary particle phenomena, it is necessary to be consistent with the principles of special relativity. The first wave equations to satisfy this requirement were given by Schrödinger (with further study by Klein and Gordon) and by Dirac. However the wavefunctions obeying these equations do not have a consistent interpretation as position probability amplitudes for a single particle. One is led to interpret them as field equations which must be 'second-quantised'. In this way the attempt to synthesise quantum mechanics and special relativity leads to quantum field theory and a framework for the description of annihilation and creation of particles. We shall not employ the concepts and methods of quantum field theory in this book, and this will prevent our giving a full discussion of the weak interaction Hamiltonian and its symmetries.

It is important to note that in a large range of elementary particle phenomena, we are concerned with interaction processes which are highly localised and not directly accessible to study. What *are* studied experimentally are the incident and final particles which are travelling freely at such large separations that their mutual interactions may be neglected. As discussed above, this makes the $\mathcal{S}$-matrix a convenient concept. It also means that we only need a relativistic description of freely travelling particles which, as we shall see, can be based on the fundamental invariance principles of space–time.

2.2 Invariance principles and conserved quantities in quantum mechanics

In this section we consider the consequences of the existence of invariance principles in quantum-mechanical terms. The discussion applies to all the invariance principles to be considered later except for time reversal symmetry, for which many of the statements made in this section must be modified. This will be done in chapter 6.

The line of development is presented in general and therefore rather abstract terms. The reader may prefer to use this section for reference when reading later chapters.

The invariance principles of elementary particle physics fall into two broad classes,

(*a*) Space–time symmetries such as uniform translations, rotations, Lorentz transformations, which arise from the existence of equivalent space–time frames of reference.

(*b*) Internal symmetries such as isospin, $SU(3)$, charge conjugation.

In the first class we have transformations of two types: those which involve the time coordinate and those which do not.

This last type is the simplest and we shall start by formulating our discussion in terms of transformations between different spatial frames of reference, and later extend the discussion to include the other kinds. We start with a description in terms of Schrödinger wave mechanics and then go on to indicate what features are more generally valid.

2.2.1 *Coordinate transformations and state vector transformations*

In setting up quantum-mechanical formalism we have implicitly chosen a frame of reference Σ with respect to which the coordinates are measured. The form of the wavefunctions and operators will in general change if we use a different frame of reference. Let us describe how this happens. The change from one frame of reference Σ to a new one Σ' is defined by the equations of transformation which give the coordinates (x', y', z') of a point P referred to the new frame Σ' in terms of the coordinates x, y, z of P referred to Σ. We symbolise this by

$$\left.\begin{aligned} x \to x' &= f(x, y, z) \\ y \to y' &= g(x, y, z) \\ z \to z' &= h(x, y, z) \end{aligned}\right\} \tag{2.20}$$

For example, if the transformation is a uniform translation by (a_x, a_y, a_z), then (2.20) reads

$$\left.\begin{aligned} x \to x' &= x + a_x \\ y \to y' &= y + a_y \\ z \to z' &= z + a_z \end{aligned}\right\} \tag{2.21}$$

A second example is a rotation in the xy-plane through an angle θ, in which case the equations of transformation are

$$\left.\begin{aligned} x \to x' &= x \cos\theta - y \sin\theta \\ y \to y' &= x \sin\theta + y \cos\theta \\ z \to z' &= z \end{aligned}\right\} \tag{2.22}$$

A third example is the inversion of the coordinate system, or parity transformation

$$\left.\begin{aligned} x \to x' &= -x \\ y \to y' &= -y \\ z \to z' &= -z \end{aligned}\right\} \tag{2.23}$$

If the system is in a state described by the wavefunction $\psi(x, y, z)$ relative to the frame of reference Σ, we define the transformed wavefunction $\psi'(x', y', z')$ as that function of the new coordinates referred to Σ', which has the same value as the old function at the same point of space. This is expressed by

$$\psi'(x', y', z') = \psi(x, y, z) \tag{2.24}$$

which simply states that the probability amplitude for finding the particle at a particular point of space is the same whichever coordinate system is used.

It is implicit that we use (2.20) in (2.24) to express both sides as functions of the same variables (primed or unprimed). Thus if we solve (2.20) for x, y, z in terms of x', y', z' which we shall express by

$$\left.\begin{aligned} x &= \bar{f}(x', y', z') \\ y &= \bar{g}(x', y', z') \\ z &= \bar{h}(x', y', z') \end{aligned}\right\} \tag{2.25}$$

then we have

$$\psi'(x', y', z') = \psi(\bar{f}(x', y', z'), \bar{g}(x', y', z'), \bar{h}(x', y', z')) \tag{2.26}$$

which serves to define ψ'. Thus for the case of the translation (2.21),

$$\psi'(x', y', z') = \psi(x' - a_x, y' - a_y, z' - a_z)$$

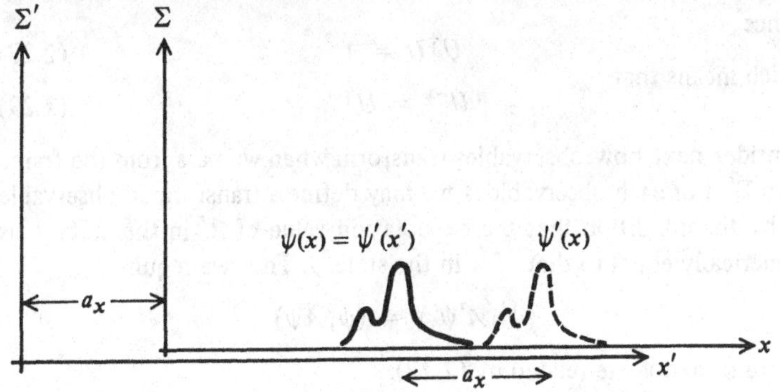

Fig. 2.1. Relation between a coordinate transformation (uniform translation by a_x) and the wavefunction transformation $\psi(x) \to \psi'(x)$. $\psi'(x)$ 'looks the same' relative to Σ as $\psi'(x')$ does relative to Σ'.

We can at this stage drop the primes from x', y' and z' throughout this equation,

$$\psi'(x, y, z) = \psi(\bar{f}(x, y, z), \bar{g}(x, y, z), \bar{h}(x, y, z)) \qquad (2.27)$$

This suggests an alternative interpretation of the transformation: we regard $\psi'(x, y, z)$ as a new wavefunction referred to the original reference frame Σ. This second interpretation of (2.27) as generating a new state ψ' in the same reference frame as ψ is called the *active* interpretation, in contrast to the original which is called the *passive* interpretation of the transformation (2.20).

Fig. 2.1 illustrates the connections for the case of a translation in one dimension. The two interpretations are not quite equivalent. The active interpretation will be very convenient in applications, but it does not exist if the transformation (2.20) is not a symmetry property of the system.

To denote that ψ' arises from ψ we write it as $U\psi$ where U is an operator which depends on the transformation (2.20) considered. If we transform a superposition of two states the result is the same as transforming the two states separately and taking the resultant superposition. It follows that U is a linear operator. Furthermore for the transformations which we shall consider ψ', is normalised if ψ is. This can be verified explicitly for the case of translation (2.21) above. It follows that U is unitary and thus

$$\int d\tau (U\psi)^* U\psi = \int d\tau \psi^* \psi$$

or

$$(U\psi, U\psi) = (\psi, \psi)$$

holds for any ψ.

Hence

$$U^{\dagger}U = 1 \tag{2.28}$$

which means that

$$U^{-1} = U^{\dagger} \tag{2.29}$$

Consider next how observables transform when we pass from the frame Σ to Σ'. For each observable A we may define a transformed observable A' by the condition that the expectation value of A' in the state ψ' is numerically equal to that of A in the state ψ. Thus we require

$$(\psi', A'\psi') = (\psi, A\psi)$$

where ψ' and ψ are related by (2.27).

Now,

$$\begin{aligned}(\psi', A'\psi') &= (U\psi, A'U\psi)\\ &= (\psi, U^{\dagger}A'U\psi)\end{aligned}$$

so the requirement is for all ψ,

$$(\psi, U^{\dagger}A'U\psi) = (\psi, A\psi)$$

and hence

$$U^{\dagger}A'U = A$$

or using (2.29)

$$A' = UAU^{\dagger} \tag{2.30}$$

A particularly important case occurs when the observable A is invariant under the transformation considered.

Then

$$A' = A$$

so that

$$UAU^{\dagger} = A \tag{2.31}$$

or

$$UA = AU$$

which states that U commutes with A.

2.2.2 *Symmetry transformations*

If Σ and Σ' are equally valid frames of reference for formulating the laws describing the behaviour of a system, we shall say that the transformation (2.20) relating Σ and Σ' is a *symmetry transformation* of the system.

For geometrical symmetry transformation not involving the time coordinate and for internal symmetry transformation it is possible to give an alternative definition. The Hamiltonian of a system gives as its eigenstates and eigenvalues the stationary states of the system and their energies, while more generally it governs the time development of the system and thus for example the scattering properties. This suggests that we define a symmetry transformation as one which leaves the Hamiltonian H of the system invariant. From (2.30) this is expressed by

$$UHU^{-1} = H \tag{2.32}$$

where U is the transformation operator corresponding to (2.20).

In the case of transformation involving the time coordinate (these comprise Lorentz transformations and time reversal) the corresponding operator U does not commute with H. However, it will be seen in chapters 4 and 6 that these symmetry transformations lead to restrictions on the form of H and of the $\mathcal{S}$-matrix.

It is characteristic of the symmetry transformations which we have to deal with, that they form families or groups of transformations. This is because we often have not just two, but several equivalent frames of reference. Thus translational invariance is expressed by the existence of infinitely many equivalent frames of reference differing only in the choice of origin. We then have a set of transformations of the form (2.21) with (a_x, a_y, a_z) taking any values. In general we have a family of transformations labelled by several parameters $\alpha_1, \alpha_2, \ldots, \alpha_n$ which we shall collectively denote by α. The general transformation will be denoted by

$$\xi \rightarrow \xi' = T_\alpha(\xi) \tag{2.33}$$

where we have denoted the coordinates collectively by ξ.

To say that a family of transformations, for example the set of all translations, forms a group simply amounts to recognising that the family possesses certain properties which from the mathematical point of view are the axioms satisfied by any group.

We shall not make explicit use of the formal apparatus of group theory in this book. However, many things which we discuss are special cases of that theory, and because of its wide use we shall from time to time remark briefly on the terminology of group theory.

Consider any three equivalent reference frames Σ, Σ' and Σ''. Let the coordinate transformation from ξ to ξ' be

$$\xi \rightarrow \xi' = T_\alpha(\xi) \tag{2.34}$$

and that from ξ' to ξ'' be

$$\xi' \rightarrow \xi'' = T_\beta(\xi') \tag{2.35}$$

On substituting (2.34) in (2.35), we obtain the equation of transformation relating ξ'' and ξ

$$\xi'' = T_\beta[T_\alpha(\xi)]$$

which is written

$$\xi'' = T_\beta T_\alpha(\xi) \tag{2.36}$$

This compound transformation is called the *product* of T_β and T_α. On the other hand because Σ and Σ'' are equivalent frames of reference there is an equation of transformation directly relating ξ and ξ'' given by

$$\xi'' = T_\gamma(\xi) \tag{2.37}$$

for some suitable values of the parameters γ. From (2.36) and (2.37) it follows that

$$T_\gamma = T_\beta T_\alpha \tag{2.38}$$

Thus the product of any two transformations of the family is itself a member of the family (*Group Axiom 1*).

The remaining group axioms are trivially satisfied in this instance. They are:

Axiom 2. There exists an identity transformation

$$\xi' = \xi$$

which is a member of the family, in the sense that for some value α_0 of the parameter α,

$$\xi' = T_{\alpha_0}(\xi) = \xi$$

Usually, it can be arranged that $\alpha_0 = 0$, as in the case of translations (2.21).

Axiom 3. For each transformation

$$\xi' = T_\alpha(\xi)$$

the inverse transformation

$$\xi = T_\alpha^{-1}(\xi')$$

is also a transformation of the family, i.e. for suitable parameters $\bar{\alpha}$

$$T_\alpha^{-1} = T_{\bar{\alpha}}$$

For example, the inverse of the translation T_{a_x, a_y, a_z}

$$T: \begin{cases} x' = x + a_x \\ y' = y + a_y \\ z' = z + a_z \end{cases}$$

is the translation $T_{-a_x, -a_y, -a_z}$.

Now to each coordinate transformation T_α there corresponds a transformation operator on wavefunctions which we denote by U_α. Consider the transformation of a wavefunction under the successive transformations from Σ to Σ' by T_α and then from Σ' to Σ'' by T_β

$$\xi' = T_\alpha(\xi) \quad \psi' = U_\alpha \psi \tag{2.39}$$

$$\xi'' = T_\beta(\xi') \quad \psi'' = U_\beta \psi' \tag{2.40}$$

Now the wavefunctions ψ'' and ψ are related on the one hand by

$$\xi'' = T_\gamma(\xi) \quad \psi'' = U_\gamma \psi$$

and on the other by combining (2.39) and (2.40) to give

$$\xi'' = T_\beta T_\alpha(\xi) \quad \psi'' = U_\beta U_\alpha \psi \tag{2.41}$$

Thus for any ψ

$$\psi'' = U_\gamma \psi = U_\beta U_\alpha \psi$$

and hence†

$$U_\gamma = U_\beta U_\alpha \tag{2.42}$$

Thus we have shown that to the product of two transformations of coordinates there corresponds the product of the unitary transformation operators. The set of operators U_α form a group of unitary operators in the Hilbert space of states, and because of the correspondence between (2.42) and (2.38) the U_α are said to be a representation of the symmetry group of the system by unitary operators.

In the case of the parity transformation (2.23) there are only two equivalent frames of reference: the original one and the inverted one. Correspondingly there is only one non-trivial symmetry transformation P (2.23). This and the identity transformation form the symmetry group. A parameter to label these transformations takes only two discrete values and such a group is called a discrete symmetry group.

When the parameters vary over a continuous range we refer to a continuous symmetry group: for example, the group of rotations.

2.2.3 *Infinitesimal transformations and constants of motion*

If U_α is a symmetry operator corresponding to a transformation not involving the time, then U_α commutes with the Hamiltonian,

† Since the wavefunctions ψ and $e^{i\alpha}\psi$ define the same state it is possible that the two sides of (2.42) can differ by a phase factor. The only case of concern to us where this happens is for rotational transformations of half-integral spin wavefunctions and there it can be explicitly allowed for. We shall proceed as if (2.42) were correct in all cases.

$$U_\alpha H = H U_\alpha \tag{2.43}$$

and is a constant of the motion for the system.

However U_α cannot be interpreted as an observable because it is not Hermitian; rather, as we have seen, it is unitary,

$$U_\alpha^\dagger U_\alpha = 1$$

Nevertheless we can obtain from a group of symmetry operators U_α depending on a continuous parameter (i.e. derived from a continuous group of symmetry transformations) an associated Hermitian observable. If U_α depends on n real parameters $\alpha_1, \ldots, \alpha_n$, then we can obtain n observables.

We assume that $\alpha = 0$ corresponds to the identity transformation; then for the corresponding operator on wavefunctions we have

$$U_0 = 1$$

where 1 denotes the unit operator. The method is to consider an infinitesimal value of the parameter: we expect ψ' will differ infinitesimally from ψ and so U will differ infinitesimally from the unit operator. We write

$$U_{\delta\alpha} = 1 + \mathrm{i}\delta\alpha G + O(\delta\alpha^2) \tag{2.44}$$

where G is an operator and i has been introduced for convenience. $O(\delta\alpha^2)$ denotes terms of higher order in $\delta\alpha$ which will be neglected in what follows. Now we impose the condition that $U_{\delta\alpha}$ be unitary. First we have

$$U_{\delta\alpha}^\dagger = 1 - \mathrm{i}\delta\alpha G^\dagger$$

so substituting in

$$U_{\delta\alpha}^\dagger U_{\delta\alpha} = 1$$

we have

$$(1 - \mathrm{i}\delta\alpha G^\dagger)(1 + \mathrm{i}\delta\alpha G) = 1$$

Thus

$$1 - \mathrm{i}\delta\alpha(G^\dagger - G) = 1$$

and

$$G^\dagger = G \tag{2.45}$$

Thus the operator G is Hermitian. It is called the *generator* of the transformation, U_α.

Since U_α commutes with H, so does G,

$$HG = GH \tag{2.46}$$

Thus corresponding to the symmetry transformation T_α, we have a Hermitian operator G which commutes with the Hamiltonian, and is therefore constant in time. This is the quantum-mechanical analogue of the classical conserved quantity. We shall find that when we consider specific examples the observable G is a familiar observable.

In the case of a discrete symmetry such as parity the above arguments are inapplicable. However, in that case we are saved by the fact that the unitary operator U_P satisfies

$$U_P^2 = 1 \tag{2.47}$$

which follows from the condition that if we perform the parity operation (2.23) twice we obtain the identity transformation, that is

$$P^2 = 1$$

Now U_P is also unitary

$$U_P^\dagger U_P = 1 \tag{2.48}$$

and it follows from (2.47) and (2.48) that

$$U_P = U_P^\dagger$$

So in this case the operator U_P is itself Hermitian and can be interpreted as an observable. We can label states by its eigenvalues.

These considerations apply to any symmetry transformation whose square is the identity transformation. Examples are charge conjugation, G-parity, operation of interchange of two identical particles (permutational symmetry).

We end this section by giving a table of some symmetry transformation groups and the associated conserved generators G to which they give rise. The first three entries will be discussed in §2.3, chapter 3 and chapter 5 respectively.

Table 2.1

Symmetry transformation	Conserved generator
Spatial translations	Linear momentum
Rotations	Angular momentum
Space inversion	Parity
Time translations	Hamiltonian (total energy)

The last example cited is a transformation involving the time coordinate which we explicitly excluded above. Let us show that for this particular case the preceding considerations apply.

Time translation invariance implies that the description of a system is independent of the choice of zero from which time is measured. The transformation from one choice of origin to another is

$$t \to t' = t + \tau$$

The associated unitary operator is defined by

$$U_\tau \psi(r, t+\tau) = \psi(r, t) \tag{2.49}$$

We have displayed one position coordinate to emphasise that it is the same on both sides.

Introducing $t' = t + \tau$, we obtain

$$U_\tau \psi(t) = \psi(t-\tau) \tag{2.50}$$

where we have omitted the prime.

The associated Hermitian observable may be identified by taking for τ the infinitesimal value $\delta\tau$, and writing

$$U_{\delta\tau} = 1 + i\delta\tau G$$

to order $\delta\tau$. Then

$$\begin{aligned} U_{\delta\psi}(t) &= (1 + i\delta\tau G)\psi(t) \\ &= \psi(t - \delta\tau) \\ &= \psi(t) - \delta\tau \frac{\partial\psi}{\partial t} \end{aligned}$$

where we have expanded by Taylor's theorem to first order in $\delta\tau$. The operator G therefore satisfies the equation

$$iG\psi = -\frac{\partial\psi}{\partial t}$$

or defining $H = \hbar G$

$$H\psi = i\hbar\frac{\partial\psi}{\partial t} \tag{2.51}$$

Now this equation has precisely the same form as the Schrödinger equation for the time development of the wavefunction, and so H may be identified with the Hamiltonian of the system.

We may say that the Hamiltonian H is the infinitesimal operator associated with the time displacement invariance of the system,

$$U_\tau = 1 + i\tau H/\hbar \tag{2.52}$$

2.3. Translational invariance

In this section we apply the concepts developed in §2.2 to the special case of translational invariance which has some simplifying features.

2.3.1 *Translations and linear momentum*

Space translation invariance requires that two frames of reference differing by a uniform translation

$$\boldsymbol{r} \to \boldsymbol{r}' = \boldsymbol{r} + \boldsymbol{a}$$

are equivalent for describing the behaviour of the system. In non-relativistic quantum mechanics we can examine the Hamiltonian of the system to determine whether it is invariant in form under the transformation (2.21) and the associated transformation of the momentum which for the case of a translation is trivial:

$$\boldsymbol{p} \to \boldsymbol{p}' = \boldsymbol{p} \tag{2.53}$$

Equation (2.53) follows from the operator expression for $\boldsymbol{p}$ in the coordinate representation

$$\boldsymbol{p} = -\mathrm{i}\hbar\boldsymbol{\nabla}$$

If several particles are involved, the equations (2.21) and (2.53) apply to the coordinates of each,

$$\boldsymbol{r}_i \to \boldsymbol{r}_i' = \boldsymbol{r}_i + \boldsymbol{a} \quad \boldsymbol{p}_i \to \boldsymbol{p}_i' = \boldsymbol{p}_i, \quad i = 1, 2, 3, \ldots, N$$

The symmetry operator $U_{\boldsymbol{a}}$ is defined by

$$U_{\boldsymbol{a}}\psi(\boldsymbol{r}_1, \boldsymbol{r}_2, \ldots, \boldsymbol{r}_N) = \psi(\boldsymbol{r}_1 - \boldsymbol{a}, \ldots, \boldsymbol{r}_N - \boldsymbol{a}) \tag{2.54}$$

Following the general discussion of the preceding section we derive the Hermitian observable by considering infinitesimal displacements. Let

$$\delta\boldsymbol{a} = (0, 0, \delta a)$$

Then

$$\begin{aligned} U(0, 0, \delta a)&\psi(x_1, y_1, z_1, \ldots, x_N, y_N, z_N) \\ &= \psi(x_1, y_1, z_1 - \delta a, \ldots, x_N, y_N, z_N - \delta a) \\ &= \psi(x_1, \ldots, z_N) - \delta a\left(\frac{\partial\psi}{\partial z_1} + \ldots + \frac{\partial\psi}{\partial z_N}\right) \end{aligned}$$

to terms of order δa.

So defining the Hermitian operator P_z by

$$U(0, 0, \delta a) = 1 - \mathrm{i}\delta a P_z/\hbar \tag{2.55}$$

we see that

$$P_z = \sum_{i=1}^{N} -\mathrm{i}\hbar\frac{\partial}{\partial z_i} \tag{2.56}$$

Hence the associated observable is the z-component of the *total* linear momentum. Similar considerations hold for displacements along the other coordinate directions, and we may write more generally

$$U_{\delta\boldsymbol{a}} = 1 - \mathrm{i}\delta\boldsymbol{a}\cdot\boldsymbol{P}/\hbar \tag{2.57}$$

where

$$P = \sum_{i=1}^{N} -\mathrm{i}\hbar\boldsymbol{\nabla}_i$$

It is possible to derive a more precise relation between the unitary operators U_a and the Hermitian operators P_x, P_y and P_z. We restrict ourselves to displacements along the z-direction and we write $U(0, 0, a)$ as $U(a)$, for brevity.

The resultant of a translation $(0, 0, a)$ followed by a translation $(0, 0, b)$ is a translation $(0, 0, a + b)$. It follows that

$$U(b)U(a) = U(b + a) \tag{2.58}$$

We now let $b = \delta a$ in (2.58) and substitute by means of (2.55)

$$U(\delta a + a) = \left(1 - \mathrm{i}\frac{\delta a}{\hbar}P_z\right) U(a)$$

which on rearrangement gives

$$\frac{1}{\delta a}[U(a + \delta a) - U(a)] = -\frac{\mathrm{i}}{\hbar}P_z U(a) \tag{2.59}$$

On taking the limit $\delta a \to 0$ on the left we obtain the derivative $\mathrm{d}U(a)/\mathrm{d}a$ of the operator $U(a)$ with respect to the scalar parameter a and (2.59) becomes a differential equation for $U(a)$:

$$\frac{\mathrm{d}U}{\mathrm{d}a} = -\mathrm{i}P_z U(a)/\hbar \tag{2.60}$$

If $U(a)$ and P_z were ordinary functions we should write the solution immediately as

$$U(a) = \mathrm{e}^{-\mathrm{i}P_z a/\hbar} \tag{2.61}$$

where we have used the condition $U(0) = 1$ to fix the (operator) constant of integration. In the case of operators the right side of (2.61) is simply an abbreviation for the series

$$1 - \frac{\mathrm{i}}{\hbar}P_z a + \frac{1}{2!}\left(-\frac{\mathrm{i}}{\hbar}P_z a\right)^2 + \ldots$$

and it may be verified that this satisfies (2.60).

Equation (2.61) shows that the operator representing a finite displacement by a in the z-direction can be expressed in terms of the operator of the z-component of momentum. The momentum operator is sometimes referred to as the *generator* of displacements.

A similar calculation can be made for displacements along the x- and y-axes, yielding the formulae

$$U(a, 0, 0) = e^{-iaP_x/\hbar}$$

$$U(0, b, 0) = e^{-ibP_y/\hbar}$$

The three operators P_x, P_y, P_z commute with one another. In Schrödinger wave mechanics this can be seen clearly from the derivation given above leading to (2.56): different spatial coordinates are involved in the three cases. However, it is of interest to show the basis for this result, in order to contrast it with other groups of symmetry transformations for which the generators do not commute.

If we consider displacements along two different directions, say a_x along the x-direction and a_z along the z-direction, then it is obvious that the order in which these are performed is immaterial. It follows that

$$U(a_x, 0, 0)U(0, 0, a_z) = U(0, 0, a_z)U(a_x, 0, 0)$$

for all a_x and a_z.

We take a_x and a_z as infinitesimal and substitute by means of (2.55) and the corresponding equation for δa_x,

$$U(\delta a_x) = 1 - i\delta a_x P_x/\hbar$$

then the condition that the resulting equation be valid to first order in δa_x and δa_z is

$$P_x P_z - P_z P_x = 0$$

Similarly, it can be shown that

$$[P_y, P_z] = 0 \quad [P_x, P_y] = 0$$

2.3.2 *Conservation of linear momentum*

Translational invariance will serve as the first simple but important illustration of how an invariance principle leads to conditions on the Hamiltonian and the $\mathcal{S}$-matrix.

We started our analysis of translational invariance in the familiar ground of Schrödinger wave mechanics and we assumed the state vector to be a function of the position coordinates. However, in a relativistic theory, position coordinates are known not to be useful concepts. In deciding what to use instead, one may take a more abstract point of view by saying there is a state vector describing fully the state of the system. But, of what is it a function? We have seen that translational invariance implies the existence of Hermitian operators P_x, P_y, P_z, which by analogy with non-relativistic theory are identified with the components of momentum. We therefore label our states by the eigenvalues of momentum.

If we have a system consisting of one particle, we are led to the momentum representation in which the complete set of basic states is the set of eigenstates of momentum denoted by ϕ_p, and satisfying

$$P\phi_p = p\phi_p$$

The general state of the particle may be written as a superposition of such states

$$\Psi = \int d^3p\, f(p)\phi_p$$

which is an integral rather than a sum because p varies over a continuous range. $f(p)$ is the momentum–space wavefunction. The probability of finding the particle with momentum in the range $(p, p + dp)$ is $|f(p)|^2 d^3p$.

We form a complete set of states for two or more particles by taking products of ϕ_p states.
Thus

$$\phi^{(12)}_{p_1, p_2} = \phi^{(1)}_{p_1}\phi^{(2)}_{p_2}$$

represents a state with particle 1 having momentum p_1 and particle 2 having momentum p_2. States such as these are a suitable basis in which to express $\mathcal{S}$-matrix elements for scattering processes as discussed above.

Consider a two particle reaction

$$a + b \rightarrow c + d$$

The amplitude for a transition from the initial state ϕ_{p_a, p_b} to a final state ϕ_{p_c, p_d} is

$$(\phi_{p_c, p_d}, \mathcal{S}\phi_{p_a, p_b})$$

Translational invariance expressed by

$$U_a\mathcal{S} = \mathcal{S}U_a$$

implies

$$P\mathcal{S} = \mathcal{S}P$$

On taking the matrix element of this equation between states ϕ_{p_a, p_b} and ϕ_{p_c, p_d} which are eigenstates of P, we obtain

$$(p_c + p_d)(\phi_{p_c, p_d}, \mathcal{S}\phi_{p_a, p_b}) = (p_a + p_b)(\phi_{p_c, p_d}, \mathcal{S}\phi_{p_a, p_b})$$

which implies that the transition amplitude between the two states considered is zero unless the total momentum in the two states is the same, i.e. unless the total momentum is conserved.

We shall meet less elementary applications of this line of argument in later chapters.

CHAPTER 3

ANGULAR MOMENTUM

In this chapter we discuss the quantum theory of angular momentum, and show how it is related to the transformation properties of systems under rotations.

The material of this chapter is of central importance because on the one hand the results have many direct applications, while on the other hand the algebra of angular momentum is the prototype of that used for any continuous group of symmetry transformations. Thus, the development of the algebra of $SU(3)$ and other higher symmetries can be better understood by analogy with angular momentum while the algebra of isospin is exactly the same as that of angular momentum.

3.1 Elementary quantum mechanics of angular momentum

It is well known from elementary quantum mechanics that the energy eigenstates of a system possessing spherical symmetry can be chosen to be eigenfunctions of angular momentum. We review these results briefly.

3.1.1 *Angular momentum operators and their commutation relations*

The components of the vector operator of angular momentum L are obtained from the classical expression

$$L = r \wedge p \tag{3.1}$$

by the substitution

$$p \rightarrow -\mathrm{i}\hbar\nabla \tag{3.2}$$

giving

$$L = -\mathrm{i}\hbar r \wedge \nabla \tag{3.3}$$

Thus,

$$\left.\begin{aligned} L_x &= -\mathrm{i}\hbar\left(y\frac{\partial}{\partial z} - z\frac{\partial}{\partial y}\right) \\ L_y &= -\mathrm{i}\hbar\left(z\frac{\partial}{\partial x} - x\frac{\partial}{\partial z}\right) \\ L_z &= -\mathrm{i}\hbar\left(x\frac{\partial}{\partial y} - y\frac{\partial}{\partial x}\right) \end{aligned}\right\} \tag{3.4}$$

The standard rules for transforming partial derivatives enable us to express these operators in spherical polar coordinates (r, θ, ϕ) as follows

$$\left.\begin{aligned} L_x &= i\hbar\left(\sin\phi\frac{\partial}{\partial\theta} + \cot\theta\cos\phi\frac{\partial}{\partial\phi}\right) \\ L_y &= i\hbar\left(-\cos\phi\frac{\partial}{\partial\theta} + \cot\theta\sin\phi\frac{\partial}{\partial\phi}\right) \\ L_z &= -i\hbar\frac{\partial}{\partial\phi} \end{aligned}\right\} \tag{3.5}$$

The square of the total angular momentum

$$\boldsymbol{L}^2 = L_x^2 + L_y^2 + L_z^2 \tag{3.6}$$

has the form

$$\boldsymbol{L}^2 = -\hbar^2\left\{\frac{1}{\sin\theta}\frac{\partial}{\partial\theta}\left(\sin\theta\frac{\partial}{\partial\theta}\right) + \frac{1}{\sin^2\theta}\frac{\partial^2}{\partial\phi^2}\right\} \tag{3.7}$$

For a Hamiltonian function of the form

$$H = -\frac{\hbar^2}{2m}\nabla^2 + V(|\boldsymbol{r}|) \tag{3.8}$$

where V is a function only of the radial distance from the origin, all three components of $\boldsymbol{L}$, and hence $\boldsymbol{L}^2$, commute with H

$$[H, \boldsymbol{L}] = 0$$

$$[H, \boldsymbol{L}^2] = 0$$

However, the components of $\boldsymbol{L}$ do not commute among themselves. One finds by direct calculation using (3.4) that

$$\left.\begin{aligned} [L_x, L_y] &= i\hbar L_z \\ [L_y, L_z] &= i\hbar L_x \\ [L_z, L_x] &= i\hbar L_y \end{aligned}\right\} \tag{3.9}$$

It follows that it is impossible to find a basis of states in which two (or more) components of $\boldsymbol{L}$ are simultaneously diagonalised (see §2.1). However (3.9) may be used to show that $\boldsymbol{L}^2$ commutes with L_x, L_y and L_z separately. Thus, the best that can be done is to diagonalise $\boldsymbol{L}^2$ and one component of $\boldsymbol{L}$, conventionally taken to be L_z, because of the simple form which it takes in spherical polar coordinates.

The result is that any energy eigenfunction of H can be written in the form

$$\psi(r, \theta, \phi) = R_{nl}(r) Y_{lm}(\theta, \phi)$$

where R_{nl} is the radial wavefunction, and $Y_{lm}(\theta, \phi)$ is a spherical harmonic function. The quantum number n distinguishes different radial functions belonging to the same l value. $Y_{lm}(\theta, \phi)$ satisfies

$$L^2 Y_{lm}(\theta, \phi) = l(l+1)\hbar^2 Y_{lm}(\theta, \phi) \tag{3.10}$$

$$L_z Y_{lm}(\theta, \phi) = m\hbar Y_{lm}(\theta, \phi) \tag{3.11}$$

where l is a positive integer and m runs over integers from $-l$ to $+l$.

The properties of the spherical harmonic functions are summarised in §3.1.2.

Although the operators L_x and L_y cannot be chosen to be diagonal, they are conserved quantities; that is they commute with H.

They are particularly useful in the linear combinations

$$L_\pm = L_x \pm iL_y \tag{3.12}$$

These, too, commute with H and have the following properties

$$\left.\begin{aligned} L_+ Y_{lm}(\theta, \phi) &= \rho_+(l, m) Y_{l, m+1}(\theta, \phi) \quad \text{for } -l \leqslant m \leqslant l-1 \\ L_+ Y_{ll}(\theta, \phi) &= 0 \end{aligned}\right\} \tag{3.13}$$

$$\left.\begin{aligned} L_- Y_{lm}(\theta, \phi) &= \rho_-(l, m) Y_{l, m-1}(\theta, \phi) \quad \text{for } -l+1 \leqslant m \leqslant l \\ L_- Y_{l, -l}(\theta, \phi) &= 0 \end{aligned}\right\} \tag{3.14}$$

where $\rho_+(l, m)$ and $\rho_-(l, m)$ are numerical factors which will be calculated below (§3.2.2), when (3.13) and (3.14) are proved. For the present we assume them to have been verified by using (3.5) and the formulae for the $Y_{lm}(\theta, \phi)$.

These equations may be used to prove the important result that the eigenfunctions

$$\psi_{nlm} = R_{nl}(r) Y_{lm}(\theta, \phi)$$

with the same n and l, and *different* m belong to the same energy eigenvalue. Of course this would come out explicitly if we had solved the Schrödinger equation with a Hamiltonian (3.8) in some particular case (e.g. the hydrogen atom).

We suppose that the Schrödinger equation is

$$H\psi_{nlm} = E\psi_{nlm}$$

and apply H to $L_+\psi_{nlm}$

$$\begin{aligned} HL_+\psi_{nlm} &= L_+H\psi_{nlm} \\ &= L_+E\psi_{nlm} \end{aligned}$$

$$= EL_+\psi_{nlm}$$

hence $L_+\psi_{nlm}$ is an eigenfunction of H belonging to the same eigenvalue E as ψ_{nlm}. Since L_+ does not act on the radial part of the wavefunction, $L_+\psi_{nlm}$ is essentially $\psi_{nl,\,m+1}$. Hence we have shown

$$H\psi_{nl,\,m+1} = E\psi_{nl,\,m+1}$$

Repeated use of this result shows that all the eigenfunctions with the same n and l and any m from $-l$ to $+l$ belong to the same energy eigenvalue. Thus for a spherically symmetrical Hamiltonian every energy level must be $(2l+1)$-fold degenerate where l is the total orbital angular momentum quantum number. The eigenstates are said to form *multiplets* of levels. It could happen for a particular potential $V(r)$, that two multiplets of different l (or even the same l) might have the same energy. This is called *accidental degeneracy*, in the sense that it does not result from angular momentum conservation.

3.1.2 *Properties of the spherical harmonic functions*

The dependence of Y_{lm} on θ and ϕ may be separated as follows

$$Y_{lm}(\theta,\phi) = \Theta_{lm}(\theta)\Phi_m(\phi) \tag{3.15}$$

where

$$\Phi_m(\phi) = (2\pi)^{-1/2}\, e^{im\phi}$$

and

$$\Theta_{lm}(\theta) = (-1)^m \left(\frac{2l+1}{2}\right)^{1/2} \left(\frac{(l-m)!}{(l+m)!}\right)^{1/2} P_l^m(\cos\theta)$$

Here $m \geqslant 0$. For $m < 0$ we define

$$\Theta_{lm}(\theta) = (-1)^m \Theta_{l,-m}(\theta)$$

$P_l^m(\cos\theta)$ is the associated Legendre function as defined in Edmonds (1957), equation (2.5.17).

The numerical coefficients ensure that Θ and Φ are separately normalised as follows

$$\int_0^{2\pi} \Phi_{m'}^*(\phi)\,\Phi_m(\phi)\,d\phi = \delta_{m'm}$$

$$\int_0^{\pi} d\theta \sin\theta\, \Theta_{l'm}(\theta)\Theta_{lm}(\theta) = \delta_{l'l}$$

Hence

$$\int_0^{\pi} d\theta \sin\theta \int_0^{2\pi} d\phi\, Y_{l'm'}^*(\theta,\phi)\, Y_{lm}(\theta,\phi) = \delta_{l'l}\delta_{m'm} \tag{3.16}$$

Table 3.1. *Spherical Harmonics* $Y_{lm}(\theta, \phi)$

Y_{00}	$\sqrt{\frac{1}{4\pi}}$
Y_{10}	$\left(\sqrt{\frac{3}{4\pi}}\right) \cos\theta$
$Y_{1\pm1}$	$\mp\left(\sqrt{\frac{3}{8\pi}}\right) \sin\theta\, e^{\pm i\phi}$
Y_{20}	$\frac{1}{2}\left(\sqrt{\frac{5}{4\pi}}\right) (3\cos^2\theta - 1)$
$Y_{2\pm1}$	$\mp\left(\sqrt{\frac{15}{8\pi}}\right) \sin\theta \cos\theta\, e^{\pm i\phi}$
$Y_{2\pm2}$	$\frac{1}{4}\left(\sqrt{\frac{15}{2\pi}}\right) \sin^2\theta\, e^{\pm 2i\phi}$

A special case is $m = 0$, when P_l^m reduces to the ordinary Legendre polynomial,

$$P_l^0(\cos\theta) = P_l(\cos\theta)$$

Table 3.1 contains expressions for the Y_{lm} for $l = 0$, 1 and 2.

3.1.3 *Spin angular momentum*

There is ample evidence from atomic physics and nuclear physics that the electron, proton and neutron possess intrinsic or spin angular momentum in addition to angular momentum arising from spatial motion.

It does not prove useful to represent this spin angular momentum in terms of internal coordinates, as we do in the case of molecules where the internal angular momentum is understood to arise from rotational motion about its centre of mass. This statement must be modified if we attempt to set up a model for 'elementary' particles as composed of 'more elementary' entities such as quarks. For the present we adopt the more conservative point of view, which is as follows.

For a particle carrying spin we postulate the existence of three spin operators S_x, S_y, S_z which obey the commutation relations

$$\left.\begin{aligned} [S_x, S_y] &= i\hbar S_z \\ [S_y, S_z] &= i\hbar S_x \\ [S_z, S_x] &= i\hbar S_y \end{aligned}\right\} \qquad (3.17)$$

formed by analogy with (3.9). This characterises them as angular momentum operators.

They are further subject to the condition

$$S_x^2 + S_y^2 + S_z^2 = s(s+1)\hbar^2 \tag{3.18}$$

where s is the spin of the particle in units of $\hbar$. For the proton, electron and neutron, $s = \frac{1}{2}$.

Further, the spin and orbital angular momentum operators commute. Writing $\boldsymbol{S} = (S_x, S_y, S_z)$ we may express this by

$$[\boldsymbol{S}, \boldsymbol{L}] = 0$$

This ensures that the spin and orbital angular momentum can in principle be measured simultaneously, even though they may be coupled by interactions such as spin–orbit coupling in atoms.

It is an empirical fact that many of the unstable elementary particles and short lived resonant states discovered in recent years carry a definite spin angular momentum. Indeed the fact that a short lived particle occurring as an intermediate state or produced as a final state resonance can be assigned a definite angular momentum independent of its mode of production or decay is part of the evidence for treating it as a particle on a similar footing to the stable particles.

We shall defer further details of the spin formalism until the next chapter. The important application of the formalism is in the partial wave analysis of scattering of particles with spin.

We shall present the partial wave analysis after discussing the Lorentz transformation properties of states. It will then be possible to see that the partial wave analysis is Lorentz invariant.

3.1.4 *Total angular momentum*

To cover cases in which angular momentum may be due to spin, or to orbital motion, or partly to each, we introduce a total angular momentum operator J, with components

$$J_x = L_x + S_x$$

$$J_y = L_y + S_y$$

$$J_z = L_z + S_z$$

and magnitude given by

$$J^2 = J_x^2 + J_y^2 + J_z^2$$

The components of J obey commutation relations (3.19) similar to those of (3.9) and (3.17) for L and S separately:

$$\left.\begin{aligned}[J_x, J_y] &= i\hbar J_z \\ [J_y, J_z] &= i\hbar J_x \\ [J_z, J_x] &= i\hbar J_y\end{aligned}\right\} \tag{3.19}$$

3.2 Matrix elements of the angular momentum operators

In this section it will be shown that the allowed eigenvalues j and m for the total angular momentum and any one of its components, and the relation between them, follow from the commutation relations (3.19) *alone*, and do not depend on the explicit differential operator expressions (3.4) or (3.5). This is important because in connection with spin we shall be led to allow half-integral values of j and m. In this case the operators cannot conveniently be represented as differential operators, since we do not introduce internal variables to describe spinning particles.

3.2.1 *Eigenvalues of angular momentum*

We introduce non-Hermitian shift operators $J_\pm$ defined by

$$J_\pm = J_x \pm iJ_y \tag{3.20}$$

which are Hermitian conjugates of one another

$$J_\pm^\dagger = J_\mp \tag{3.21}$$

The commutation relations (3.19) may be rewritten in terms of $J_\pm$ as follows:

$$[J_z, J_\pm] = \pm\hbar J_\pm \tag{3.22}$$

$$[J_+, J_-] = 2\hbar J_z \tag{3.23}$$

The total angular momentum J^2 may be rewritten in terms of $J_\pm$ and J_z as

$$J^2 = J_-J_+ + J_z^2 + \hbar J_z \tag{3.24}$$

$$= J_+J_- + J_z^2 - \hbar J_z \tag{3.25}$$

In so far as J^2 commutes with J_x and J_y it also commutes with $J_\pm$

$$[J^2, J_\pm] = 0$$

and, of course,

$$[J^2, J_z] = 0$$

We consider a system with angular momentum operators J. We choose to diagonalise J^2 and J_z denoting their eigenvalues in the general state as $\kappa^2\hbar^2$ and $m\hbar$. The eigenvalue of J^2 has been written as $\kappa^2\hbar^2$ to take account of the fact that because J^2 is a sum of the squares of Hermitian operators its eigenvalues must be non-negative (positive or zero).

(*Proof*. For any state ψ, eigenstate of Q or not,

$$(\psi, Q^2\psi) = (\psi, Q^\dagger Q\psi) = \Sigma(\psi, Q^\dagger\phi_n)(\phi_n, Q\psi)$$
$$= \sum_n |(\phi_n, Q\psi)|^2 \geqslant 0$$

where we introduced a complete set of states ϕ_n satisfying

$$\sum_n \phi_n\phi_n^* = 1)$$

Such an operator is called *positive semi-definite.*

Out of the operators of the system we choose others, symbolised by A with eigenvalues a, which commute with J_x, J_y and J_z and hence with J^2 and J_z and such that a state of the system is completely specified by the eigenvalues a, κ^2 and m. We denote a state by $\psi(a, \kappa^2, m)$ where the eigenvalues are written as arguments rather than subscripts for clarity in what follows. A, J^2 and J_z are what Dirac calls a *complete commuting set of observables* for the system.

We have not specified the system in detail because the argument is of very general validity.

We focus attention on states of fixed values of a and of κ^2. We first show that the possible m values of such states are restricted.

The operator

$$J_x^2 + J_y^2 = J^2 - J_z^2$$

has the states $\psi(a, \kappa^2, m)$ as eigenstates with eigenvalue $(\kappa^2 - m^2)\hbar^2$; On the other hand $J_x^2 + J_y^2$ is a positive semi-definite operator and has only non-negative eigenvalues. Hence

$$\kappa^2\hbar^2 - m^2\hbar^2 \geqslant 0$$

or

$$|m| \leqslant \kappa \tag{3.26}$$

The state vectors $J_\pm\psi(a, \kappa^2, m)$ are eigenstates of J_z with eigenvalues $m \pm 1$, because with the aid of (3.22) we calculate

$$\begin{aligned} J_z J_\pm\psi(a, \kappa^2, m) &= (J_\pm J_z \pm \hbar J_\pm)\psi(a, \kappa^2, m) \\ &= (J_\pm m\hbar \pm \hbar J_\pm)\psi(a, \kappa^2, m) \\ &= (m \pm 1)\hbar J_\pm\psi(a, \kappa^2, m) \end{aligned} \tag{3.27}$$

On the other hand $J_\pm$ commute with J^2 and with A, hence the states $J_\pm\psi(a,\kappa^2,m)$ belong to the same eigenvalues κ^2 and a of those operators,

$$AJ_\pm\psi(a,\kappa^2,m) = aJ_\pm\psi(a,\kappa^2,m) \tag{3.28}$$

and similarly

$$J^2J_\pm\psi(a,\kappa^2,m) = \hbar^2\kappa^2J_\pm\psi(a,\kappa^2,m) \tag{3.29}$$

So by applying J_+ repeatedly to a state $\psi(a,\kappa^2,m)$ we obtain new states belonging to a, $\kappa^2\hbar^2$, but with new J_z eigenvalues $(m+1)\hbar$, $(m+2)\hbar, \ldots$ Now after a certain number of steps this process must stop because of condition (3.26). This can only be so if for some value of m which we denote by m_0, we have

$$J_+\psi(a,\kappa^2,m_0) = 0 \tag{3.30}$$

Similarly, since a repeated application of J_- gives states with eigenvalues $(m-1)\hbar$, $(m-2)\hbar, \ldots$, and since there must be a minimum value of m, which we denote by m_1, we must have

$$J_-\psi(a,\kappa^2,m_1) = 0 \tag{3.31}$$

Now from (3.30) we have

$$J_-J_+\psi(a,\kappa^2,m_0) = 0$$

but by (3.24) this may be written

$$(J^2 - J_z^2 - \hbar J_z)\psi(a,\kappa^2,m_0) = 0$$

and hence

$$(\kappa^2 - m_0^2 - m_0)\hbar^2\psi(a,\kappa^2,m_0) = 0$$

i.e.

$$\kappa^2 = m_0(m_0+1) \tag{3.32}$$

which relates m_0 to κ.

A similar calculation based on (3.31) and using (3.25) gives

$$\kappa^2 = m_1(m_1-1) \tag{3.33}$$

From (3.32) and (3.33) we have

$$m_0(m_0+1) = m_1(m_1-1)$$

which has solutions $m_1 = -m_0$ or $m_1 = m_0+1$. The second must be rejected, since $m_0\hbar$ was the largest eigenvalue of J_z by hypothesis. We now replace m_0 by j to be in accord with standard notation.

We have therefore shown that the possible J_z eigenvalues associated with a fixed J^2 eigenvalue of $j(j+1)$ run from $+j$ by integer steps down to $-j$. Since the total number of steps $(2j+1)$ is a positive integer or zero, the restriction on j is

$$j = \text{integer or half integer or zero}$$

We in future label our states by j rather than $j(j+1)$. The set of states $\psi(a, j, m)$ with $m = -j, -j+1, \ldots, j-1, +j$ is called a *multiplet* of angular momentum j.

We note that although $j(j+1)\hbar^2$ is the eigenvalue of J^2, $j\hbar$ itself has further significance as the largest eigenvalue of J_z in a multiplet.

3.2.2 *Matrix elements of angular momentum operators*

The next task is to find the matrix elements of the operators J in the basis of states $\psi(a, j, m)$.

We showed in the preceding section that the state $J_+\psi(a, j, m)$ belongs to the eigenvalues a, $j(j+1)\hbar^2$, and $(m+1)\hbar$ of the operators A, J^2, and J_z. Now by the assumption on the way the operators A were chosen to complement J^2 and J_z, any state with these eigenvalues must be the state $\psi(a, j, m+1)$ apart from a multiplicative constant.

We therefore write

$$J_+\psi(a, j, m) = \rho_+(j, m)\psi(a, j, m+1) \tag{3.34}$$

Similarly we have

$$J_-\psi(a, j, m) = \rho_-(j, m)\psi(a, j, m-1) \tag{3.35}$$

This may be re-expressed by

$$(\psi(a, j, m+1), J_+\psi(a, j, m)) = \rho_+(j, m)$$

or since $J_- = J_+^\dagger$

$$(J_-\psi(a, j, m+1), \psi(a, j, m)) = \rho_+(j, m)$$

and therefore

$$(\psi(a, j, m), J_-\psi(a, j, m+1)) = \rho_+(j, m)^*$$

but from (3.35) with $m+1$ replacing m

$$(\psi(a, j, m), J_-\psi(a, j, m+1)) = \rho_-(j, m+1)$$

Therefore

$$\rho_-(j, m+1) = \rho_+(j, m)^* \tag{3.36}$$

It follows easily that

$$J_+J_-\psi(a, j, m) = |\rho_+(j, m-1)|^2\,\psi(a, j, m)$$

and

$$J_-J_+\psi(a, j, m) = |\rho_+(j, m)|^2\,\psi(a, j, m)$$

and then on recalling the commutation relation

$$[J_+, J_-] = 2\hbar J_z$$

we obtain

$$|\rho_+(j, m-1)|^2 - |\rho_+(j, m)|^2 = 2m\hbar \qquad (3.37)$$

This equation enables us to compute $|\rho_+(j, m)|^2$ recursively starting from $m = +j$. Looking at (3.34) for $m = +j$ we see that we must have

$$\rho_+(j, +j) = 0 \qquad (3.38)$$

in order to be consistent with (3.30). Equations (3.37) and (3.38) then give, after using the formula for the sum of an arithmetic progression,

$$|\rho_+(j, m)|^2 = [j(j+1) - m(m+1)]\hbar^2$$

The standard choice of phase (called the Condon and Shortley phase convention) is to take the real positive square root so that

$$\begin{aligned} \rho_+(j, m) &= \hbar[j(j+1) - m(m+1)]^{1/2} \\ &= \hbar[(j-m)(j+m+1)]^{1/2} \end{aligned} \qquad (3.39)$$

Hence from (3.36)

$$\begin{aligned} \rho_-(j, m) &= \hbar[j(j+1) - m(m-1)]^{1/2} \\ &= \hbar[(j+m)(j-m+1)]^{1/2} \end{aligned} \qquad (3.40)$$

Equation (3.40) shows that

$$\rho_-(j, -j) = 0$$

as is necessary for (3.35) to be consistent with (3.31).

To summarise, omitting the label a,

$$\left.\begin{aligned} J^2\psi(j,m) &= j(j+1)\hbar^2\psi(j, m) \\ J_z\psi(j, m) &= m\hbar\psi(j, m) \\ J_\pm\psi(j, m) &= \hbar[(j \mp m)(j \pm m + 1)]^{1/2}\psi(j, m \pm 1) \end{aligned}\right\} \qquad (3.41)$$

It should be emphasised that the derivation rests solely on the commutation relations of J_x, J_y and J_z, and on no other property.

It can be verified that for integer j and m the spherical harmonics $Y_{jm}(\theta, \phi)$ satisfy (3.41).

The mathematical derivation has shown the possibility of j and m being half-integral, and Nature takes advantage of this. In this case the operators $\boldsymbol{J}$ are represented by matrices.

The matrix elements of J^2, J_z and $J_\pm$ with respect to the $\psi(j, m)$ are easily obtained. The operators we are dealing with when applied to a state of given j, gives states of the same j, possibly with different m (or a linear combination of such states). Therefore we need only consider matrix elements between states of the same j. The states $\psi(j, m)$ are assumed to be individually normalised to 1 while states with different j or m are orthogonal since they belong to different eigenvalues of a Hermitian operator (J^2 or J_z). Thus we have

$$(\psi(j', m'), \psi(j, m)) = \delta_{j'j}\delta_{m'm}$$

Taking the scalar product of the equations (3.41) with $\psi(j, m')$ we obtain the matrix elements of J^2, J_z and $J_\pm$. From the latter we obtain the matrix elements of J_x and J_y, with results as follows,

$$(\psi(j, m), J^2\psi(j, m)) = j(j+1)\hbar^2$$

$$(\psi(j, m), J_z\psi(j, m)) = m\hbar$$

$$(\psi(j, m \pm 1), J_x\psi(j, m)) = \tfrac{1}{2}\hbar[(j \mp m)(j \pm m + 1)]^{1/2}$$

$$(\psi(j, m \pm 1), J_y\psi(j, m)) = \tfrac{1}{2}i\hbar[(j \mp m)(j \pm m + 1)]^{1/2}$$

For the important case of $j = \frac{1}{2}$ the matrix elements of J_x, J_y and J_z with respect to the basic states may be written as 2×2 matrices as follows

$$(\psi(\tfrac{1}{2}, m'), J_x\psi(\tfrac{1}{2}, m)) = (S_x)_{m'm} \text{ etc.}$$

It is conventional to use the Pauli matrices defined by

$$\sigma_x = \begin{pmatrix} 0 & 1 \\ 1 & 0 \end{pmatrix} \quad \sigma_y = \begin{pmatrix} 0 & -i \\ i & 0 \end{pmatrix} \quad \sigma_z = \begin{pmatrix} 1 & 0 \\ 0 & -1 \end{pmatrix}$$

Then we find

$$S_i = \tfrac{1}{2}\hbar\sigma_i$$

where i represents x, y, or z.

In addition to the commutation relations

$$[\sigma_x, \sigma_y] = 2i\sigma_z, \text{ etc.} \qquad (3.42)$$

which are satisfied in virtue of (3.19), the Pauli matrices also obey the multiplication rules

$$\sigma_x\sigma_y = i\sigma_z, \text{ etc.} \qquad (3.43)$$

from which (3.42) follows.

3.3 Rotational invariance

We now turn to the connection between angular momentum and rotational invariance. For purposes of orientation we again start by treating the wave-mechanical case.

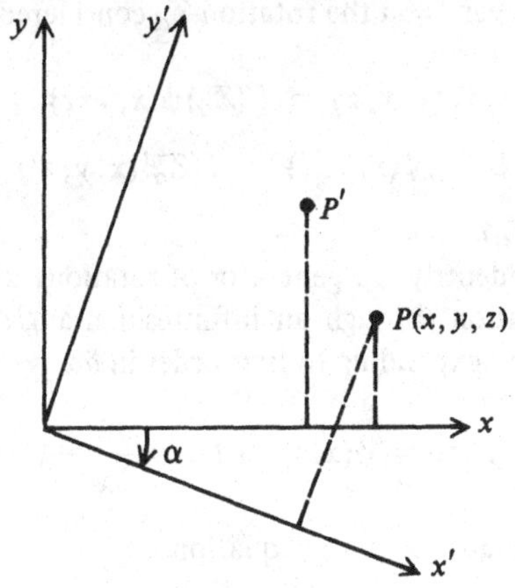

Fig. 3.1. Definition of the rotation Z_α.

3.3.1 *Operators of finite rotations*

Consider a clockwise rotation through an angle α about the z-axis. This is defined by the coordinate transformation

$$\left.\begin{aligned} x' &= x\cos\alpha - y\sin\alpha \\ y' &= x\sin\alpha + y\cos\alpha \\ z' &= z \end{aligned}\right\} \qquad (3.44)$$

We denote this rotation by Z_α. We shall sometimes refer to it as the standard rotation.

The situation in the xy-plane is shown in fig. 3.1. On the passive interpretation as described in §2.2, (3.44) gives the coordinates of P as referred to the new frame. If we adopt the active point of view (3.44) gives the coordinates of a new point P' obtained from P but referred to the same (unprimed) coordinate system. On the active point of view, the effect of Z_α on a given wavefunction $\psi(x, y, z)$ is to generate a new wavefunction $\psi'(x, y, z)$ defined by

$$\psi'(x, y, z) = \psi(Z_\alpha^{-1}(x, y, z))$$

The active rotation of ψ so defined is anticlockwise with respect to the x, y, z-axes. The vector symbolised $Z_\alpha^{-1}(x, y, z)$ has components

$$(x\cos\alpha + y\sin\alpha, -x\sin\alpha + y\cos\alpha, z) \qquad (3.45)$$

obtained by inverting (3.44) and dropping the primes.

According to our general analysis, ψ' and ψ are related by a unitary operator which depends on the rotation Z_α considered. We denote it by $U(Z_\alpha)$,

$$\psi'(x, y, z) = U(Z_\alpha)\psi(x, y, z)$$

Then we have

$$U(Z_\alpha)\psi(x, y, z) = \psi(Z_\alpha^{-1}(x, y, z)) \tag{3.46}$$

which defines $U(Z_\alpha)$.

We proceed to identify the generator of rotations about the z-axis by considering a rotation through an infinitesimal angle $\delta\alpha$. On inserting (3.45) in (3.46) and expanding to first order in $\delta\alpha$, we obtain

$$\psi'(x, y, z) = \psi(x, y, z) + y\delta\alpha\frac{\partial\psi}{\partial x} - x\delta\alpha\frac{\partial\psi}{\partial y} \tag{3.47}$$

We define the generator Q by the equation

$$U(Z_{\delta\alpha}) = 1 - \mathrm{i}\frac{\delta\alpha}{\hbar}Q \tag{3.48}$$

On substituting (3.48) in (3.46) and comparing with (3.47) we find

$$Q = -\mathrm{i}\hbar\left(x\frac{\partial}{\partial y} - y\frac{\partial}{\partial x}\right) = L_z$$

and (3.48) becomes

$$U(Z_\alpha) = 1 - \mathrm{i}\delta\alpha L_z/\hbar \tag{3.49}$$

Thus the generator of rotations about the z-axis is the z-component of angular momentum. It follows from similar calculations that L_x and L_y are generators of rotations about the x- and y-axes.

As in the case of translational invariance the operator $U(Z_\alpha)$ corresponding to a finite rotation can be expressed in terms of the corresponding generator by

$$U(Z_\alpha) = \mathrm{e}^{-\mathrm{i}\alpha L_z/\hbar} \tag{3.50}$$

Similarly if Y_β denotes a rotation through β about the y-axis

$$U(Y_\beta) = \mathrm{e}^{-\mathrm{i}\beta L_y/\hbar}$$

A similar formula exists for a rotation about the x-axis.

It may be shown that the rotation through an angle ω about a general axis defined by a unit vector $\hat{n}$ is represented by the operator

$$U(\hat{n}, \omega) = \mathrm{e}^{-\mathrm{i}\omega\hat{n}\cdot\boldsymbol{L}/\hbar}$$

In contrast to the case of translations, rotations about different axes do not commute; consequently the operators representing them do not in general commute. An examination of the multiplication rules for finite rotations can be shown to lead to the commutation rules between the generators L_x, L_y and L_z already given in (3.9).

For this the reader is referred to books on the group theory of angular momentum.

As an illustration of the properties of the rotation operators and the formulae developed above, consider the application of $U(Z_\alpha)$ to a momentum eigenstate

$$\psi = e^{ipx/\hbar}$$

corresponding to momentum $p = (p, 0, 0)$.

Using (3.46) one finds the rotated state

$$U(Z_\alpha)\psi = e^{i(px \cos \alpha + py \sin \alpha)/\hbar}$$

describing a state with momentum $(p \cos \alpha, p \sin \alpha, 0)$ as expected.

As another application of the rotation operators we show how an eigenstate of the z-component of angular momentum may be characterised by its rotational properties.

We first note that if we go over to spherical polar coordinates, (r, θ, ϕ) the defining equation for $U(Z_\alpha)$, (3.46), takes the simple form

$$U(Z_\alpha)\psi(r, \theta, \phi) = \psi(r, \theta, \phi - \alpha) \tag{3.51}$$

Now let ψ_{nlm} denote an eigenstate of angular momentum (n denotes other quantum numbers) of the form

$$\psi_{nlm}(r, \theta, \phi) = R_{nl}(r) Y_{lm}(\theta, \phi)$$

Then it follows from the form of the spherical harmonic function (and in particular its ϕ dependence) that

$$\psi_{nlm}(r, \theta, \phi - \alpha) = e^{-im\alpha}\psi_{nlm}(r, \theta, \phi)$$

hence (3.51) becomes

$$U(Z_\alpha)\psi_{nlm}(r, \theta, \phi) = e^{-im\alpha}\psi_{nlm}(r, \theta, \phi) \tag{3.52}$$

This also follows by using (3.50). So for an eigenstate of J_z the effect of a rotation about the z-axis is to multiply the state vector by a phase factor, i.e. ψ_{nlm} is an eigenstate of $U(Z_\alpha)$ with eigenvalue $e^{-im\alpha}$, which has unit modulus in accordance with the fact that $U(Z_\alpha)$ is a unitary operator.

If (3.52) holds for a state ψ then we can say ψ is an eigenstate of J_z with eigenvalue m. We shall have occasion to use this more general method of characterising an angular momentum eigenstate in chapter 4.

All that we have said so far has referred to a single particle. If we have several particles, then the wavefunction will contain the coordinates of all of them, and under a rotation of the frame of reference the coordinates of all the particles will undergo a rotation. It is clear that (3.46) defining the rotation operator should be replaced by

$$U(Z_\alpha)\psi(x_1, y_1, z_1; x_2, y_2, z_2, \ldots)$$
$$= \psi(Z_\alpha^{-1}(x_1, y_1, z_1), Z_\alpha^{-1}(x_2, y_2, z_2), \ldots)$$

The calculation leading to (3.49) now gives

$$U(Z_\alpha) = 1 - \mathrm{i}\frac{\alpha}{\hbar}(L_z^{(1)} + L_z^{(2)} + \ldots) + O(\alpha^2) \tag{3.53}$$

where $L_z^{(n)}$ denotes the z-component of angular momentum, of the nth particle,

$$L_z^{(n)} = -\mathrm{i}\hbar\left(x_n\frac{\partial}{\partial y_n} - y_n\frac{\partial}{\partial x_n}\right)$$

so the generator of rotations about the z-axis now becomes the sum of the z-components of angular momentum, in other words the z-component of the total angular momentum.

In the case of a particle with spin the rotation operator for a rotation Z is

$$U(Z_\alpha) = \mathrm{e}^{-\mathrm{i}\alpha(L_z + S_z)/\hbar} \tag{3.54}$$

and that for a rotation Y is

$$U(Y_\beta) = \mathrm{e}^{-\mathrm{i}\beta(L_y + S_y)/\hbar} \tag{3.55}$$

Because L and S commute, these may also be written as

$$U(Z_\alpha) = \mathrm{e}^{-\mathrm{i}\alpha L_z/\hbar}\,\mathrm{e}^{-\mathrm{i}\alpha S_z/\hbar}$$
$$U(Y_\beta) = \mathrm{e}^{-\mathrm{i}\beta L_y/\hbar}\,\mathrm{e}^{-\mathrm{i}\beta S_y/\hbar}$$

The forms of these operators may be understood by observing that to obtain the correctly rotated state from a given state we must rotate both the spatial and spin parts of the state vector. We shall illustrate the use of these operators in §3.4.

3.3.2 *Rotational invariance and angular momentum conservation*

In elementary quantum mechanics we can examine the Hamiltonian operator to see whether it is invariant under rotations. In nuclear or particle physics the problem is often to find the Hamiltonian or scattering operator by working back from experimental data. We therefore assume rotational invariance and see what restrictions this imposes on for example the scattering operator. Let us illustrate briefly both these approaches.

Invariance of a Hamiltonian H under rotations is expressed by

$$U(R)H = HU(R) \tag{3.56}$$

where R is any rotation of coordinates. By taking for R an infinitesimal rotation about the z-axis one may show that

$$J_z H = HJ_z$$

and a similar calculation shows

$$J_x H = HJ_x$$

$$J_y H = HJ_y$$

Altogether we have

$$[J, H] = JH - HJ = 0 \tag{3.57}$$

We may therefore say that rotational invariance implies that the three angular momentum operators J commute with the Hamiltonian, H. We may then take the eigenstates of H to be eigenstates of J^2 and J_z too, and all the usual consequences of angular momentum follow. Note that J here is the sum of the orbital angular momenta and spins of all the particles of the system.

Let us next consider a scattering or reaction process. According to the general discussion in §2.1.3, (3.56) implies that $U(R)$ commutes with the scattering operator $\mathcal{S}$,

$$U(R)\mathcal{S} = \mathcal{S}U(R) \tag{3.58}$$

Alternatively we may take this as the basic postulate of rotational invariance. (3.58) likewise implies

$$J\mathcal{S} = \mathcal{S}J \tag{3.59}$$

and hence in particular

$$J^2\mathcal{S} = \mathcal{S}J^2 \tag{3.60}$$

Consider a particular reaction symbolised by $a \to b$. If the initial state is ϕ_a and the final state is ϕ_b, the amplitude for this process is given by the $\mathcal{S}$-matrix element $(\phi_b, \mathcal{S}\phi_a)$.

What does rotational invariance imply for the $\mathcal{S}$-matrix element? On taking the ϕ_b–ϕ_a matrix element of (3.59) we obtain

$$(\phi_b, J\mathcal{S}\phi_a) = (\phi_b, \mathcal{S}J\phi_a)$$

If the states ϕ_a and ϕ_b are eigenstates of J^2 and J_z, we can conclude that the amplitude for the process $a \to b$ is zero unless ϕ_a and ϕ_b have equal total angular momenta and equal z-projection of the total angular momentum. If we display these eigenvalues explicitly by writing our states as ϕ_{ajm}, we can state this result as

$$(\phi_{bj'm'}, \mathcal{S}\phi_{ajm}) = \delta_{jj'}\delta_{m'm}(\phi_{bjm}, \mathcal{S}\phi_{ajm})$$

It is a further consequence of rotational invariance that $(\phi_{bjm}, \mathcal{S}\phi_{ajm})$ must be independent of m. In semi-classical terms, m describes the orientation of the angular momentum vector relative to a conventionally chosen z-direction but rotational invariance prohibits any dependence of the transition probability $|(\phi_b, \mathcal{S}\phi_a)|^2$ on such a quantity.

We may prove this formally by noting that $J_\pm$ commute with $\mathcal{S}$ so in particular

$$J_+\mathcal{S} = \mathcal{S}J_+$$

and on taking the ϕ_{bjm}–ϕ_{ajm-1} matrix element of this equation between eigenstates of angular momentum we find

$$(\phi_{bjm}, J_+\mathcal{S}\phi_{aj,m-1}) = (\phi_{bjm}, \mathcal{S}J_+\phi_{aj,m-1})$$

and since $J_+^\dagger = J_-$

$$(J_-\phi_{bjm}, \mathcal{S}\phi_{aj,m-1}) = (\phi_{bjm}, \mathcal{S}J_+\phi_{aj,m-1})$$

or using (3.47)

$$[(j+m)(j-m+1)]^{1/2}(\phi_{bj,m-1}, \mathcal{S}\phi_{aj,m-1})$$
$$= [(j-m+1)(j+m)]^{1/2}(\phi_{bjm}, \mathcal{S}\phi_{ajm})$$

which shows that the matrix element has the same value for different values of m, with the same j. To summarise we can write

$$(\phi_{bj'm'}, \mathcal{S}\phi_{ajm}) = \delta_{j'j}\delta_{m'm}\mathcal{S}^j_{ba} \tag{3.61}$$

We have assumed here that the states a and b were eigenstates of the total angular momentum. However, the initial state and the detected state in a scattering experiment are not eigenstates of total angular momentum: they are usually eigenstates of linear momentum (plane waves).

The process of relating the plane wave states to angular momentum eigenstates and thereby exploiting the consequences of rotational invariance in scattering and reaction processes is called the partial wave analysis. We shall discuss it in the next chapter. One of the tools we shall require there is a description of the effect of a finite rotation R on an angular momentum eigenstate ϕ_{jm}. We therefore discuss this next.

3.4 Representation of finite rotations

We have introduced the unitary operators $U(R)$ which represents the effect of a finite rotation R on the state vectors of a system. We now consider in detail how this operator acts on eigenstates of J^2 and J_z.

3.4.1 *The matrix elements of finite rotations*

The first step is to find a suitable set of parameters to describe the most general rotation of coordinates. This is the same problem as describing the motion of a rigid body with one point fixed. The Euler angles used in that problem prove convenient here.

We are going to use the rotations to rotate the wavefunction according to the active point of view. It is useful to visualise this process as analogous to rotating a rigid body: the wavefunction may be pictured as a solid filling space with variable density. The active rotation of the wavefunction corresponds to the rigid rotation of the solid in space around the origin. We imagine a set of cartesian axes fixed in the body, and a set fixed in space, which are initially coincident. A positive rotation about Ox moves Oy towards Oz, and cyclically for the other two axes.

The rotation with Euler angles α, β, γ is the transformation from Σ to Σ_f illustrated in fig. 3.2. It is defined by

(*a*) A positive rotation through an angle α about the z-axis sending $\Sigma(x, y, z)$ into $\Sigma'(x', y', z')$;

(*b*) A positive rotation through an angle β about the y'-axis sending $\Sigma'(x', y', z')$ into $\Sigma''(x'', y'', z'')$;

(*c*) A positive rotation through an angle γ about the z''-axis sending $\Sigma''(x'', y'', z'')$ into $\Sigma_f(x_f, y_f, z_f)$.

Other definitions of Euler angles exist, but the above is universal in quantum mechanics. α, β, γ constitute the three parameters specifying the final orientation of the body-fixed coordinate system Σ_f with respect to the space-fixed axes. We denote the rotation by $R(\alpha, \beta, \gamma)$.

The ranges of the angles are

$$0 \leqslant \alpha \leqslant 2\pi$$
$$0 \leqslant \beta \leqslant \pi$$
$$0 \leqslant \gamma \leqslant 2\pi$$

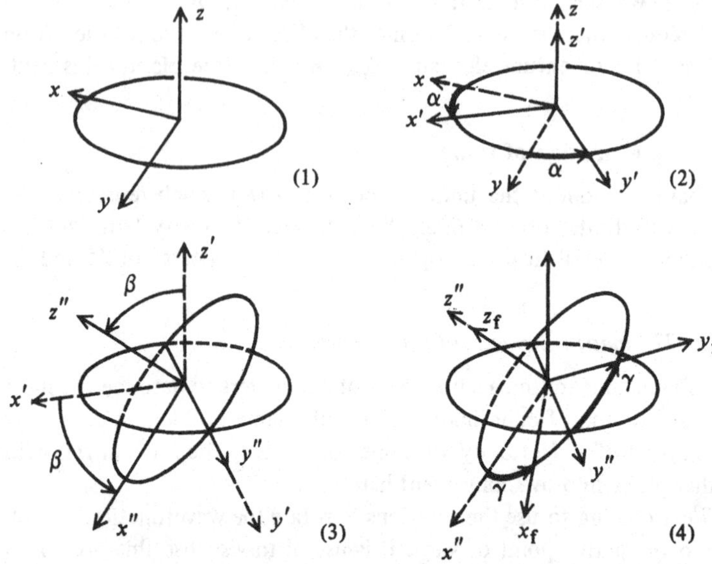

Fig. 3.2. Sequence of rotations used to build the general rotation of coordinates with Euler angles (α, β, γ). (1) shows the coordinate system Σ. (2), (3) and (4) show the operations (*a*), (*b*) and (*c*) described in the text.

If $\beta = 0$, the net rotation is a rotation about the original z-axis by an angle $\alpha + \gamma$. So in this case only the sum of α and γ has significance.

The operator corresponding to the rotation $R(\alpha, \beta, \gamma)$ is given by

$$\begin{aligned} U(R(\alpha, \beta, \gamma)) &= U(Z''_\gamma)U(Y'_\beta)U(Z_\alpha) \\ &= e^{-iJ''_z\gamma}\, e^{-iJ'_y\beta}\, e^{-iJ_z\alpha} \end{aligned} \tag{3.62}$$

where the primes indicate that the second and third rotations refer to rotated reference frames.

The second step in the calculation is to rewrite this formula so that everything is referred to the original unprimed space-fixed set of axes.

This is done as follows. J'_y is the generator of rotations about the axis y', which is the axis resulting from y by the rotation Z_α. Thus J'_y and its companions J'_z and J'_x have exactly the same properties with respect to the primed frame Σ', as J_y, J_z and J_x have for the frame Σ. It follows that in particular,

$$J'_y = U(Z_\alpha) J_y U(Z_\alpha)^\dagger \tag{3.63}$$

This is an instance of (2.30).

It follows from (3.63) that

$$e^{-i\beta J'_y} = e^{-i\alpha J_z}\, e^{-i\beta J_y}\, e^{+i\alpha J_z} \tag{3.64}$$

Similarly since J''_z refers to the axis obtained from the z-axis by successive rotations Z_α and Y'_β, we have

$$e^{-i\gamma J''_z} = e^{-i\beta J'_y}\, e^{-i\alpha J_z}\, e^{-i\gamma J_z}\, e^{+i\alpha J_z}\, e^{+i\beta J'_y} \tag{3.65}$$

Substituting (3.65) and (3.64) into (3.62) and making use of (3.64) once more, we find

$$U(R(\alpha, \beta, \gamma)) = e^{-i\alpha J_z}\, e^{-i\beta J_y}\, e^{-i\gamma J_z} \tag{3.66}$$

This important simple result is interpreted as follows: the sequence of rotations prescribed before has the same effect as (a') a rotation γ about the z-axis, followed by (b') a rotation β about the y-axis, followed by (c') a rotation α about the z-axis. In this prescription all the rotations refer to the axes in the *original space-fixed frame* $\Sigma(x, y, z)$.

We now apply the unitary rotation operator $U(R(\alpha, \beta, \gamma))$ to an angular momentum eigenfunction $\psi(j, m)$ whose properties with respect to the operators J_x, J_y and J_z were derived in §3.2.

Since $\psi(j, m)$ is an eigenstate of J_z we have

$$\begin{aligned} U(R(\alpha, \beta, \gamma))\psi(j, m) &= e^{-i\alpha J_z}\, e^{-i\beta J_y}\, e^{-i\gamma J_z}\, \psi(j, m) \\ &= e^{-i\alpha J_z}\, e^{-i\beta J_y}\, e^{-im\gamma}\, \psi(j, m) \end{aligned} \tag{3.67}$$

Now J_y is not diagonal with respect to the basis $\psi(j, m)$; however, the J_y has non-vanishing matrix elements only between states of the same j, so the most general possibility is that $e^{-i\beta J_y}\psi(j, m)$ is a linear combination of the states $\psi(j, m)$. The coefficients of this linear combination will be denoted by $d^j_{m'm}(\beta)$, so that

$$e^{-i\beta J_y}\, \psi(j, m) = \sum_{m'=-j}^{j} \psi(j, m')\, d^j_{m'm}(\beta) \tag{3.68}$$

The coefficients have been written last to emphasise the order of the indices. Combining (3.67) and (3.68), we obtain

$$U(R(\alpha, \beta, \gamma))\psi(j, m) = \sum_{m'=-j}^{j} \psi(j, m')\mathfrak{D}^{(j)}_{m'm}(\alpha, \beta, \gamma) \tag{3.69}$$

where we have put

$$\mathfrak{D}^{(j)}_{m'm}(\alpha, \beta, \gamma) = e^{-im'\alpha}\, d^j_{m'm}(\beta)\, e^{-im\gamma} \tag{3.70}$$

These functions are called the Wigner $\mathfrak{D}$-functions (or the 'curly $\mathfrak{D}$-functions'). They are most conveniently calculated by group-theoretical methods, and tables of them are available.

If we take the scalar product of both sides of (3.69) with the state $\psi(j, m')$, we obtain

$$(\psi(j, m'), U(R(\alpha, \beta, \gamma))\psi(j, m)) = \mathfrak{D}^{(j)}_{m'm}(\alpha, \beta, \gamma)$$

which shows that the $\mathfrak{D}$-functions are the matrix elements of the rotation operator with respect to the basis $\psi(j, m)$.

The $d^j_{m'm}(\beta)$ functions are the matrix elements in the basis of angular momentum eigenstates of the operator $e^{-i\beta J_y}$. Now we showed in §3.2 that J_y has purely imaginary matrix elements in this basis, so iJ_y and hence $e^{-i\beta J_y}$ have only purely real matrix elements. For a given j, $d^j_{m'm}(\beta)$ is a $(2j+1)$ by $(2j+1)$ matrix with real coefficients which are functions of β.

For example, for $j = \frac{1}{2}$ and $j = 1$

$$\begin{array}{cc} m = +\frac{1}{2} & -\frac{1}{2} \end{array}$$

$$d^{1/2}_{m'm}(\beta) = \begin{bmatrix} \cos\frac{1}{2}\beta & -\sin\frac{1}{2}\beta \\ \sin\frac{1}{2}\beta & \cos\frac{1}{2}\beta \end{bmatrix} \begin{array}{l} m' = +\frac{1}{2} \\ -\frac{1}{2} \end{array} \tag{3.71}$$

$$\begin{array}{ccc} m = +1 & 0 & -1 \end{array}$$

$$d^1_{m'm}(\beta) = \begin{bmatrix} \dfrac{1+\cos\beta}{2} & \dfrac{-\sin\beta}{2^{1/2}} & \dfrac{1-\cos\beta}{2} \\ \dfrac{\sin\beta}{2^{1/2}} & \cos\beta & \dfrac{-\sin\beta}{2^{1/2}} \\ \dfrac{1-\cos\beta}{2} & \dfrac{\sin\beta}{2^{1/2}} & \dfrac{1+\cos\beta}{2} \end{bmatrix} \begin{array}{l} m' = +1 \\ 0 \\ -1 \end{array} \tag{3.72}$$

To illustrate the use of these quantities by a simple example let us consider the transformation of the spin eigenfunctions of a particle of spin $\frac{1}{2}$. We assume the particle to be fixed at rest at the origin so that we need not consider any orbital motion.

Let us introduce a special notation for the spin eigenfunctions previously denoted by $\psi(j, m)$

$$\phi_+ = \psi(\tfrac{1}{2}, +\tfrac{1}{2}) \quad \phi_- = \psi(\tfrac{1}{2}, -\tfrac{1}{2})$$

These satisfy

$$\left.\begin{aligned} S_z\phi_+ &= +\tfrac{1}{2}\hbar\phi_+ \\ S_z\phi_- &= -\tfrac{1}{2}\hbar\phi_- \end{aligned}\right\} \tag{3.73}$$

The matrix of S_z with respect to the basis $\phi_\pm$ is simply the Pauli matrix σ_z multiplied by $\frac{1}{2}\hbar$. For this reason one also writes

$$\phi_+ = \begin{pmatrix} 1 \\ 0 \end{pmatrix} \quad \phi_- = \begin{pmatrix} 0 \\ 1 \end{pmatrix}$$

so that (3.73) may be regarded as explicit matrix equations. The general spin state of an $S = \frac{1}{2}$ particle may be written

$$\psi = a\phi_+ + b\phi_- = \begin{pmatrix} a \\ b \end{pmatrix}$$

and is called a spinor. It is normalised if $|a|^2 + |b|^2 = 1$.

We may find the spin eigenfunctions corresponding to a particle with spin component of $\pm\frac{1}{2}\hbar$ along an arbitrary direction in space, defined by polar angles (θ, ϕ), by applying a suitable rotation operator $U(R)$ to $\phi_\pm$. A rotation which rotates the z-direction into the direction (θ, ϕ) is conveniently given by $R(\phi, \theta, 0)$ in the Euler angle notation: a rotation through θ about the y-axis followed by a rotation through ϕ about the z-axis. Note that other choices are possible.

Then we have for $\lambda = \pm\frac{1}{2}$

$$\begin{aligned} U(R(\phi, \theta, 0))\phi_\lambda &= \sum_{\mu = \pm 1/2} \phi_\mu \mathfrak{D}^{(1/2)}_{\mu\lambda}(\phi, \theta, 0) \\ &= \sum_{\mu} \phi_\mu \, \mathrm{e}^{-\mathrm{i}\mu\phi} d^{1/2}_{\mu\lambda}(\theta) \end{aligned}$$

Making use of the explicit formulae (3.71) and denoting the rotated states by ϕ'_λ, we have

$$\begin{aligned} \phi'_+ &= \cos\tfrac{1}{2}\theta \, \mathrm{e}^{-\mathrm{i}\phi/2}\phi_+ + \sin\tfrac{1}{2}\theta \, \mathrm{e}^{\mathrm{i}\phi/2}\phi_- \\ \phi'_- &= -\sin\tfrac{1}{2}\theta \, \mathrm{e}^{-\mathrm{i}\phi/2}\phi_+ + \cos\tfrac{1}{2}\theta \, \mathrm{e}^{\mathrm{i}\phi/2}\phi_- \end{aligned}$$

In the column vector notation we may write

$$\phi'_+ = \begin{bmatrix} \cos\tfrac{1}{2}\theta \, \mathrm{e}^{-\mathrm{i}\phi/2} \\ \sin\tfrac{1}{2}\theta \, \mathrm{e}^{\mathrm{i}\phi/2} \end{bmatrix} \tag{3.74}$$

$$\phi'_- = \begin{bmatrix} -\sin\tfrac{1}{2}\theta \, \mathrm{e}^{-\mathrm{i}\phi/2} \\ \cos\tfrac{1}{2}\theta \, \mathrm{e}^{\mathrm{i}\phi/2} \end{bmatrix} \tag{3.75}$$

which are the normalised eigenfunctions of spin along the direction (θ, ϕ).

The appearance of the half angles is characteristic of a spin-$\frac{1}{2}$ system. It has the consequence that if we rotate through an angle of 2π about any axis (say the y-axis: corresponding to putting $\theta = 2\pi$, $\phi = 0$ in the above formulae) then

$$\phi'_{\pm} = -\phi_{\pm}$$

Although a rotation through 2π corresponds geometrically to 'no change', the rotation of spin-$\frac{1}{2}$ eigenfunctions through 2π leads to a change of sign. This phase factor is common to both spin states and does not lead to any observable consequences. One does not usually keep track of rotations through 2π, and if two or more rotations are applied to a spin-$\frac{1}{2}$ eigenfunction the effective rotation may amount to more than 2π radians in which case the state vector will pick up a minus sign. One speaks of the double-valuedness of the eigenfunctions for spin $\frac{1}{2}$.

A rotation through 4π leaves the sign unchanged.

These remarks on double valuedness apply to all states of half-integral j because of the form of the rotation matrices $\mathfrak{D}^{(j)}_{m'm}(\alpha, \beta, \gamma)$. For half-integral j the α and γ dependence is through $\frac{1}{2}\alpha$ and $\frac{1}{2}\gamma$, and the $d^j_{m'm}(\beta)$ involve odd or even powers of cos $\frac{1}{2}\beta$ and sin $\frac{1}{2}\beta$ according as j is half-integral or integral respectively.

3.4.2 *Group theory and rotations*

The purpose of this section is to introduce some terminology of group theory in the context of rotations.

We defined a rotation by its Euler angles but it is more convenient here to use the 3 by 3 matrix of coefficients giving the coordinates of the transformed point $\boldsymbol{r}'$ in terms of those of the original point $\boldsymbol{r}$. Using (x_1, x_2, x_3) instead of (x, y, z) we have

$$x'_i = \sum_j R_{ij} x_j$$

where all indices run over 1 to 3 as do all sums. The standard rotation (3.48) is denoted by the matrix

$$\begin{bmatrix} \cos\alpha & -\sin\alpha & 0 \\ \sin\alpha & \cos\alpha & 0 \\ 0 & 0 & 1 \end{bmatrix}$$

This matrix is orthogonal, that is it satisfies

$$RR^{\mathrm{T}} = R^{\mathrm{T}}R = 1 \tag{3.76}$$

where R^{T} denotes the transpose of R, with components

$$(R^{\mathrm{T}})_{ij} = R_{ji}$$

and 1 denotes the unit matrix δ_{ij}. R also has positive determinant

$$\det(R) = +1 \tag{3.77}$$

It is the content of Euler's theorem (of rigid body mechanics) that any matrix satisfying (3.76) and (3.77) corresponds to a rotation about some axis through some angle. Equation (3.76) expresses the facts that distances from the origin are preserved and orthogonal directions remain orthogonal under a rotation. Equation (3.77) excludes the transformation

$$P = \begin{bmatrix} -1 & 0 & 0 \\ 0 & -1 & 0 \\ 0 & 0 & -1 \end{bmatrix}$$

which is the space inversion operation, and all products of P with rotations. These may be considered separately without loss of generality.

The collection of all 3 by 3 rotation matrices forms a group called the rotation group. This is proved by verifying that the axioms given in §2.2 are satisfied in this case. If we perform one rotation R taking x into x'

$$x'_j = \sum_i R_{ji} x_i$$

followed by a second rotation S taking x' into x''

$$x''_k = \sum_j S_{kj} x'_j$$

then the resultant transformation may be written

$$x''_k = \sum_i T_{ki} x_i$$

where the transformation matrix is given by

$$T_{ki} = \sum_j S_{kj} R_{ji} \tag{3.78}$$

But T is an orthogonal matrix if R and S are, for remembering the transposition rule

$$(SR)^{\mathrm{T}} = R^{\mathrm{T}} S^{\mathrm{T}}$$

we have

$$T^{\mathrm{T}} T = (SR)^{\mathrm{T}} SR = R^{\mathrm{T}} S^{\mathrm{T}} SR$$

$$= R^{\mathrm{T}} R = 1$$

Similarly

$$T T^{\mathrm{T}} = 1$$

Likewise the useful theorem on the determinant of a product matrix

$$\det(SR) = \det(S)\cdot\det(R)$$

shows that

$$\det(T) = +1$$

So the rotation R followed by a rotation S is equivalent to a single rotation T whose parameters are in principle obtainable from those of R and S by means of (3.78). Equation (3.76) shows that the matrix of the rotation inverse to a rotation R is simply given by R^{T} since their product is the unit matrix 1 corresponding to no rotation (the identity of the rotation group).

Following the discussion of §2.2 the unitary operators $U(R)$ corresponding to the rotations R satisfy

$$U(S)U(R) = U(T)$$

if

$$SR = T$$

The $U(R)$ are therefore said to form a *representation* of the rotation group by unitary operators.

These operators act in the Hilbert space of states which is infinite-dimensional. If we choose a complete orthonormal basis of states in Hilbert space, then referred to this basis $U(R)$ is defined by an infinite matrix (as with any operator in the matrix representation of quantum mechanics).

We now introduce the ideas of invariant subspace and irreducible subspace for the operators $U(R)$. These general definitions apply to any group.

It may happen that by suitably choosing the basis, that one can pick out a subset V of the basis functions with the property that if an arbitrary rotation operator $U(R)$ is applied to any one of the basis functions ψ of V, then the resulting function $U(R)\psi$ is a linear combination of the basis functions of the set. Geometrically the set of basis functions of V span a subspace of the whole Hilbert space of states, and the condition just described means that if $U(R)$ (for any R) is applied to any vector lying in this subspace the resulting vector also lies wholly within this subspace. Such a subspace is called an *invariant subspace* for the operators $U(R)$ of the group. One may go on to see whether one can find a smaller subspace of the subspace V which has this property. If so, the original subspace V is said to be *reducible*. If not the subspace V is said to be *irreducible*.

Let us illustrate this terminology using the results of the rotation group. Suppose we choose the basis of states to be eigenstates of total angular momentum of its z-component, and denote them by $\psi(a, j, m)$. Here a represents the other labels. We have seen that if an arbitrary rotation is applied to $\psi(a, j, m)$, the resulting state is a linear combination of states of the same a and j

$$U(R)\psi(a, j, m) = \sum_{m'} \psi(a, j, m')\mathfrak{D}^{(j)}_{m'm}(R) \qquad (3.79)$$

Thus the set of $(2j + 1)$ states $\psi(a, j, m)$, $-j \leqslant m \leqslant +j$, spans an invariant subspace for the operators $U(R)$. This is true for every j. It is possible to show using group-theoretical techniques that such a set of states spans an irreducible subspace. This means it is not possible to pick new combinations of them less than $(2j + 1)$ in number, which transform only among themselves under all rotations.

As a concrete illustration, consider all the bound state eigenfunctions for a spinless electron in a Coulomb potential. The set of all eigenfunctions $\psi(n, l, m)$ belonging to a fixed value of n all have the same energy. The set of such states form an invariant subspace for the group of rotations but it is not irreducible. On the other hand the $(2l + 1)$ states $\psi(n, l, m)$ corresponding to fixed values of n and l spans an irreducible subspace.

Let us next define an irreducible representation. On applying a rotation operator $U(R)$ to each member of the basis $\psi(a, j, m)$ spanning an irreducible subspace, we obtain a matrix of coefficients $\mathfrak{D}^{(j)}_{m'm}(R)$ as in (3.79).

If we make a further rotation S on (3.69) then we have

$$\begin{aligned} U(S)U(R)\psi(a, j, m) &= U(S) \sum_{m'} \psi(a, j, m')\mathfrak{D}^{(j)}_{m'm}(R) \\ &= \sum_{m'} U(S)\psi(a, j, m')\mathfrak{D}^{(j)}_{m'm}(R) \\ &= \sum_{m'} \sum_{m''} \psi(a, j, m'')\mathfrak{D}^{(j)}_{m''m'}(S)\mathfrak{D}^{(j)}_{m'm}(R) \end{aligned} \qquad (3.80)$$

where we have introduced the matrix $\mathfrak{D}^{(j)}_{m''m'}(S)$ corresponding to the rotation S. But the operator on the left is simply $U(T)$ where $T = SR$, and

$$U(T)\psi(a, j, m) = \sum_{m''} \psi(a, j, m'')\mathfrak{D}^{(j)}_{m''m}(T) \qquad (3.81)$$

Comparing the last two equations we have

$$\mathfrak{D}^{(j)}_{m''m}(T) = \sum_{m'} \mathfrak{D}^{(j)}_{m''m'}(S)\mathfrak{D}^{(j)}_{m'm}(R) \qquad (3.82)$$

This shows that for a given j, the matrix corresponding to the product of two rotations is equal to the product of the matrices corresponding to the two separate rotations. For this reason the matrices $\mathfrak{D}^{(j)}_{m'm}(R)$ are said to be from a *representation* of the group of rotations, on the states $\psi(a, j, m)$ as basis. When as in this case the basis spans an irreducible subspace, the representation is said to be *irreducible*.

If R and S are both rotations about the z-axis one may see from (3.70) (with $\beta = \gamma = 0$) that (3.82) is satisfied. If R and S are both rotations about the y-axis say by β and β' respectively we obtain from (3.82) an identity satisfied by the $d^j_{m'm}$ matrices

$$d^j_{m''m}(\beta' + \beta) = \sum_{m'} d^j_{m''m'}(\beta')d^j_{m'm}(\beta) \qquad (3.83)$$

3.4.3 *Properties of the $d^j_{m'm}(\beta)$ functions*

For future convenience we tabulate here some of the properties of the rotation functions $d^j_{m'm}(\beta)$ which as we have seen are the only non-trivial ingredient in the definition of the functions $\mathfrak{D}^{(j)}_{m'm}(\alpha, \beta, \gamma)$. Additional properties and derivations of the relations given here may be found in the standard texts on the quantum theory of angular momentum, such as Rose (1957) or Edmonds (1957).

Symmetries

$$d^j_{\mu\lambda}(\beta) = d^j_{-\lambda,-\mu}(\beta) \qquad (3.84a)$$

$$= (-1)^{\mu-\lambda}d^j_{\lambda\mu}(\beta) \qquad (3.84b)$$

$$= (-1)^{\mu-\lambda}d^j_{-\mu,-\lambda}(\beta) \qquad (3.84c)$$

$$d^j_{\mu\lambda}(\pi - \beta) = (-1)^{j+\mu}d^j_{\mu,-\lambda}(\beta) \qquad (3.85a)$$

$$d^j_{\mu\lambda}(-\beta) = d^j_{\lambda\mu}(\beta) \qquad (3.85b)$$

$$d^j_{\mu\lambda}(\pi + \beta) = (-1)^{j+\mu}d^j_{-\mu,\lambda}(\beta) \qquad (3.85c)$$

Note that $j \pm \mu$ and $\mu \pm \lambda$ are always integers.

Special values

$$d^j_{\mu\lambda}(0) = \delta_{\mu\lambda} \qquad (3.86a)$$

$$d^j_{\mu\lambda}(\pi) = (-1)^{j-\lambda}\delta_{\mu,-\lambda} \qquad (3.86b)$$

$$d^j_{\mu\lambda}(-\pi) = (-1)^{j+\lambda}\delta_{\mu,-\lambda} \qquad (3.86c)$$

$$d^j_{\mu\lambda}(2\pi) = (-1)^{2j} \qquad (3.86d)$$

For $j = l =$ an integer,

$$d^{l}_{m0}(\beta) = (-1)^m \left(\frac{(l-m)!}{(l+m)!}\right)^{1/2} P_l^m(\cos\beta) \tag{3.87a}$$

$$d^{l}_{0m}(\beta) = \left(\frac{(l-m)!}{(l+m)!}\right)^{1/2} P_l^m(\cos\beta) \tag{3.87b}$$

$$d^{l}_{00}(\beta) = P_l(\cos\beta) \tag{3.87c}$$

where P_l and P_l^m denote the Legendre polynomial and associated Legendre function respectively.

From these formulae and (3.15) we have

$$\mathfrak{D}^{(l)*}_{m0}(\alpha, \beta, 0) = \left(\frac{4\pi}{2l+1}\right)^{1/2} Y_{lm}(\beta, \alpha) \tag{3.87d}$$

Orthogonality and completeness

$$\int_0^{\pi} d^{j_1}_{\mu\lambda}(\beta)\, d^{j_2}_{\mu\lambda}(\beta) \sin\beta\, d\beta = \frac{2}{2j_1+1}\delta_{j_1 j_2} \tag{3.88a}$$

$$\sum_j \frac{2j+1}{2} d^{j}_{\mu\lambda}(\beta)\, d^{j}_{\mu\lambda}(\beta') = \delta(\cos\beta - \cos\beta') \tag{3.88b}$$

3.5 Vector addition of angular momenta

3.5.1 *Clebsch–Gordan coefficients*

Suppose that we have a system in two parts a and b, each with angular momentum described by an angular momentum operator J_a or J_b. a and b might for example be the spins of two particles, or the spin and orbital angular momentum of a single particle. The angular momenta are assumed to be kinematically independent. This is expressed by

$$[J_a, J_b] = 0$$

which means that we can prescribe the eigenvalues of J_a^2 and J_{az}, and J_b^2 and J_{bz} separately. A possible state of the whole system which is an eigenstate of these four operators is then of the form

$$\psi_a(j_a, m_a)\psi_b(j_b, m_b)$$

We may also define the total angular momentum

$$J = J_a + J_b \tag{3.89}$$

the z-component of which is

$$J_z = J_{az} + J_{bz}$$

and we see that $\psi_a(j_a, m_a)\psi_b(j_b, m_b)$ is an eigenstate of J_z with eigenvalue $m_a + m_b$. If we introduce the square of the total angular momentum

$$J^2 = (J_a + J_b)^2 = J_a^2 + J_b^2 + 2J_a\ J_b \tag{3.90}$$

then we see that because of the last term, $\psi_a\psi_b$ may not be an eigenfunction of J^2: in general it will be a linear combination of eigenfunctions $\psi(j, m)$ of J^2 and J_z, with different values of j, but all having $m = m_a + m_b$. This is expressed by

$$\psi_a(j_a, m_a)\psi_b(j_b, m_b) = \sum_j G^{jm}_{j_a m_a j_b m_b}\psi(j, m) \tag{3.91}$$

Conversely a state vector $\psi(j, m)$ for the total system which is an eigenfunction of J^2 and J_z will contain components in which m_a and m_b have all pairs of values which add to give the required total m. Thus

$$\psi(j, m) = \sum_{m_a} C^{jm}_{j_a m_a j_b m_b}\psi_a(j_a, m_a)\psi_b(j_b, m_b) \tag{3.92}$$

Here m_b is fixed by m and m_a and no separate summation over m_b is required. If all the wavefunctions are assumed to be normalised and orthogonal, then we must have

$$\sum_j |G|^2 = 1 \quad \text{for given } j_a, j_b, m_a, m_b$$

and

$$\sum_{m_a} |C|^2 = 1 \quad \text{for given } j_a, j_b, j, m$$

C may be alternatively expressed as an overlap integral in the space of wavefunctions of the whole system. Taking the scalar product of (3.91) with $\psi(j, m)$ we have

$$G^{jm}_{j_a m_a j_b m_b} = (\psi(j, m), \psi_a(j_a, m_a)\psi_b(j_b, m_b))$$

Similarly

$$C^{jm}_{j_a m_a j_b m_b} = (\psi_a(j_a, m_a)\psi_b(j_b, m_b), \psi(j, m))$$

It follows from the general rule

$$(\psi_1, \psi_2) = (\psi_2, \psi_1)^*$$

that

$$G = C^*$$

We shall later show by construction that the C can be arranged to be real, and hence

$$G = C$$

We therefore drop the symbol G. The $C^{jm}_{j_am_aj_bm_b}$ are called Clebsch–Gordan coefficients or vector coupling coefficients.

Substitution of (3.92) into (3.91) and of (3.91) into (3.92) gives two relations which will be useful later

$$\sum_j C^{jm}_{j_am_aj_bm_b} C^{jm}_{j_am'_aj_bm'_b} \begin{cases} = 1 \text{ if } m_a = m'_a \\ = 0 \text{ if } m_a \neq m'_a \end{cases} \tag{3.93}$$

$$\sum_{m_a} C^{jm}_{j_am_aj_bm_b} C^{j'm}_{j_am_aj_bm_b} \begin{cases} = 1 \text{ if } j = j' \\ = 0 \text{ if } j \neq j' \end{cases} \tag{3.94}$$

3.5.2 *The method of calculating Clebsch–Gordan coefficients*

We show how to calculate the C–G. coefficients from first principles. This helps one to understand their significance. This process becomes laborious for large values of j. However tables of C–G. coefficients are available, and Wigner has derived a formula for the general C–G. coefficient by group-theoretical methods.

There are two keys to the calculation. The first is the use of a step operator to relate states of the same j and different m. We recall that in the treatment of a single angular momentum we made use of the shift operators $J_\pm$. In particular we note

$$\left.\begin{aligned} J_-\psi(j,m) &= \rho_-(j,m)\psi(j,m-1) \\ \rho_-(j,m) &= \hbar[(j+m)(j-m+1)]^{1/2} \end{aligned}\right\} \tag{3.95}$$

We therefore define for the total angular momentum J of (3.89), the shift operator which is simply related to the shift operator of the constituent parts,

$$J_- = J_{a-} + J_{b-} \tag{3.96}$$

The second key is the choice of starting state. Consider the state $\psi(j,m)$ with the largest possible $m = m_a + m_b$. This occurs when $m_a = j_a$, $m_b = j_b$ which are their maximum possible values. To allow this value of m, but no larger ones, j must be $j_a + j_b$.

On examining (3.92) we see that in this case the sum can contain only one term

$$\psi(j_a+j_b, j_a+j_b) = C^{j_a+j_b,\, j_a+j_b}_{j_aj_aj_bj_b}\, \psi(j_a, j_a)\psi_b(j_b, j_b) \tag{3.97}$$

Normalisation requires $|C|^2 = 1$, and by a choice of phase $C = 1$, hence

$$\psi(j_a+j_b, j_a+j_b) = \psi_a(j_a, j_a)\psi_b(j_b, j_b) \tag{3.98}$$

We now apply the shift operator (3.96) to both sides of (3.98). Using (3.95) which in this case gives

$$\rho_-(j, j) = (2j)^{1/2}$$

we have

$$[2(j_a + j_b)]^{1/2}\psi(j_a + j_b, j_a + j_b - 1)$$
$$= (2j_b)^{1/2}\psi(j_a, j_a)\psi_b(j_b, j_b - 1) + (2j_a)^{1/2}\psi_a(j_a, j_a - 1)\psi_b(j_b, j_b) \tag{3.99}$$

Comparing this with the defining equation (3.92) for $j = j_a + j_b, m = j_a + j_b - 1$, we may read off the Clebsch–Gordan coefficients

$$C^{j_a+j_b,\, j_a+j_b-1}_{j_a j_a,\, j_b j_b-1} = \left(\frac{j_b}{j_a + j_b}\right)^{1/2}, \quad C^{j_a+j_b,\, j_a+j_b-1}_{j_a j_a-1,\, j_b j_b} = \left(\frac{j_a}{j_a + j_b}\right)^{1/2}$$

We may verify that (3.94) is satisfied.

This process may be used to work right down the multiplet $j = j_a + j_b$, ending with

$$\psi(j_a + j_b, -j_a - j_b) = \psi_a(j_a, -j_a)\psi_b(j_b, -j_b) \tag{3.100}$$

which is simple again.

How do we find combinations of the $\psi_a\psi_b$ with $j = j_a + j_b - 1$? Equation (3.99) has furnished a single state with $m = j_a + j_b - 1$ as a linear combination of two states. Since the total number of states must be conserved in passing from the (m_a, m_b) labelling scheme to the (j, m) scheme, there must be an independent linear combination of those two states which again has $m = j_a + j_b - 1$.

Consider the expression

$$[2(j_a + j_b)]^{1/2}\psi(j_a + j_b - 1, j_a + j_b - 1)$$
$$= (2j_a)^{1/2}\psi_a(j_a, j_a)\psi_b(j_b, j_b - 1) - (2j_b)^{1/2}\psi_a(j_a, j_a - 1)\psi_b(j_b, j_b) \tag{3.101}$$

The coefficients on the right and the sign ensure that this state is orthogonal to (3.99), while the coefficient on the left ensures that ψ is normalised.

We shall not go through the labour of verifying that this state indeed has $j = j_a + j_b - 1$. Application of the shift operator J_- to this state produces a 'string' of states with $j = j_a + j_b - 1$ and at each stage the C–G. coefficients may be read off.

A sign convention was involved in the placing of the minus sign in (3.101). We shall not attempt to define the phase conventions, but

(3.101) is in accord with the Condon and Shortley phase conventions which are now almost universally used in the quantum theory of angular momentum. The commonly used 'Review of particle properties' also agrees with Condon and Shortley.

We shall not carry the general discussion further as the formulae would become unmanageable but it is possible to obtain in this way the eigenfunctions of total angular momentum with j-values down to the smallest value which can occur, $|j_a - j_b|$. Let us note finally that it is clear that all the C–G. coefficients will be real because they arise essentially from the matrix elements $\rho_-(j, m)$ of the shift operators J_{a-}, J_{b-} and J_- which were determined to be real in §3.2.

3.5.3 *An example:* $j_a = \frac{1}{2}$, $j_b = 1$

We carry through the method in detail for this case. The resulting C–G. coefficients may be tabulated in a square array with one column corresponding to each pair of values of j and m of the total quantum numbers, and one row for each pair of values of m_a and m_b. This is shown in table 3.2.

Table 3.2. *Clebsch–Gordan coefficients for* $j_a = \frac{1}{2}$, $j_b = 1$

		j:	$\frac{3}{2}$	$\frac{3}{2}$	$\frac{3}{2}$	$\frac{3}{2}$	$\frac{1}{2}$	$\frac{1}{2}$
m_a	m_b	m:	$+\frac{3}{2}$	$+\frac{1}{2}$	$-\frac{1}{2}$	$-\frac{3}{2}$	$+\frac{1}{2}$	$-\frac{1}{2}$
$+\frac{1}{2}$	$+1$		1					
$+\frac{1}{2}$	0			$(\frac{2}{3})^{1/2}$			$(\frac{1}{3})^{1/2}$	
$+\frac{1}{2}$	-1				$(\frac{1}{3})^{1/2}$			$(\frac{2}{3})^{1/2}$
$-\frac{1}{2}$	$+1$			$(\frac{1}{3})^{1/2}$			$-(\frac{2}{3})^{1/2}$	
$-\frac{1}{2}$	0				$(\frac{2}{3})^{1/2}$			$-(\frac{1}{3})^{1/2}$
$-\frac{1}{2}$	-1					1		

In this case (3.98) becomes

$$\psi(\tfrac{3}{2}, +\tfrac{3}{2}) = \psi_a(\tfrac{1}{2}, +\tfrac{1}{2})\psi_b(1, +1)$$

and (3.99) gives

$$3^{1/2}\psi(\tfrac{3}{2}, +\tfrac{1}{2}) = 2^{1/2}\psi_a(\tfrac{1}{2}, +\tfrac{1}{2})\psi_b(1, 0) + 1^{1/2}\psi_a(\tfrac{1}{2}, -\tfrac{1}{2})\psi_b(1, +1) \tag{3.102}$$

We recommend the rule of always adding a sign to the ms (unless zero) which helps to distinguish them from the js.

Applying $J_- = J_{a-} + J_{b-}$ to (3.102) we obtain on the right four states in principle, but $J_{a-}\psi(\frac{1}{2}, -\frac{1}{2})$ is zero, so we have

Table 3.3. *Clebsch–Gordan coefficients for* $j_a = j_b = \frac{1}{2}$

		j:	1	1	1	0
m_a	m_b	m:	$+1$	0	-1	0
$+\frac{1}{2}$	$+\frac{1}{2}$		1			
$+\frac{1}{2}$	$-\frac{1}{2}$			$(\frac{1}{2})^{1/2}$		$(\frac{1}{2})^{1/2}$
$-\frac{1}{2}$	$+\frac{1}{2}$			$(\frac{1}{2})^{1/2}$		$-(\frac{1}{2})^{1/2}$
$-\frac{1}{2}$	$-\frac{1}{2}$				1	

$$2\cdot 3^{1/2}\psi(\tfrac{3}{2},-\tfrac{1}{2}) = (1\cdot 2)^{1/2}\psi_a(\tfrac{1}{2},-\tfrac{1}{2})\psi_b(1,0) + (2\cdot 2)^{1/2}\psi_a(\tfrac{1}{2},+\tfrac{1}{2})\psi_b(1,-1) + (1\cdot 2)^{1/2}\psi_a(\tfrac{1}{2},-\tfrac{1}{2})\psi_b(1,0) \tag{3.103}$$

and hence

$$\psi(\tfrac{3}{2},-\tfrac{1}{2}) = (\tfrac{1}{3})^{1/2}\psi_a(\tfrac{1}{2},+\tfrac{1}{2})\psi_b(1,-1) + (\tfrac{2}{3})^{1/2}\psi_a(\tfrac{1}{2},-\tfrac{1}{2})\psi_b(1,0) \tag{3.104}$$

Finally as a check we apply J_- yet again to (3.103) or (3.104) to obtain

$$\psi(\tfrac{3}{2},-\tfrac{3}{2}) = \psi_a(\tfrac{1}{2},-\tfrac{1}{2})\psi_b(1,-1).$$

The respective orthogonal combinations of the pairs of states in (3.102) and (3.103) give the $j = \frac{1}{2}$ states

$$3^{1/2}\psi(\tfrac{1}{2},+\tfrac{1}{2}) = 1^{1/2}\psi_a(\tfrac{1}{2},+\tfrac{1}{2})\psi_b(1,0) - 2^{1/2}\psi_a(\tfrac{1}{2},-\tfrac{1}{2})\psi_b(1,+1) \tag{3.105}$$

$$3^{1/2}\psi(\tfrac{1}{2},-\tfrac{1}{2}) = 2^{1/2}\psi_a(\tfrac{1}{2},+\tfrac{1}{2})\psi_b(1,-1) - 1^{1/2}\psi_a(\tfrac{1}{2},-\tfrac{1}{2})\psi_b(1,0) \tag{3.106}$$

which completes the entries in table 3.2.

The calculation may be checked by means of the orthogonality relations (3.94) and (3.93).

Since $[\rho_-(j,m)]^2$ is an integer even if j is a half-integer, the C–G. coefficients are square roots of ratios of integers. Hence the convention is sometimes adopted of tabulating the square of the C–G. coefficient, as in the 'Review of particle properties'. This review also tabulates primarily by m and secondarily by j, so that the non-zero entries occur in square blocks down the diagonal.

It should be noted the C–G. coefficients depend on the order in which j_a and j_b are coupled; if we had taken $j_a = 1$, $j_b = \frac{1}{2}$ some signs are changed because of the steps leading to (3.105) and (3.106).

It can be shown that in general

Table 3.4. *Clebsch–Gordan coefficients for* $j_a = j_b = 1$

		j:	2	2	2	2	2	1	1	1	0
m_a	m_b	m:	+2	+1	0	−1	−2	+1	0	−1	0
+1	+1		1								
+1	0			$(\frac{1}{2})^{1/2}$				$(\frac{1}{2})^{1/2}$			
0	+1			$(\frac{1}{2})^{1/2}$				$-(\frac{1}{2})^{1/2}$			
+1	−1				$(\frac{1}{6})^{1/2}$				$(\frac{1}{2})^{1/2}$		$(\frac{1}{3})^{1/2}$
0	0				$(\frac{2}{3})^{1/2}$				0		$-(\frac{1}{3})^{1/2}$
−1	+1				$(\frac{1}{6})^{1/2}$				$-(\frac{1}{2})^{1/2}$		$(\frac{1}{3})^{1/2}$
0	−1					$(\frac{1}{2})^{1/2}$				$(\frac{1}{2})^{1/2}$	
−1	0					$(\frac{1}{2})^{1/2}$				$-(\frac{1}{2})^{1/2}$	
−1	−1						1				

$$C^{jm}_{j_b m_b j_a m_a} = (-1)^{j_a + j_b - j} C^{jm}_{j_a m_a j_b m_b} \qquad (3.107)$$

3.5.4 $j_a = j_b = \frac{1}{2}$ *or* 1

Tables 3.3 and 3.4 arise from calculations similar to those of the preceding example, for the addition of two equal angular momenta of magnitude $\frac{1}{2}$ or 1.

In cases where $j_a = j_b$, symmetry considerations can reduce the labour of calculation or alternatively serve as a check on the method described. The following rule of symmetry emerges from the procedure described in §3.5.2. When $j = 2j_a$ the signs are all positive and hence $\psi(2j_a, m)$ is a *symmetrical* combination of the constituent state vectors $\psi_a^{(1)}$ and $\psi_a^{(2)}$.

When we move to $\psi(2j_a - 1, m)$ minus signs are introduced (cf. (3.101)) and this state is *antisymmetric* in the constituents.

The symmetry character then alternates as we go through the possible values of j

$$\text{Even: } 2j_a,\ 2j_a - 2,\ 2j_a - 4, \ldots$$

$$\text{Odd: } 2j_a - 1,\ 2j_a - 3,\ 2j_a - 5, \ldots$$

This rule is exemplified in tables 3.3 and 3.4. It will be particularly useful when the whole algebra of coupling is applied to isospin, and used to discuss decays of mesons in chapter 8.

CHAPTER 4

LORENTZ INVARIANCE

It is supposed that the reader is familiar with the general ideas of the special theory of relativity as applied to macroscopic phenomena. Important among these ideas is the concept of Lorentz invariance by which we mean invariance under relativistic transformations between frames of reference in uniform relative motion. The basic postulates of the theory are assumed to be valid for microscopic phenomena and no evidence has as yet been found to challenge this assumption.

We start by reviewing the mathematical formalism of Lorentz transformations and four-vectors. The remainder of the chapter is concerned with Lorentz invariance in quantum mechanics.

We shall adopt units in which $\hbar$ and c are equal to 1.

4.1 Lorentz transformations and four-vector algebra

4.1.1 *Lorentz transformation equations*

We start by recalling the Lorentz transformation formulae relating the space and time coordinates of an event, for example the collision of two point particles, as measured with respect to two inertial frames of reference in relative motion. An inertial frame is one in which Galileo's law of inertia is valid.

When the two frames, denoted by Σ and Σ', are such that the (x, y, z)- and (x', y', z')-axes are respectively parallel and the relative motion is along the common z-axis with velocity u, the Lorentz transformation equations are

$$\left.\begin{aligned} x' &= x \\ y' &= y \\ z' &= (1-u^2)^{-1/2}(z+ut) \\ t' &= (1-u^2)^{-1/2}(t+uz) \end{aligned}\right\} \tag{4.1}$$

This transformation has the property that the quadratic combination

$$s^2 \equiv t^2 - x^2 - y^2 - z^2 \tag{4.2}$$

has the same value in both the primed and unprimed coordinate systems.

If (t_A, x_A, y_A, z_A) and (t_B, x_B, y_B, z_B) are the space–time coordinates of two events, the quantity

$$\Delta s = [(t_B - t_A)^2 - (x_B - x_A)^2 - (y_B - y_A)^2 - (z_B - z_A)^2]^{1/2} \tag{4.3}$$

is called the *interval* between the two events. It has the same value in any inertial frame. A quantity which has the same value in any inertial frame is called a *Lorentz scalar*.

The equations (4.1) will sometimes be referred to as the standard Lorentz transformation. Many of the results of relativistic invariance may be understood with the aid of this special case. However for a broader understanding we want to define a general Lorentz transformation and then the Lorentz group. The interval squared (4.2) may be used to give a compact definition of the general Lorentz transformation, as follows:

Any linear transformation of t, x, y, z among themselves which leaves s^2 invariant is a Lorentz transformation between two frames of reference. (4.4)

That this is reasonable can be seen as follows. If we have two frames Σ and Σ' whose relative velocity v lies in an arbitrary direction, the equations of Lorentz transformation between them take a more general form than (4.1). We shall not write it down since it is cumbersome and can be found in the standard text-books. However, the general transformation equations can be reduced to the form (4.1) by rotating the two frames so that their respective space axes are parallel and so that v lies along the common z-axis. This sort of consideration shows that Lorentz transformations and rotations have to be considered together.

It is characteristic of a rotation of coordinates that the quantity $x^2 + y^2 + z^2$ remains invariant while the time coordinate is unaffected (see the discussion of §3.4.2). Thus rotations of coordinates are included in the class of transformations which leave the interval invariant.

We now introduce a more compact notation as follows. We set

$$x_0 = t, \quad x_1 = x, \quad x_2 = y, \quad x_3 = z$$

and the set of four coordinates (x_0, x_1, x_2, x_3) will be collectively denoted by x_μ. Here and elsewhere Greek indices run from 0 to 3.

The quantity s^2 may be denoted compactly by

$$s^2 = \sum_{\mu\nu} g_{\mu\nu} x_\mu x_\nu$$

where the symbol $g_{\mu\nu}$ called the metric tensor takes the following values

$$g_{\mu\nu} = 0 \qquad \text{if} \quad \mu \neq \nu$$

$$g_{00} = +1, \quad g_{11} = g_{22} = g_{33} = -1$$

The reader should note that other index conventions exist. A common one is to use the imaginary time coordinate $x_4 = it$ and the interval $\sigma = is$ so that (4.2) is replaced by

$$\sigma^2 = x_1^2 + x_2^2 + x_3^2 + x_4^2$$

in which all the signs are positive. However, we have preferred to work with purely real quantities.

With this notation the square of the interval between two events (4.3) becomes

$$\begin{aligned} \Delta s^2 &= \sum_{\mu\nu} g_{\mu\nu}(x_{B\mu} - x_{A\mu})(x_{B\nu} - x_{A\nu}) \\ &= \sum_{\mu\nu} g_{\mu\nu}(x_{B\mu}x_{B\nu} + x_{A\mu}x_{A\nu} - 2x_{B\mu}x_{A\nu}) \end{aligned}$$

Introducing the notations

$$\begin{aligned} x_B \cdot x_A &= \sum_{\mu\nu} g_{\mu\nu} x_{B\mu} x_{A\nu} \\ x_A^2 &= x_A \cdot x_A \end{aligned}$$

we have

$$\Delta s^2 = x_B^2 + x_A^2 - 2x_B \cdot x_A$$

A general Lorentz transformation (4.4) may be written

$$x'_\mu = \sum_\nu \Lambda_{\mu\nu} x_\nu \tag{4.5}$$

The coefficients $\Lambda_{\mu\nu}$ form a 4×4 real matrix acting on the column vector x_ν. Thus the standard Lorentz transformation (4.1) is represented by the matrix

$$\begin{bmatrix} (1-u^2)^{-1/2} & 0 & 0 & u(1-u^2)^{-1/2} \\ 0 & 1 & 0 & 0 \\ 0 & 0 & 1 & 0 \\ u(1-u^2)^{-1/2} & 0 & 0 & (1-u^2)^{-1/2} \end{bmatrix}$$

The transformation (4.5) is a Lorentz transformation (4.4) provided

$$s^2 = x' \cdot x' = x \cdot x \tag{4.6}$$

This implies a condition on the matrix $\Lambda_{\mu\nu}$. On substituting (4.5) in (4.6) we have

$$\sum_{\mu\nu}\sum_{\rho\sigma} g_{\mu\nu}(\Lambda_{\mu\rho}x_\rho)(\Lambda_{\nu\sigma}x_\sigma) = \sum_{\alpha\beta} g_{\alpha\beta}x_\alpha x_\beta \tag{4.7}$$

Comparing coefficients of $(x_1)^2, x_1x_2$, etc., we obtain a set of conditions which may be compactly written

$$\sum_{\mu\nu} g_{\mu\nu}\Lambda_{\mu\rho}\Lambda_{\nu\sigma} = g_{\rho\sigma} \tag{4.8}$$

This is the condition on Λ that it should represent a Lorentz transformation. Equation (4.8) may be written in a matrix notation. We let Λ represent the matrix $\Lambda_{\nu\sigma}$, and introduce the matrix

$$G = g_{\mu\nu} = \begin{bmatrix} 1 & 0 & 0 & 0 \\ 0 & -1 & 0 & 0 \\ 0 & 0 & -1 & 0 \\ 0 & 0 & 0 & -1 \end{bmatrix}$$

Then (4.8) is the $\rho\sigma$ matrix element of the equation

$$\Lambda^{\mathrm{T}}G\Lambda = G \tag{4.9}$$

where Λ^{T} denotes the transposed matrix of Λ.

4.1.2 *The Lorentz group*

The Lorentz transformations between inertial frames form a group called the *Lorentz group.* Following the general argument of §2.2.2, if Σ, Σ' and Σ'' are three frames and $\Lambda^{(1)}_{\nu\sigma}$ denotes the transformation relating coordinates in Σ and Σ', and $\Lambda^{(2)}_{\mu\nu}$ that relating Σ' and Σ'', then the transformation matrix relating coordinates in Σ and Σ'' is

$$\Lambda^{(3)}_{\mu\sigma} = \sum_{\nu} \Lambda^{(2)}_{\mu\nu}\Lambda^{(1)}_{\nu\sigma} \tag{4.10}$$

and is called the product of $\Lambda^{(1)}_{\nu\sigma}$ and $\Lambda^{(2)}_{\mu\nu}$. Now if Σ and Σ' are equivalent frames of reference for physical laws, and if Σ' and Σ'' are similarly equivalent, then Σ and Σ'' must be equivalent, and so the transformation relating them is a Lorentz transformation. Mathematically, one shows that $\Lambda^{(3)}_{\mu\sigma}$ satisfies (4.8) and hence the Lorentz transformations form a group.

As we observed above, simple rotations of spatial coordinates have been included as transformations of the Lorentz group, in the mathematical sense. The criterion (4.8) for a Lorentz transformation has also included more general transformations of coordinates corresponding to space inversion and time reversal. In the space–time notation the operation of space inversion is

$$\left.\begin{aligned} x_0 \to x_0' &= +x_0 \\ x_1 \to x_1' &= -x_1 \\ x_2 \to x_2' &= -x_2 \\ x_3 \to x_3' &= -x_3 \end{aligned}\right\} \tag{4.11}$$

while time reversal is defined by

$$\left.\begin{aligned} x_0 \to x_0' &= -x_0 \\ x_1 \to x_1' &= +x_1 \\ x_2 \to x_2' &= +x_2 \\ x_3 \to x_3' &= +x_3 \end{aligned}\right\} \tag{4.12}$$

Both of these transformations clearly leave $(x_0)^2-(x_1)^2-(x_2)^2-(x_3)^2$ invariant.

We wish to exclude these transformations for the present. Their consequences as invariance transformations will be dealt with separately in chapters 5 and 6. There is no loss of generality in doing this.

When we wish to refer to a Lorentz transformation between spatial frames with no rotation of coordinates, we speak of a pure Lorentz transformation or a boost. Equation (4.1) is a boost.

4.1.3 *Four-vectors*

Let us now define a four-vector. The definition is exactly parallel to that for 'ordinary' three-vectors (see Feynman, Leighton and Sands, 1963). A four-vector denoted by V_μ is specified in one frame Σ by its four components $V_0V_1V_2V_3$, and on passing to a second frame Σ' its components transform in exactly the same way as the coordinates x_μ, thus

$$V_\mu' = \sum_\nu \Lambda_{\mu\nu} V_\nu \tag{4.13}$$

We shall also use the notation $V_\mu = (V_0, \boldsymbol{V})$.

Quantities (tensors) with more complicated transformation properties exist, but we shall not require them.

It follows from (4.8) that the 'square' of the four-vector V defined by

$$V \cdot V = (V_0)^2 - (V_1)^2 - (V_2)^2 - (V_3)^2 \tag{4.14}$$

has the same value in all Lorentz frames. Similarly it may be shown that if V_μ and W_μ are the two four-vectors the quantity

$$\begin{aligned} V \cdot W &= V_0 W_0 - V_1 W_1 - V_2 W_2 - V_3 W_3 \\ &= V_0 W_0 - \boldsymbol{V} \cdot \boldsymbol{W} \end{aligned} \tag{4.15}$$

has the same value in all Lorentz frames. It is called the scalar product of V_μ and W_μ.

The value of recognising the Lorentz transformation properties of a quantity, e.g. scalar, four-vector, or tensor, is that the Lorentz invariant formulation of physical laws such as those of mechanics may be done systematically. Furthermore, the algebra of four-vectors is a powerful calculational tool in relativistic kinematics, which is our next topic.

4.2 Relativistic kinematics

As a classical particle moves it traces out a world line in space–time. If we consider two infinitesimally close points on that world line the *interval* $\mathrm{d}s$ between them is related to the coordinate differences by

$$\mathrm{d}s^2 = (\mathrm{d}x_0)^2 - (\mathrm{d}x_1)^2 - (\mathrm{d}x_2)^2 - (\mathrm{d}x_3)^2 \tag{4.16}$$

$\mathrm{d}s$ is a Lorentz scalar and thus has a common value in all frames of reference. The increment $\mathrm{d}s$ in s along the world line is therefore a natural variable to label the points (events) along it.

The four-velocity of the particle at any point is defined as the rate of change of coordinate with respect to s along the world line

$$w_\mu = \frac{\mathrm{d}x_\mu}{\mathrm{d}s} \tag{4.17}$$

Because s is a scalar, w_μ is a four-vector like $\mathrm{d}x_\mu$. On dividing (4.16) by $\mathrm{d}s^2$ we see that the components of w_μ satisfy

$$w \cdot w = 1 \tag{4.18}$$

The relation to the ordinary velocity $\boldsymbol{v}$ with components

$$\frac{\mathrm{d}x_1}{\mathrm{d}x_0}, \quad \frac{\mathrm{d}x_2}{\mathrm{d}x_0}, \quad \frac{\mathrm{d}x_3}{\mathrm{d}x_0}$$

is

$$w_0 = \frac{1}{(1-v^2)^{1/2}}, \quad w = \frac{v}{(1-v^2)^{1/2}} \tag{4.19}$$

or

$$v = \left(\frac{w_1}{w_0}, \frac{w_2}{w_0}, \frac{w_3}{w_0}\right)$$

The four-momentum is defined by

$$p_\mu = mw_\mu \tag{4.20}$$

where m is a Lorentz scalar called the rest-mass of the particle.

From (4.18) the four-momentum satisfies

$$p^2 = p \cdot p = m^2 \tag{4.21}$$

The relativistic energy E and three-momentum $\boldsymbol{p}$ are contained in p_μ, thus

$$p_\mu = (E, \boldsymbol{p})$$

and from (4.21) we have the well-known relation

$$E^2 = \boldsymbol{p}^2 + m^2$$

In terms of the three-velocity

$$E = \frac{m}{(1-v^2)^{1/2}}, \quad \boldsymbol{p} = \frac{m\boldsymbol{v}}{(1-v^2)^{1/2}} \tag{4.22}$$

from which we may relate E and $\boldsymbol{p}$ with their non-relativistic counterparts by making the approximation $|\boldsymbol{v}| \ll 1$, in a well-known way. We note that the velocity $\boldsymbol{v}$ can be recovered from E and $\boldsymbol{p}$ by

$$\boldsymbol{v} = \frac{\boldsymbol{p}}{E} = \frac{\boldsymbol{p}}{p_0} \tag{4.23}$$

In relativistic mechanics the two separate conservation laws of energy and of three-momentum are united into a single conservation law of four-momentum. Thus, for an isolated system of N particles, we have

$$\sum_{r=1}^{N} p_\mu^{(r)} = \text{constant} \tag{4.24}$$

Just as in the case of classical mechanics, this conservation law can be derived from an invariance principle: in this case, invariance against displacements of the origin of coordinates in space, and of displacements in time. The important application of the conservation of four-momentum is to the kinematics of scattering, reaction and decay processes. The reason that purely classical considerations can be used is that in a typical experiment, as we have noted already, the particles observed before and after the interaction are in almost sharp eigenstates of momentum and energy. So the kinematical problem is equivalent to the classical one.

The application of Lorentz invariance to kinematics has been thoroughly treated elsewhere and we shall not consider it further. Instead we turn our attention to Lorentz invariance in quantum mechanics.

4.3 Lorentz invariance in quantum mechanics: spinless particles

We wish to formulate a quantum mechanical description of elementary particle processes which is consistent with the principles of Lorentz invariance. Until recently, the almost universally used method of doing this was by means of specific relativistic wave equations such as the Klein–Gordon equation for spin zero particles or the Dirac equation for spin one-half particles. The Dirac equation gives an excellent description of the interactions of the classical elementary particles (electron, photon, neutron) with one another via the electromagnetic field, and is an essential ingredient in the more accurate treatment by quantum electrodynamics. The study of such wave equations remains important for these purposes.

However, the last few years have seen the discovery of the rich spectrum of hadron resonances with high spins, which it is often desired to treat phenomenologically on an equal footing with the stable particles, e.g. the Ω^- with spin $\frac{3}{2}$ and stable against strong decay. Now the method of wave equations becomes cumbersome for high spin. On the other hand, within an $\mathcal{S}$-matrix formulation, only a detailed quantum description of freely moving particles is required. This can be wholly based on invariance arguments, and the resulting formalism which will be described in the remainder of this chapter treats in a uniform manner particles of any mass (including zero) and any spin.

The approach has its foundation in a mathematical paper by Wigner (1939) on the representations of the inhomogeneous Lorentz group (the group of Lorentz transformations and displacements in space and time). However it received an immense impetus from the work of Jacob and Wick (1959), who established a set of standard conventions which transformed the generality and rigour of Wigner's analysis into a form suitable for practical application. The resulting theory is often called the helicity formalism. We shall try to present sufficient detail for the reader to understand how it is used in some simple cases. For a more rigorous treatment we shall refer the reader elsewhere.

We want to work within the basis of momentum eigenstates and since these are characterised by their simple properties under translations, we start by considering relationships between the symmetry operations of displacements in space–time and the Lorentz transformations.

A translation of a reference frame in space–time is specified by the displacement four-vector a_μ. The corresponding coordinate transformation is

$$T_{a_\mu}: x_\mu \to x'_\mu = x_\mu + a_\mu \tag{4.25}$$

Thus a_0 is the displacement of the origin of the time coordinate, while $\boldsymbol{a}$ is the vector of spatial displacement.

The combination of a special Lorentz transformation with matrix $\Lambda_{\mu\nu}$ followed by a displacement by a_μ corresponds to the coordinate transformation

$$x'_\mu = a_\mu + \sum_\nu \Lambda_{\mu\nu} x_\nu \tag{4.26}$$

between reference frames Σ and Σ'. This is the most general change of reference frame with respect to which all the laws of physics remain invariant. The transformation (4.26) will be denoted by $\{a, \Lambda\}$.

The collection of all transformations (4.26) forms a group which is called the inhomogeneous Lorentz group or the *Poincaré group*. Space and time inversions are still excluded.

According to the general theory, there corresponds to each transformation (4.26) a unitary operator acting on the state vectors of the system. The unitary operator corresponding to the translation a_μ will be denoted by $U(a_\mu)$, and that corresponding to the Lorentz transformation $\Lambda_{\mu\nu}$ by $U(\Lambda_{\mu\nu})$. To the general transformation $\{a, \Lambda\}$, the operator

$$U(\{a, \Lambda\}) = U(a)\, U(\Lambda)$$

corresponds, since the operators U must multiply together in the same way as the corresponding coordinate transformations, i.e. the Us form a unitary operator representation of the Poincaré group.

We start by considering the space–time translations $U(a_\mu)$. We have already dealt with spatial translations in §2.3 and with time displacements at the end of §2.2. It remains to combine these results in a relativistic notation.

It is clear from (4.25) that a purely spatial translation by $\boldsymbol{a}$ and a pure time translation a_0 commute since they act on different coordinates. It follows that the corresponding unitary operators denoted by $U(\boldsymbol{a})$ and $U(a_0)$ also commute

$$U(a_0)\,U(a) = U(a)\,U(a_0) \tag{4.27}$$

We now introduce the generator of infinitesimal space–time displacement transformations by writing

$$U(a_\mu) = 1 + \mathrm{i}a\cdot P \tag{4.28}$$

to first order in a_μ. Here

$$a\cdot P = \Sigma\, g_{\mu\nu} a_\mu P_\nu = a_0 P_0 - \boldsymbol{a}\cdot\boldsymbol{P} \tag{4.29}$$

and on comparing with (2.57) and (2.52) we identify $\boldsymbol{P}$ as the momentum operator and P_0 as the Hamiltonian or energy operator H. On substituting (4.28) in (4.27) we have

$$(1+\mathrm{i}a_0 H)(1-\mathrm{i}\boldsymbol{a}\cdot\boldsymbol{P}) = (1-\mathrm{i}\boldsymbol{a}\cdot\boldsymbol{P})(1+\mathrm{i}a_0 H)$$

and hence

$$H\boldsymbol{P} = \boldsymbol{P}H \tag{4.30}$$

which states that the Hamiltonian and momentum operators commute. It follows that our basic states may be chosen to be simultaneous eigenstates of energy and momentum. We write the basic states as $\phi_{E\boldsymbol{p}}$; then we have

$$H\phi_{E\boldsymbol{p}} = E\phi_{E\boldsymbol{p}} \tag{4.31a}$$

$$\boldsymbol{P}\phi_{E\boldsymbol{p}} = \boldsymbol{p}\phi_{E\boldsymbol{p}} \tag{4.31b}$$

If we are dealing with a compound system, for example a deuteron, H and $\boldsymbol{P}$ are the operators of *total* energy and *total* momentum, and E and $\boldsymbol{p}$ are their eigenvalues. In this case the states might carry additional labels describing the internal properties of the system.

Now (4.31*a*) and (4.31*b*) are true both in non-relativistic and in relativistic quantum theory. We shall now show that if our theory is to be invariant under Lorentz transformations, then the eigenvalues E and $\boldsymbol{p}$ of each state of our system must satisfy

$$E^2 - \boldsymbol{p}^2 = m^2 \tag{4.32}$$

where m is a constant of the system which we shall identify as the mass of the particle, or for compound systems the total energy in the centre of momentum frame. Consider first a special case of the standard Lorentz transformations (4.1) between two frames Σ and Σ', in the form

$$\left.\begin{aligned} x'_0 &= (1-v^2)^{-1/2}\{x_0 + vx_3\} \\ x'_1 &= x_1 \\ x'_2 &= x_2 \\ x'_3 &= (1-v^2)^{-1/2}\{x_3 + vx_0\} \end{aligned}\right\} \quad (4.1a)$$

which we denote by Λ_0.

Now consider a state of the system which is assigned the state vector ϕ_{Ep} for the observer Σ. By analogy with a classical free particle we would expect that for the observer Σ' this state is described as having energy E' and momentum p' transformed in the same way as the coordinates,

$$\left.\begin{aligned} E' &= (1-v^2)^{-1/2}(E + vp_3) \\ p'_1 &= p_1 \\ p'_2 &= p_2 \\ p'_3 &= (1-v^2)^{-1/2}(p_3 + vE) \end{aligned}\right\} \quad (4.32)$$

and Σ' would assign the state vector $\phi_{E'p'}$ to the system. Accepting this argument for the moment, if we go over to the active point of view, then the state $\phi_{E'p'}$ is a *new* state of the system referred to the *original* frame Σ, and the relation between ϕ_{Ep} and $\phi_{E'p'}$ is

$$\phi_{E'p'} = U(\Lambda_0)\phi_{Ep}$$

In proving this result, the general case of an arbitrary Lorentz transformation is as easy to treat as any particular one. We require the formula

$$U(\Lambda)^{-1}P_\mu U(\Lambda) = \sum_\nu \Lambda_{\mu\nu}P_\nu \quad (4.33)$$

which expresses the fact that the four operators P_μ transform as a four-vector under Lorentz transformations. Equation (4.33) is obtained by observing that the general Poincaré transformation (4.26) may be accomplished by *first* performing a displacement

$$a'_\mu = \sum_\nu \Lambda^{-1}_{\mu\nu}a_\nu$$

followed by a Lorentz transformation $\Lambda_{\mu\nu}$, for

$$\begin{aligned} x'_\mu &= \sum_\sigma \Lambda_{\mu\sigma}\left(\sum_\nu (\Lambda^{-1})_{\sigma\nu}a_\nu + x_\sigma\right) \\ &= a_\mu + \sum_\sigma \Lambda_{\mu\sigma}x_\sigma \end{aligned}$$

Hence
$$T_a\Lambda = \Lambda T_{a'}$$
and so for operators
$$U(a)\,U(\Lambda) = U(\Lambda)\,U(a') \tag{4.34}$$
On substituting
$$U(a) = 1 + \mathrm{i}a\cdot P$$
$$U(a') = 1 + \mathrm{i}a'\cdot P$$

and making use of the Lorentz invariance of the dot product to write
$$a'\cdot P = (\Lambda a')\cdot(\Lambda P)$$
$$= a\cdot\Lambda P,$$
we can compare coefficients of a_μ in (4.34) we arrive at (4.33).

With the aid of (4.33) we can now show that the state obtained by applying $U(\Lambda)$ to $\phi_{E\boldsymbol{p}}$ is an eigenstate of H and $\boldsymbol{P}$ with eigenvalues E' and $\boldsymbol{p}'$ which are related to E and $\boldsymbol{p}$ by the Lorentz transformation $\Lambda_{\mu\nu}$.

Equation (4.33) can be rewritten
$$P_\mu U(\Lambda) = U(\Lambda)\,\{\sum_\nu \Lambda_{\mu\nu}P_\nu\}$$
and applying this to $\phi_{E\boldsymbol{p}}$ we have
$$P_\mu U(\Lambda)\,\phi_{E\boldsymbol{p}} = U(\Lambda)\,\{\sum_\nu \Lambda_{\mu\nu}p_\nu\}\,\phi_{E\boldsymbol{p}}$$
$$= \{\sum_\nu \Lambda_{\mu\nu}p_\nu\}\,U(\Lambda)\phi_{E\boldsymbol{p}} \tag{4.35}$$
since the number $\{\ldots\}$ can be commuted past the operator $U(\Lambda)$.

Equation (4.35) shows that $U(\Lambda)\,\phi_{E\boldsymbol{p}}$ is essentially the state $\phi_{E'\boldsymbol{p}'}$ where $(E'\boldsymbol{p}') = p'_\mu = \sum_\nu \Lambda_{\mu\nu}p_\nu$. On putting a possible normalisation constant equal to 1, we have
$$U(\Lambda)\,\phi_{E\boldsymbol{p}} = \phi_{E'\boldsymbol{p}'} \tag{4.36}$$
Since E' and $\boldsymbol{p}'$ are related to E and $\boldsymbol{p}$ by the Lorentz transformation matrix $\Lambda_{\mu\nu}$ it follows from the properties of a Lorentz transformation that
$$E'^2 - \boldsymbol{p}'^2 = E^2 - \boldsymbol{p}^2 = m^2 \tag{4.37}$$
where m is identified with the mass of the particle.

There is nothing to prevent m taking the value zero, in which case we obtain a relativistic quantum description of massless particles. We defer further consideration of massless particles to §4.5.

We now claim that if we take the set of states $\phi_{E\boldsymbol{p}}$ for all E and $\boldsymbol{p}$ subject to
$$E^2 - \boldsymbol{p}^2 = m^2 \tag{4.38}$$

as the basis states, then we have a quantum description of a particle of mass m consistent with Lorentz invariance. We shall show below that this particle has spin zero. The general state of the particle is of course, described by superposition of these basic states.

That the description proposed is Lorentz invariant can be seen as follows. Consider a frame of reference Σ and a state of the system to which the observer Σ assigns the state vector $\phi_{E\boldsymbol{p}}$. In a second frame Σ' related to the first by a Lorentz transformation matrix $\Lambda_{\mu\nu}$, the system will be assigned the state vector $\phi_{E'\boldsymbol{p}'}$, given by (4.36). If the two frames of reference are to be completely equivalent, $\phi_{E'\boldsymbol{p}'}$ must be a *possible* state of the system for the first observer Σ. But under the description proposed, $\phi_{E\boldsymbol{p}}$ for all E and $\boldsymbol{p}$ (subject only to (4.38)) are admissible states; hence such a state does exist for Σ and Lorentz invariance is proved.

Equation (4.36) may be used to relate states with different values of E and $\boldsymbol{p}$ (all referred to the same frame Σ). In particular we may relate all states of the basis to a reference state which can be taken as the rest state with $E = m, \boldsymbol{p} = 0$, i.e. ϕ_{m0}.

Before illustrating this process it is convenient to change our notation. Since E is determined in terms of $\boldsymbol{p}$ and the mass it can be omitted from the state vector. We then ought to carry the mass as a label, but it will be omitted. Also we shall sometimes specify the momentum vector in polar coordinate form $\boldsymbol{p} = (p, \theta, \phi)$, so we may write either $\phi_{\boldsymbol{p}}$ or $\phi_{p\theta\phi}$.

Now we may generate a state in which the particle is moving along the $+z$-direction with momentum p by applying a boost to the rest state now denoted by ϕ_{000}. Thus

$$\phi_{p00} = U(\Lambda)\,\phi_{000} \tag{4.39}$$

For Λ we must take a pure Lorentz transformation along the z-direction. Such a transformation is given by (4.32) with the substitutions $(E, \boldsymbol{p}) \to (m, 0)$ and $(E'\boldsymbol{p}') \to (E, \boldsymbol{p})$. The velocity v must be taken equal to p/E. Such a Lorentz transformation will be frequently referred to in what follows and we shall denote it by $\mathfrak{Z}_p$. So (4.39) becomes

$$\phi_{p00} = U(\mathfrak{Z}_p)\,\phi_{000} \tag{4.40}$$

To obtain a state in which the momentum $\boldsymbol{p}$ has direction (θ, ϕ) we simply apply a suitable pure rotation to the state ϕ_{p00}. The effect of rotation on momentum eigenstates was mentioned in §3.3.1. The rotation must take the momentum vector along z-axis into the direction (θ, ϕ). A suitable rotation is $R(\phi, \theta, 0)$ in the Euler angle notation.

A pure rotation does not change the magnitude of p, hence we have

$$\begin{aligned}\phi_{p\theta\phi} &= U(R(\phi,\theta,0))\,\phi_{p00} \\ &= U(R(\phi,\theta,0))\,U(\mathfrak{Z}_p)\,\phi_{000}\end{aligned} \tag{4.41}$$

Note that we could have taken a different rotation: for example, $R(\phi,\theta,\gamma)$ with an arbitrary γ would give the same result because the state ϕ_{p00} is invariant under rotations about the z-direction. This will no longer be true when we deal with particles with spin.

Let us now show that our description is appropriate to spinless particles. The assumption that we are dealing with a particle of spin zero was made when we stated that the particle at rest was in a unique state ϕ_{000}. For a particle of spin s however, there are $(2s+1)$ rest-states distinguished by the eigenvalue of S_z, and which transform among themselves under rotations. For the spinless particle rest-state we have

$$U(R)\,\phi_{000} = \phi_{000} \tag{4.42}$$

for all R.

We shall consider the description of a particle with spin in §4.4.

We have not checked that the operators $U(\Lambda)$ are unitary. The exact form of the unitarity condition is related to the normalisation of the states ϕ_{Ep}. The orthogonality and normalisation of the set of basic momentum states will be taken to be

$$(\phi_{p'},\phi_p) = (2\pi)^3 2E_p \delta^{(3)}(p'-p) \tag{4.43}$$

where we have introduced the abbreviation

$$E_p = (p^2+m^2)^{1/2}$$

As expected, the normalisation condition involves a δ-function because the momentum eigenvalue has a continuous range of values.

In non-relativistic theory the symbol on the left would be interpreted as the integral over the configuration space of the product of two wavefunctions (one of them complex-conjugated) which are functions of the configuration space variables. In the present case we have not given such a meaning, nor is it necessary; however, one may continue to refer to such a quantity as the overlap integral.

Let us show that with the orthonormality condition (4.43) the operator with the property (4.36) is unitary. We again consider the standard Lorentz transformation (4.32) but the result is valid for any Lorentz transformation $\Lambda_{\mu\nu}$.

We take two states ϕ_p and ϕ_q. Then

$$U(\Lambda)\phi_p = \phi_{p'}, \quad U(\Lambda)\phi_q = \phi_{q'}$$

where

$$p_1' = p_1, \quad p_2' = p_2, \quad p_3' = \Lambda_{33}p_3 + \Lambda_{30}E_p$$

$$E_{p'} = \Lambda_{00}E_p + \Lambda_{03}p_3$$

and

$$\Lambda_{33} = \Lambda_{00} = (1-v^2)^{-\frac{1}{2}}, \quad \Lambda_{30} = \Lambda_{03} = v(1-v^2)^{-\frac{1}{2}}$$

with a similar set of equations for q'.

$U(\Lambda)$ is unitary if

$$(U(\Lambda)\phi_q, U(\Lambda)\phi_p) = (\phi_q, \phi_p) \tag{4.44}$$

Now the left-hand side is simply

$$(\phi_{q'}, \phi_{p'}) = (2\pi)^3 2E_{p'}\delta^{(3)}(q'-p')$$

$$\delta^{(3)}(q'-p') = \delta(q_1'-p_1')\delta(q_2'-p_2')\delta(q_3'-p_3')$$

$$= \delta(q_1-p_1)\delta(q_2-p_2)\cdot\delta([E_q-E_p]\Lambda_{30}+[q_3-p_3]\Lambda_{33})$$

The last δ-function simplifies by using the rule for a δ-function of a function: if $f(x)$ is any continuous function of x which vanishes for $x=a, f(a)=0$, then

$$\delta(f(x)) = \frac{1}{\left|\left(\frac{\mathrm{d}f}{\mathrm{d}x}\right)_{x=a}\right|}\delta(x-a)$$

The argument of the last δ-function vanishes for $p_3 = q_3$, and

$$\frac{\partial}{\partial p_3}\{(E_q-E_p)\Lambda_{30}+(q_3-p_3)\Lambda_{33}\}$$

$$= -\frac{p_3}{E_p}\Lambda_{30}-\Lambda_{33}$$

$$= -\frac{E_{p'}}{E_p}$$

so that the last δ-function is simply

$$\frac{E_p}{E_{p'}}\delta(q_3-p_3)$$

Hence

$$(\phi_{q'}, \phi_{p'}) = (2\pi)^3 2E_{p'}\cdot\frac{E_p}{E_{p'}}\delta^{(3)}(q-p)$$

which is equal to the right side of (4.44) and so the operator $U(\Lambda)$ satisfying (4.36) is indeed unitary provided we normalise our states according to (4.43).

We end this section by establishing the normalisation and overlap integrals for an arbitrary one particle state. A general state Ψ is a superposition of the basic states ϕ_p, and this will be written as

$$\Psi = \frac{1}{(2\pi)^{3/2}} \int \frac{d^3p}{(2E_p)^{1/2}} a(p)\phi_p \tag{4.45}$$

Here again the 2π and the $2E_p$ factors are for convenience, and lead to simplifications in other formulae as we shall immediately see.

The complex conjugate of (4.45) is

$$\Psi^* = \frac{1}{(2\pi)^{3/2}} \int \frac{d^3p}{(2E_p)^{1/2}} a^*(p)\phi_p^* \tag{4.46}$$

and the normalisation integral is

$$(\Psi, \Psi) = \frac{1}{(2\pi)^3} \int \frac{d^3p}{(2E_p)^{1/2}} a^*(p) \int \frac{d^3p'}{(2E_{p'})^{1/2}} a(p')(\phi_p, \phi_{p'})$$

On substituting (4.43) on the right, we obtain

$$(\Psi, \Psi) = \int d^3p\, |a(p)|^2 \tag{4.47}$$

If the integral on the right is normalised to unity, we may interpret $|a(p)|^2 d^3p$ as the probability for the state Ψ of finding the particle with momentum in the range $(p, p + d^3p)$.

4.4 Particles with spin

All that has been done in the preceding section remains valid for particles with spin. However, the state vector for a particle with spin must carry an additional label λ describing the orientation of the spin.

The label which should be present to indicate the magnitude of the spin will be suppressed when there is no ambiguity. Thus we write our basic states as $\phi_{p\lambda}$ or $\phi_{p\theta\phi\lambda}$.

We start by considering the rest-states with $p = 0$ (we are still dealing with a massive particle). For our basic rest-states $\phi_{000\lambda}$, λ is taken to be the component of spin along the z-direction of the frame Σ in which we are working. We introduce the total angular momentum operator $\boldsymbol{J}$.

In the rest-state there is no orbital contribution and so $\phi_{000\lambda}$ satisfies the equations

$$J^2\phi_{000\lambda} = s(s+1)\phi_{000\lambda}$$

$$J_3\phi_{000\lambda} = \lambda\phi_{000\lambda} \tag{4.48}$$

The shift operators $J_\pm$ relate the states with different values of λ

$$J_\pm\phi_{000\lambda} = [(s\mp\lambda)(s\pm\lambda+1)]^{1/2}\phi_{000\lambda\pm 1} \tag{4.49}$$

and the standard phase convention described in chapter 3 is adopted.

The $(2s+1)$ states $\phi_{000\lambda}$ form a complete set of rest-states.

We now define states $\phi_{p\theta\phi\lambda}$ with arbitrary momentum following the procedure used in §4.3 for a spinless particle: a state $\phi_{p00\lambda}$ in which the momentum is directed along the z-direction is obtained by applying a pure Lorentz transformation $\mathfrak{Z}_p$.

So

$$\phi_{p00\lambda} = U(\mathfrak{Z}_p)\phi_{000\lambda} \tag{4.50}$$

This defines the state $\phi_{p00\lambda}$, and λ still denotes the eigenvalue of the z-component of spin. To prove this we observe that a rotation Z_α *about* the z-axis commutes with a Lorentz transformation $\mathfrak{Z}_p$ *along* the z-axis,

$$Z_\alpha\mathfrak{Z}_p = \mathfrak{Z}_p Z_\alpha \tag{4.51}$$

because Z_α acts only on the x- and y-coordinates while $\mathfrak{Z}_p$ acts on the z- and t-coordinates. In the now familiar way it follows that

$$U(Z_\alpha)U(\mathfrak{Z}_p) = U(\mathfrak{Z}_p)U(Z_\alpha)$$

Now

$$U(Z_\alpha) = e^{-i\alpha J_3}$$

and hence

$$J_3U(\mathfrak{Z}_p) = U(\mathfrak{Z}_p)J_3 \tag{4.52}$$

Using (4.48) we have

$$\begin{aligned} J_3\phi_{p00\lambda} &= J_3U(\mathfrak{Z}_p)\phi_{000\lambda} \\ &= U(\mathfrak{Z}_p)J_3\phi_{000\lambda} \\ &= \lambda U(\mathfrak{Z}_p)\phi_{000\lambda} \end{aligned}$$

Hence

$$J_3\phi_{p00\lambda} = \lambda\phi_{p00\lambda} \tag{4.53}$$

Note that a state with p along Oz has no component of orbital angular momentum along Oz so the J_3 eigenvalue is the same as that of the spin component.

The *helicity* of a particle is defined as the projection of the total angular momentum of the particle along the direction of motion; that is, as the eigenvalue of

$$\frac{\boldsymbol{J}\cdot\boldsymbol{P}}{|\boldsymbol{P}|} \tag{4.54}$$

which for eigenstates of $\boldsymbol{P}$, is the same thing as $\boldsymbol{J}\cdot\hat{\boldsymbol{p}}$ where $\hat{\boldsymbol{p}}$ is a unit vector along the momentum direction. Thus the eigenstate $\phi_{p00\lambda}$ as defined above has helicity λ. We note that the helicity operator is invariant under rotations, being the scalar product of two vector operators.

To define the $(2s+1)$ spin states of the particle with momentum in an arbitrary direction (θ, ϕ), we simply apply a rotation $U(R(\phi, \theta, 0))$ to the states $\phi_{p00\lambda}$. Thus we have

$$\phi_{p\theta\phi\lambda} = U(R(\phi, \theta, 0))\, U(\mathcal{Z}_p)\, \phi_{000\lambda} \tag{4.55}$$

Because the helicity operator is a scalar, $\phi_{p\theta\phi\lambda}$ so defined has the same eigenvalue of helicity as does the unrotated state $\phi_{p00\lambda}$, viz. λ. Thus the label λ on the state vector $\phi_{p\theta\phi\lambda}$ can be identified as the component of spin along the direction of motion.

We have defined a complete set of states $\phi_{p\theta\phi\lambda}$ for a particle of mass $m \neq 0$ and spin s. For each momentum λ is the component of spin along the direction of motion, and takes values in the range $-s$ to $+s$. For a particle at rest, λ is the spin projection along the z-axis. The normalisation of the basic states is

$$(\phi_{p'\lambda'}, \phi_{p\lambda}) = (2\pi)^3 2E_p \delta^{(3)}(\boldsymbol{p}' - \boldsymbol{p})\,\delta_{\lambda'\lambda} \tag{4.56}$$

and the general state of the particle is

$$\Psi = \frac{1}{(2\pi)^{3/2}} \int \frac{d^3p}{(2E_p)^{1/2}} \sum_{\lambda=-s}^{s} a_\lambda(\boldsymbol{p})\, \phi_{p\lambda} \tag{4.57}$$

$a_\lambda(\boldsymbol{p})$ is the amplitude for finding the particle with momentum $\boldsymbol{p}$ and helicity λ in the state Ψ.

The identification of the quantum number λ as the helicity for the basis state $\phi_{p\lambda}$ rests on the sequence of operations ($\mathcal{Z}_p$ followed by $R(\phi, \theta, 0)$) used to 'generate' the general state from the rest-state. For example we could have chosen to generate the state $\phi_{p\lambda}$ by applying a pure Lorentz transformation in the direction $\boldsymbol{P}$ to the state $\phi_{000\lambda}$. In this case λ does not have the simple interpretation as the helicity although we obtain a perfectly good set of $(2s+1)$ states of the particle with momentum $\boldsymbol{P}$. The method we have used is due to Jacob and Wick

(1959) and the resulting description of particle states is referred to as the 'helicity formalism'.

We have deviated slightly from the original conventions of Jacob and Wick who used a rotation $R(\phi, \theta, -\phi)$. We have interpreted the transformations (4.55) in the active sense to generate the state $\phi_{p\theta\phi\lambda}$ from $\phi_{000\lambda}$, but given a particle in the state $\phi_{p\theta\phi\lambda}$ we may imagine going to the rest-frame Σ_0 of the particle. This is accomplished by inverting (4.55). λ is unchanged by this transformation and is therefore sometimes referred to as the z-projection of spin in the rest-frame of the particle.

We end this section with some remarks on spin which will serve to illustrate the formalism developed above. The detailed application to the description of scattering and decay will be presented in subsequent sections.

We sometimes require to describe a particle in an eigenstate of spin along a direction other than its direction of motion. This can be done in the helicity formalism by writing such states as superpositions of helicity states.

Consider a particle of spin $s = \frac{1}{2}$. For the particle at rest, there are two states $\phi_{0\pm 1/2}$ which we write as $\phi_{0\pm}$.

The spin operators S_x, S_y, S_z referred to these states are given in terms of the Pauli matrices

$$S = \tfrac{1}{2}\hbar\boldsymbol{\sigma}$$

The eigenstates of S_x with eigenvalues $\pm\frac{1}{2}\hbar$ are given by the superpositions

$$2^{-1/2}(\phi_{0+} \pm \phi_{0-})$$

of the basic states. On applying the standard helicity transformation (4.55) we obtain

$$2^{-1/2}(\phi_{p+} \pm \phi_{p-}) \tag{4.58}$$

which, following the remarks above, are interpreted as the states of the particle which *in its rest-frame* are eigenstates of S_x. Alternatively we may imagine a set of x, y, z-axes attached to the particle state ('body fixed axes') which are boosted by $\mathcal{Z}_p$ and then rotated by $R(\phi, \theta, 0)$ to produce the frame x', y', z'. The states (4.58) are then eigenstates of $S_{x'}$. This is illustrated in fig. 4.1.

4.5 Massless particles

The study of the Lorentz transformation properties of particle states discussed above can be extended very easily to particles of zero mass, and enables one to understand the special features of this case.

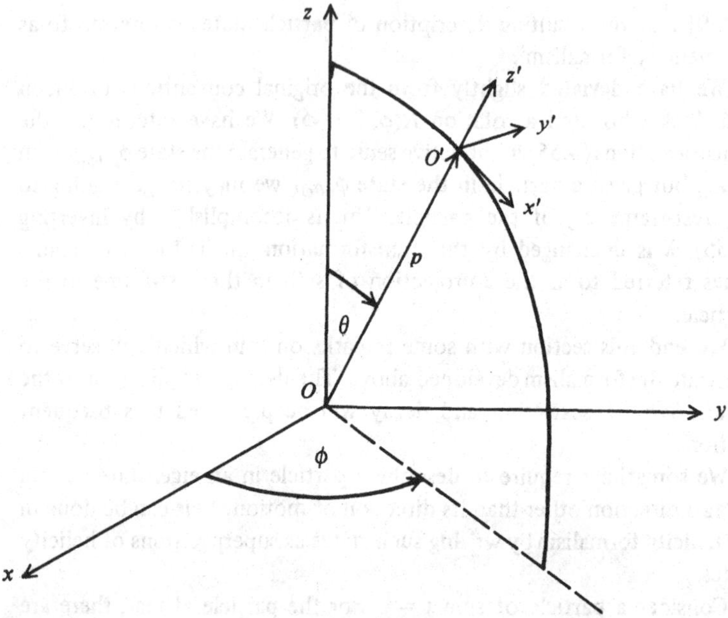

Fig. 4.1. Definition of the helicity frame *Ox'y'z'* for a particle of momentum *p*. For clarity the helicity frame is located at the tip of the corresponding momentum vector.

The first point of difference is that there is no rest-frame for a massless particle; in every frame of reference its velocity is equal to c. Hence in enumerating the states of a massless particle we cannot take the rest-state as the reference state as we did for a massive particle. It is necessary to choose some other simple state, for example, that in which the particle has its momentum directed along the z-axis with magnitude equal to a reference value, $\mathring{p}$. States with other momentum values, but still directed along the z-axis are obtained by applying a Lorentz boost to this state.

Thus

$$\phi_{p00} = U(\mathfrak{Z})\,\phi_{\mathring{p}00}$$

Then a state with momentum in the direction (θ, ϕ) is obtained by rotation

$$\phi_{p\theta\phi} = U(R(\phi, \theta, 0))\,U(\mathfrak{Z})\,\phi_{\mathring{p}00} \qquad (4.59)$$

The parameter in the Lorentz transformation $\mathfrak{Z}$ must be chosen to transform the four-momentum $(\mathring{p}, 0, 0, \mathring{p})$ into the four-momentum $(p, 0, 0, p)$.

We have not yet mentioned spin. The only known massless particles are the photon and the neutrino and both of these carry spin angular momentum. The most important feature of massless particles for the present discussion is the following: a massless particle 'of spin s' has instead of the expected $(2s+1)$ substates, only *one*, or under special circumstances, *two* substates of angular momentum. In order to understand the reason for this we follow Wigner and ask the question: why does a massive particle have $(2s+1)$ angular momentum substates? We can answer this on the basis of the discussion given in the last section. Lorentz invariance requires that there be a physically equivalent frame in which the particle is at rest, and rotational invariance applied in turn in that frame requires that there be $(2s+1)$ substates (this follows from the considerations of chapter 3). Now, when we turn to a massless particle there is no rest-state so this argument is not applicable. Instead if we agree to start with our reference state $\phi^{\circ}_{\mathfrak{p}00\lambda}$ and label the spin substates by the helicity λ, that is, the eigenvalue of the helicity operator

$$\frac{\boldsymbol{J}\cdot\boldsymbol{P}}{|\boldsymbol{P}|}$$

then the states obtained by boosting this state by $\mathfrak{Z}$ will have the same helicity, and any state obtained by rotating the momentum into a general direction will again have the same helicity eigenvalue

$$\phi_{p00\lambda} = U(\mathfrak{Z})\,\phi^{\circ}_{\mathfrak{p}00\lambda} \tag{4.60}$$

$$\phi_{p\theta\phi\lambda} = U(R(\phi,\theta,0))\,U(\mathfrak{Z})\,\phi^{\circ}_{\mathfrak{p}00\lambda} \tag{4.61}$$

Thus for a massless particle *the helicity is a Lorentz invariant*: it has the same value for all observers. This means that there is no requirement from Lorentz and rotational invariance for the existence of $(2s+1)$ spin states in the case of a massless particle. The argument we have just given may be placed on a sound mathematical basis, but we shall not do so here.

It should be noted that for a massless particle moving along the $+z$-axis in a reference frame Σ, we cannot move to a new frame Σ' which has 'overtaken' the particle. For a massive particle on the other hand this is possible and has the effect that from Σ' the particle appears to be moving along the negative z-axis. If the spin is aligned along $+z$, then it remains aligned along $+z$ for Σ', and so a positive helicity in Σ appears as a negative helicity $-\lambda$ in Σ', when Σ and Σ' are related in this way. Although we cannot use this device to change the helicity of a massless particle, there is another argument which implies the existence of more than one helicity state for a massless particle. This involves the space inversion or parity operation, and hence it can be invoked only if the interactions in which the particle participates conserve parity.

The photon interacts only via the electromagnetic interaction, which appears to conserve parity, and correspondingly we find two helicity states of the photon, $\lambda = \pm 1$, which have equal interactions with charged particles. Because $|\lambda| = 1$, we say that the photon has spin 1. The photon states

$$\phi^{\gamma}_{p00,\pm 1} \tag{4.62}$$

describe the quanta of the two polarisation states of a plane electromagnetic wave with vector potential

$$A_{\pm} = \mp 2^{-1/2}(e_x \pm ie_y)e^{ipz-ipt} \tag{4.63}$$

where e_x and e_y are unit vectors along the coordinate axes. Equation (4.63) with the upper sign describes a wave in which the electric vector rotates around the direction of motion according to the right-hand rule, and is therefore called a right circularly polarised wave. Correspondingly $\phi^{\gamma}_{p00,+1}$ is called a right circularly polarised photon state. This convention is opposite to that used in classical optics, where the wave described by A_+ is called left circularly polarised.

The neutrinos appear to interact only via the weak interaction which does not conserve parity. Correspondingly we find only one helicity state of the neutrino. There is a complication in that the neutrino carries a generalised charge (lepton number), so the above argument applies only to a neutrino of definite lepton number. This situation will be discussed further in §5.4.3.

4.6 Expansion of two particle helicity states in eigenstates of angular momentum

We wish to apply the helicity formalism to reactions such as

$$a + b \to c + d$$

and decays

$$A \to a + b$$

in which the particles may have spin. We have seen that as a consequence of rotational invariance the $\mathcal{S}$-matrix is diagonal in the angular momentum representation.

To exploit this we must be able to expand the momentum eigenstates of two particles in eigenstates of total angular momentum.

4.6.1 *Two particle states*

A state of two free particles a and b with momenta $\boldsymbol{p}_a$ and $\boldsymbol{p}_b$, and helicities λ_a and λ_b, may be written as a product

$$\phi^a_{\boldsymbol{p}_a\lambda_a}\phi^b_{\boldsymbol{p}_b\lambda_b} \tag{4.64}$$

of two constituent states of the type we have already discussed.

We shall be particularly interested in states of zero total linear momentum, that is, centre of momentum (CM) states of the two particles. CM states are the analogue for the two particle system of the rest-states for a single particle, and more general states of motion can be generated from CM states by applying Lorentz transformation operators.

A CM state may be labelled by the magnitude and polar angles of the CM momentum of say, particle a,

$$\boldsymbol{p} = \boldsymbol{p}_a = -\boldsymbol{p}_b$$

together with the helicities λ_a and λ_b, and is denoted by

$$\phi^{ab}_{p\theta\phi\lambda_a\lambda_b}$$

Let us relate such states to the product states (4.64).

For p directed along the positive z-axis we take

$$\phi^{ab}_{p00\lambda_a\lambda_b} = \phi^a_{p00\lambda_a}\hat{\phi}^b_{p\pi0\lambda_b} \tag{4.65}$$

The states on the right are essentially those defined in (4.55), but some attention to phase factors is necessary.

For particle a travelling along $+Oz$, we take $\theta = \phi = 0$ in (4.55), but for particle b travelling along $-Oz$, we first boost using $U(\mathcal{Z}_p)$ and then rotate the momentum vector into the negative z-direction. We must set $\theta = \pi$, but the value of ϕ must be chosen by convention. Different values of ϕ only change the state by a phase factor. Following Wick, we take

$$\hat{\phi}^b_{p\pi0\lambda_b} = \mathrm{e}^{-i\pi s_b}U(R(\pi,\pi,0))\,U(\mathcal{Z}_p)\,\phi^b_{000\lambda_b} \tag{4.66}$$

This defines $\hat{\phi}$. The reason for this choice is that if we take the limit $p \to 0$, it can be shown that the state $\hat{\phi}^b_{p\pi0\lambda_b}$ reduces to $\phi^b_{000,-\lambda_b}$,

$$\lim_{p\to 0}\hat{\phi}^b_{p\pi0\lambda_b} = \phi^b_{000,-\lambda_b} \tag{4.67}$$

The minus sign can be understood by reference to fig. 4.2. The caret serves as a reminder that when $\boldsymbol{p}$ points along the negative z-direction we do not take (4.55) with $\theta = \pi$, $\phi = 0$ but instead use (4.66).

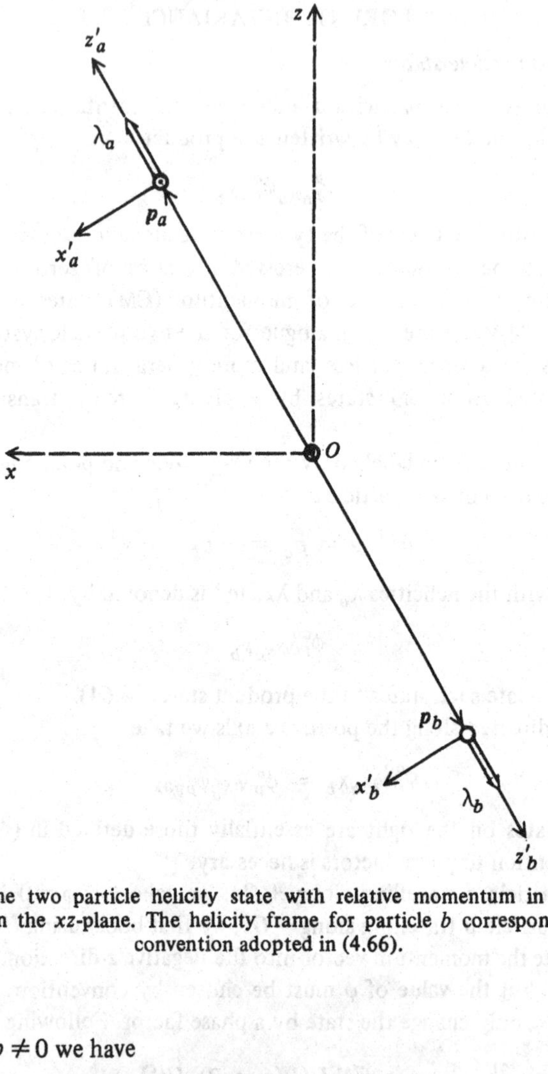

Fig. 4.2. The two particle helicity state with relative momentum in a general direction in the xz-plane. The helicity frame for particle b corresponds to the convention adopted in (4.66).

When $p \neq 0$ we have

$$J_z \phi^{ab}_{p00\lambda_a\lambda_b} = (\lambda_a - \lambda_b)\, \phi^{ab}_{p00\lambda_a\lambda_b} \tag{4.68}$$

A CM state with momentum p along a direction (θ, ϕ) is obtained with the aid of a rotation $R(\phi, \theta, 0)$. Thus

$$\phi^{ab}_{p\theta\phi\lambda_a\lambda_a} = U(R(\phi, \theta, 0))\, \phi^{ab}_{p00\lambda_a\lambda_b} \tag{4.69}$$

It should be noted that in this formalism the two particles are treated asymmetrically so in practice one must specify which particle is the 'first' (a) and which is the 'second' (b). Of course, no physical result can depend on the choice made.

4.6.2 *Expansion in eigenstates of angular momentum*

We now consider the expansion of a two particle CM momentum eigenstate in eigenstates of total angular momentum. The key observation here is that λ_a and λ_b, like p, are scalar quantities, invariant under rotations, and it is therefore possible to have simultaneous eigenstates of magnitude of CM momentum, total angular momentum and its component, and helicities of a and b.

These eigenstates will be denoted by

$$\psi^{ab}_{pJM\lambda_a\lambda_b} \tag{4.70}$$

They satisfy

$$J^2\psi^{ab}_{pJM\lambda_a\lambda_b} = J(J+1)\,\psi^{ab}_{pJM\lambda_a\lambda_b} \tag{4.71}$$

$$J_z\psi^{ab}_{pJM\lambda_a\lambda_b} = M\psi^{ab}_{pJM\lambda_a\lambda_b} \tag{4.72}$$

For fixed p, λ_a, λ_b they form a complete set of CM states, and so it must be possible to expand an eigenstate of linear momentum in terms of them.

We therefore write

$$\phi^{ab}_{p\theta\phi\lambda_a\lambda_b} = \sum_{JM} C_{JM\lambda_a\lambda_b}(\theta,\phi)\,\psi^{ab}_{pJM\lambda_a\lambda_b} \tag{4.73}$$

and seek to determine the expansion coefficients $C_{JM\lambda_a\lambda_b}(\theta,\phi)$. Consider first the special case $\theta = \phi = 0$. Omitting labels which do not change, we have

$$\phi_{00\lambda_a\lambda_b} = \sum_{JM} C_{JM\lambda_a\lambda_b}(0,0)\,\psi_{JM\lambda_a\lambda_b} \tag{4.74}$$

Let us apply to both sides the operator

$$U(Z_\alpha) = \mathrm{e}^{-\mathrm{i}\alpha J_z}$$

corresponding to a rotation by α about the z-axis. Making use of (4.68) and (4.72) we obtain

$$\mathrm{e}^{-\mathrm{i}(\lambda_a-\lambda_b)\alpha}\phi_{00\lambda_a\lambda_b} = \sum_{JM} C_{JM\lambda_a\lambda_b}(0,0)\,\mathrm{e}^{-\mathrm{i}M\alpha}\psi_{JM\lambda_a\lambda_b} \tag{4.75}$$

Since this must hold for all α we conclude that

$$C_{JM\lambda_a\lambda_b}(0,0) = 0 \quad \text{unless} \quad M = \lambda_a - \lambda_b \tag{4.76}$$

We write

$$C_{JM\lambda_a\lambda_b}(0,0) = C_J\delta_{M,\lambda_a-\lambda_b}$$

C_J will be determined by normalisation considerations to be

$$C_J = \left(\frac{2J+1}{4\pi}\right)^{1/2} \tag{4.77}$$

So for the case $\theta = \phi = 0$ we have

$$\phi^{ab}_{p00\lambda_a\lambda_b} = \sum_J C_J \psi^{ab}_{pJ,\lambda_a-\lambda_b,\lambda_a\lambda_b} \tag{4.78}$$

Now according to (4.69) the general CM state is obtained from the left-hand side of (4.78) by means of a rotation $U(R(\phi, \theta, 0))$, thus omitting labels p and ab we have

$$\begin{aligned}\phi_{\theta\phi\lambda_a\lambda_b} &= U(R(\phi, \theta, 0))\,\phi_{00\lambda_a\lambda_b} \\ &= \Sigma\, C_J U(R(\phi, \theta, 0))\,\psi_{J,\lambda_a-\lambda_b,\lambda_a\lambda_b}\end{aligned}$$

and so

$$\phi_{\theta\phi\lambda_a\lambda_b} = \sum_J C_J \sum_M \psi_{JM\lambda_a\lambda_b} \mathfrak{D}^J_{M,\lambda_a-\lambda_b}(\phi, \theta, 0) \tag{4.79}$$

where we have used the formula (3.69) for the effect of a finite rotation on an eigenfunction of total angular momentum.

Comparing (4.79) with (4.73) we find

$$C_{JM\lambda_a\lambda_b}(\theta, \phi) = C_J \mathfrak{D}^J_{M,\lambda_a-\lambda_b}(\phi, \theta, 0) \tag{4.80}$$

This result fixes the range of values of J and M. The properties of $\mathfrak{D}^J_{M'M}(\alpha, \beta, \gamma)$ require that J and M take integer or half-integer values according as $\lambda_a - \lambda_b$ is integer or half-integer, and hence according as $s_a - s_b$ is integer or half-integer. Furthermore, J cannot be less than M or M', and hence in the expansion $J \geqslant |\lambda_a - \lambda_b|$.

We have shown that

$$\phi^{ab}_{p\theta\phi\lambda_a\lambda_b} = \sum_{J \geqslant |\lambda_a-\lambda_b|} \sum_{M=-J}^{J} C_J \mathfrak{D}^J_{M,\lambda_a-\lambda_b}(\phi, \theta, 0)\psi^{ab}_{pJM\lambda_a\lambda_b} \tag{4.81}$$

where

$$\mathfrak{D}^J_{M,\lambda_a-\lambda_b}(\phi, \theta, 0) = e^{-iM\phi} d^J_{M,\lambda_a-\lambda_b}(\theta)$$

Equation (4.81) is the key result of the helicity formalism for two particle states.

If both a and b have zero spin

$$\lambda_a - \lambda_b = 0$$

and we can use the identity (3.87*d*)

$$\mathfrak{D}^L_{M0}(\phi, \theta, 0) = \left(\frac{4\pi}{2L+1}\right)^{1/2} Y^*_{LM}(\theta, \phi)$$

to write (4.81) in the form

$$\phi^{ab}_{p\theta\phi} = \sum_{LM} Y^*_{LM}(\theta, \phi)\, \psi^{ab}_{pLM} \tag{4.82}$$

Since J is an integer we write L for J. This is essentially the plane wave expansion formula of non-relativistic scattering theory which is usually written as the mathematical identity

$$e^{i\boldsymbol{p}\cdot\boldsymbol{r}} = 4\pi \sum_{LM} i^L Y^*_{LM}(\theta, \phi) j_L(pr) Y_{LM}(\theta_r, \phi_r) \tag{4.83}$$

where (θ, ϕ) and (θ_r, ϕ_r) are the polar angles of $\boldsymbol{p}$ and $\boldsymbol{r}$ respectively.

Apart from the occurrence of the less familiar $\mathfrak{D}^J_{M'M}$-function, (4.81) is hardly more complicated and is valid for arbitrary spins s_a and s_b. In the older treatments, when the spins of all particles were referred to a common z-axis, the generalisation of (4.83) to two spinning particles would involve two Clebsch–Gordan coefficients to combine the spins s_a and s_b with the orbital angular momentum ℓ to give a state of total angular momentum j.

4.6.3 *Normalisation of states*

It follows from the normalisation (4.43) adopted for the one particle states that the two particle states satisfy the normalisation condition

$$(\phi^{ab}_{q_a q_b \mu_a \mu_b}, \phi^{ab}_{p_a p_b \lambda_a \lambda_b})$$
$$= (2\pi)^6 2E_a 2E_b \delta^{(3)}(\boldsymbol{q}_a - \boldsymbol{p}_a)\delta^{(3)}(\boldsymbol{q}_b - \boldsymbol{p}_b)\delta_{\mu_a\lambda_a}\delta_{\mu_b\lambda_b} \tag{4.84}$$

where

$$E_a = (p_a^2 + m_a^2)^{1/2}$$

and

$$E_b = (p_b^2 + m_b^2)^{1/2}$$

It is convenient to describe the two particle state by the total four-momentum P and relative three-momentum $\boldsymbol{p}$, defined by

$$P_\mu = p_{a\mu} + p_{b\mu}$$

$$(m_a + m_b)\boldsymbol{p} = m_b \boldsymbol{p}_a - m_a \boldsymbol{p}_b$$

For the most part we are interested in the centre of momentum system in which the total three-momentum is zero so that

$$P_\mu = (W, 0)$$

In this case the magnitude of the CM relative momentum is uniquely given by the CM total energy so that a complete set of variables specifying a CM state is: W and two polar angles θ, ϕ to define the orientation of the relative momentum.

It can be shown that CM states defined by

$$\Phi_{W\theta\phi\lambda_a\lambda_b} = \frac{1}{2\pi}\left(\frac{p}{4W}\right)^{1/2} \phi_{p_a=p,p_b=-p,\lambda_a\lambda_b} \tag{4.85}$$

are normalised as follows

$$(\Phi_{W\theta'\phi'\lambda_a'\lambda_b'}, \Phi_{W\theta\phi\lambda_a\lambda_b}) = \delta(\cos\theta' - \cos\theta)\,\delta(\phi' - \phi)\,\delta_{\lambda_a'\lambda_a}\delta_{\lambda_b'\lambda_b} \tag{4.86}$$

In parallel with (4.85) we redefine our two particle angular momentum eigenstates by putting

$$\Psi_{WJM\lambda_a\lambda_b} = \frac{1}{2\pi}\left(\frac{p}{4P_0}\right)^{1/2} \psi_{WJM\lambda_a\lambda_b}$$

So the partial wave expansion formula retains its form

$$\Phi_{W\theta\phi\lambda_a\lambda_b} = \sum_{JM} C_J \mathfrak{D}^J_{M,\lambda_a-\lambda_b}(\phi, \theta, 0)\,\Psi_{WJM\lambda_a\lambda_b} \tag{4.87}$$

It may be verified with the aid of $\mathfrak{D}$-function identities that the standard normalisation for the angular momentum eigenstates

$$(\Psi_{WJ'M'\lambda_a'\lambda_b'}, \Psi_{WJM\lambda_a\lambda_b}) = \delta_{J'J}\delta_{M'M}\delta_{\lambda_a'\lambda_a}\delta_{\lambda_b'\lambda_b} \tag{4.88}$$

is consistent with the normalisation (4.86) and the expansion (4.87) provided we choose

$$C_J = \left(\frac{2J+1}{4\pi}\right)^{1/2}$$

We finally note that the orthonormality formula (3.88a) may be used to invert (4.87). On multiplying both sides by $\mathfrak{D}^{J*}_{M,\lambda_a-\lambda_b}(\phi, \theta, 0)$ and integrating over ϕ and θ, we obtain

$$\Psi_{WJM\lambda_a\lambda_b} = \left(\frac{2J+1}{4\pi}\right)^{1/2} \int_0^{2\pi} d\phi \int_0^{\pi} d\theta \sin\theta \times \mathfrak{D}^{J*}_{M,\lambda_a-\lambda_b}(\phi, \theta, 0)\,\Phi_{W\theta\phi\lambda_a\lambda_b} \tag{4.89}$$

which expresses an eigenstate of total angular momentum as a (continuous) superposition of plane wave helicity states. This is the original form of the angular momentum expansion presented by Jacob and Wick.

4.7 Partial wave analysis of two particle scattering

It is shown in appendix A that the CM differential cross-section for a binary reaction

$$a+b\to c+d$$

between definite helicity states is given in terms of the matrix elements of the transition operator $\mathcal{T}$ by

$$\frac{\mathrm{d}\sigma}{\mathrm{d}\Omega}=\frac{1}{(8\pi W)^2}\frac{p_{cd}}{p_{ab}}|\mathcal{T}_{\lambda_c\lambda_d,\lambda_a\lambda_b}(W,\theta,\phi)|^2 \tag{4.90}$$

where the $\mathcal{T}$-matrix element is defined by

$$(\phi^{cd}_{p_cp_d\lambda_c\lambda_d},\mathcal{T}\phi^{ab}_{p_ap_b\lambda_a\lambda_b})$$
$$=(2\pi)^4\delta^{(4)}(p_c+p_d-p_a-p_b)\mathcal{T}_{\lambda_c\lambda_d,\lambda_a\lambda_b}(W,\theta,\phi) \tag{4.91}$$

In (4.91) the initial and final momenta are as follows

$$p_{ab}=p_a=-p_b \quad \text{with polar angles } (0,0)$$

$$p_{cd}=p_c=-p_d \quad \text{with polar angles } (\theta,\phi)$$

If we use instead the states Φ given by (4.85) which have the simpler normalisation (4.86), then the expression for the CM differential cross-section becomes

$$\frac{\mathrm{d}\sigma}{\mathrm{d}\Omega}=\left(\frac{2\pi}{p_{ab}}\right)^2|(\Phi_{W\theta\phi\lambda_c\lambda_d},\mathcal{T}\Phi_{W00\lambda_a\lambda_b})|^2 \tag{4.92}$$

The partial wave expansion for the $\mathcal{T}$-matrix element is made by introducing the angular momentum expansion (4.87) for the initial and final states

$$(\Phi^{cd}_{W\theta\phi\lambda_c\lambda_d},\mathcal{T}\Phi^{ab}_{W00\lambda_a\lambda_b})$$
$$=\sum_{JM}\sum_{J'M'}C^*_J\mathcal{D}^{J*}_{M,\lambda_c-\lambda_d}(\phi,\theta,0)\,C_{J'}\mathcal{D}^{J'}_{M',\lambda_a-\lambda_b}(0,0,0)$$
$$\times(\Psi^{cd}_{WJM\lambda_c\lambda_d},\mathcal{T}\Psi^{ab}_{WJ'M'\lambda_a\lambda_b})$$

As we showed in §3.3.2 it follows from rotational invariance that the $\mathcal{T}$-matrix element in the angular momentum basis has the form

$$(\Psi^{cd}_{WJM\lambda_c\lambda_d},\mathcal{T}\Psi^{ab}_{WJ'M'\lambda_a\lambda_b})=\mathcal{T}^J_{\lambda_c\lambda_d,\lambda_a\lambda_b}(W)\,\delta_{J'J}\delta_{MM'} \tag{4.93}$$

where $\mathcal{T}^J$ is independent of M.

On introducing this expression into the preceding expansion and noting that

$$\mathfrak{D}^{J'}_{M',\lambda_a-\lambda_b}(0,0,0) = \delta_{M',\lambda_a-\lambda_b}$$

we obtain

$$(\Phi^{cd}_{W\theta\phi\lambda_c\lambda_d}, \mathcal{T}\Phi^{ab}_{W00\lambda_a\lambda_b})$$

$$= \sum_J \frac{2J+1}{4\pi} \mathfrak{D}^{J*}_{\lambda_a-\lambda_b,\lambda_c-\lambda_d}(\phi,\theta,0)\mathcal{T}^J_{\lambda_c\lambda_d,\lambda_a\lambda_b}(W) \quad (4.94)$$

Since $\mathfrak{D}^J_{mn}$ is only defined for $|m|$ and $|n|$ less than or equal to J, the sum over J starts at a value equal to the lesser of $|\lambda_a - \lambda_b|$ or $|\lambda_c - \lambda_d|$.

It is convenient to define the helicity scattering amplitude by

$$f_{\lambda_c\lambda_d,\lambda_a\lambda_b}(W,\theta,\phi)$$

$$= \frac{1}{2p_{ab}} \sum_J (2J+1)\mathfrak{D}^{J*}_{\lambda_a-\lambda_b,\lambda_c-\lambda_d}(\phi,\theta,0)\mathcal{T}^J_{\lambda_c\lambda_d,\lambda_a\lambda_b} \quad (4.95)$$

in terms of which the cross-section is given by

$$\frac{d\sigma}{d\Omega} = |f_{\lambda_c\lambda_d,\lambda_a\lambda_b}(W,\theta,\phi)|^2 \quad (4.96)$$

Equations (4.95) and (4.96) are the basic formulae of the helicity partial wave expansion for a reaction between particles of arbitrary spin.

From the form of the $\mathfrak{D}^J$ functions it follows that the ϕ-dependence of f can be factored out, and we can write

$$f_{\lambda_c\lambda_d,\lambda_a\lambda_b}(W,\theta,\phi) = e^{i(\lambda_a-\lambda_b)\phi} f_{\lambda_c\lambda_d,\lambda_a\lambda_b}(\theta,0)$$

The ϕ-dependence disappears from the expression for the differential cross-section as is expected as grounds of rotational invariance. It is usual to take the xz-plane ($\phi = 0$) as the scattering plane. $d\sigma/d\Omega$ given by (4.96) is the differential cross-section for particles in definite helicity states. If the states of polarisation of the final particles are not detected it is necessary to sum (4.96) over λ_c and λ_d, while if the initial beam and target are unpolarised it is necessary to take the average of (4.96) over the possible values of λ_a and λ_b respectively.

The 'unpolarised' cross-section is therefore

$$\overline{\frac{d\sigma}{d\Omega}} = \frac{1}{(2s_a+1)(2s_b+1)} \sum_{\lambda_c\lambda_d\lambda_a\lambda_b} |f_{\lambda_c\lambda_d,\lambda_a\lambda_b}(W,\theta,\phi)|^2 \quad (4.97)$$

For a general discussion of polarisation in the helicity formalism we refer the reader to Jacob and Wick (1959).

In considering the consequences of the discrete symmetries of space inversion and time reversal for reactions and decays, it is necessary to know the effect of these operations on the helicity states. This will be discussed in the chapters on parity and time reversal.

We finally note that any of the particles a, b, c or d may be massless, and the partial wave expansion remains valid, but the possible values of the helicity are restricted as discussed in §4.5.

4.8 Pion–nucleon scattering

This section is devoted to the illustration of the helicity formalism by the important case of elastic scattering of a spin-0 and a spin-$\frac{1}{2}$ particle. This includes πN and KN scattering as well as reactions of the type

$$\bar{K} + N \to \pi + \Lambda$$

$$\bar{K} + N \to \pi + \Sigma$$

The formulae of the helicity analysis will be related to the formulae developed in older analyses which were made in analogy with non-relativistic scattering theory. This will show that the older analyses were indeed compatible with the requirements of Lorentz invariance although this was not always apparent.

For definiteness we shall deal with the case of πN scattering. We take particle $a(c)$ to be the nucleon and $b(d)$ to be the pion, so that

$$s_a = s_c = \tfrac{1}{2}, \quad s_b = s_d = 0$$

The indices λ_b and λ_d are correspondingly suppressed and the values $\pm\frac{1}{2}$ for λ_a and λ_c are denoted by $\pm$.

The scattering geometry in the CM is shown in fig. 4.3. The incident nucleon (rather than the pion) is directed along the positive z-axis, and the direction of the final nucleon has polar angles (θ, ϕ).

The CM differential cross-section between definite helicity states is

$$\frac{d\sigma}{d\Omega} = |f_{\mu\lambda}(\theta, \phi)|^2$$

where

$$f_{\mu\lambda}(\theta, \phi) = \frac{1}{2p}\sum_J (2J+1)\, e^{i\lambda\phi} d^J_{\lambda\mu}(\theta) \mathcal{T}^J_{\mu\lambda}(W) \tag{4.98}$$

The ϕ-dependence factors out as noted above, but it is convenient to carry it for future discussion of some polarisation questions. The scattering amplitude for $\phi = 0$ is denoted briefly by $f(\theta)$ so we have

$$f_{\mu\lambda}(\theta, \phi) = e^{i\lambda\phi} f_{\mu\lambda}(\theta)$$

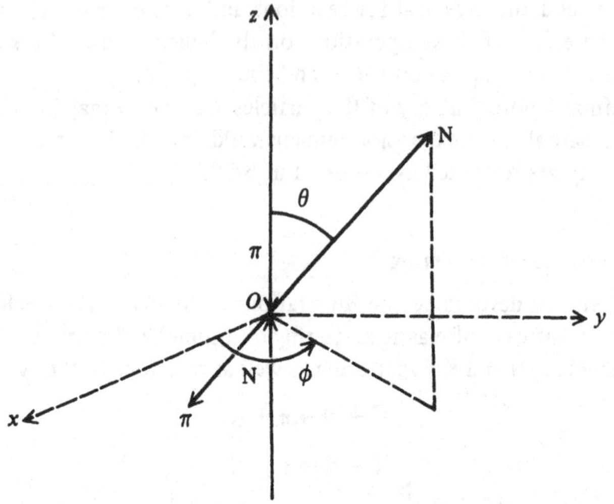

Fig. 4.3. Scattering geometry for πN scattering.

There are four helicity amplitudes *a priori*, but conservation of parity leads to relations between them. It will be shown in the next chapter that as a consequence of parity conservation

$$f_{\mu\lambda}(\theta) = (-1)^{\mu-\lambda} f_{-\mu,-\lambda}(\theta)$$

Hence

$$\left.\begin{aligned} f_{--}(\theta) &= f_{++}(\theta) \\ f_{+-}(\theta) &= -f_{-+}(\theta) \end{aligned}\right\} \tag{4.99}$$

These relations reduce the number of independent amplitudes to two.

We shall see in chapter 6 that time reversal leads to no additional restrictions on the amplitudes for this process.

The two amplitudes f_{-+} and f_{++} are called the *helicity flip* and *non-flip* amplitudes respectively.

For $\theta = 0$, $d^J_{\lambda\mu}(\theta)$ is zero unless $\mu = \lambda$, so that the forward scattering amplitude is pure non-flip; for $\theta = \pi$, $d^J_{\lambda\mu}(\theta)$ is zero unless $\mu = -\lambda$ and so backward scattering is pure helicity flip. In both cases the direction of the nucleon spin remains unchanged.

The consequences of parity conservation for the partial wave amplitudes are

$$\left.\begin{aligned} \mathcal{T}^J_{--} &= \mathcal{T}^J_{++} \\ \mathcal{T}^J_{+-} &= \mathcal{T}^J_{-+} \end{aligned}\right\} \tag{4.100}$$

and again time reversal invariance can be shown to lead to no further restrictions.

The helicity amplitudes may be written as a matrix

$$f = \begin{bmatrix} f_{++} & f_{+-} \\ f_{-+} & f_{--} \end{bmatrix} = \begin{bmatrix} f_{++} & -f_{-+} \\ f_{-+} & f_{++} \end{bmatrix}$$

where the second form results from parity conservation. f is a matrix in the helicity representation which means that column index refers to the initial nucleon spin projection along the z-axis, while its row refers to the final nucleon spin projection along the direction with polar angles (θ, ϕ).

In treatments based on the non-relativistic partial wave analysis, the final nucleon spin is referred to the same axis as the initial nucleon spin which in this case is the z-axis. Let us now consider the connection between the two treatments.

$f_{\mu\lambda}(\theta, \phi)$ may be regarded as the matrix element

$$f_{\mu\lambda} = (\chi_\mu^{(\mathrm{h}')}, f\chi_\lambda^{(\mathrm{h})})$$

of a transition operator between a helicity state of the initial nucleon

$$\chi_+^{(\mathrm{h})} = \begin{pmatrix} 1 \\ 0 \end{pmatrix} \quad \text{or} \quad \chi_-^{(\mathrm{h})} = \begin{pmatrix} 0 \\ 1 \end{pmatrix}$$

i.e. a spinor with z-axis as quantisation axis; and a helicity spinor for the final nucleon

$$\chi_+^{(\mathrm{h}')} = \begin{pmatrix} 1 \\ 0 \end{pmatrix} \quad \text{or} \quad \chi_-^{(\mathrm{h}')} = \begin{pmatrix} 0 \\ 1 \end{pmatrix}$$

for which the quantisation axis is the direction (θ, ϕ) of the final nucleon.

We now ask for the amplitude for scattering into the state in which the final nucleon has spin projection m along Oz. Such a spin state for the final nucleon denoted by $\chi_m^{(\mathrm{f})}$ is obtained from the helicity state by the rotation

$$R(\phi, \theta, 0)^{-1} = R(0, -\theta, -\phi)$$

(see fig. 4.3). Hence

$$\begin{aligned} \chi_{m'}^{(\mathrm{f})} &= U(R(\phi, \theta, 0)^{-1}\chi_{m'}^{(\mathrm{h}')} \\ &= U(Y_{-\theta})\,U(Z_{-\phi})\,\chi_{m'}^{(\mathrm{h}')} \\ &= \sum_\mu \chi_\mu^{(\mathrm{h}')} d_{\mu m'}^{(1/2)}(-\theta)\, \mathrm{e}^{im'\phi} \end{aligned}$$

Inserting the explicit expressions for the elements of the d-matrix we have

$$\chi_+^{(f)} = \chi_+^{(h')} \cos\tfrac{1}{2}\theta\, e^{i\phi/2} - \chi_-^{(h')} \sin\tfrac{1}{2}\theta\, e^{i\phi/2}$$
$$\chi_-^{(f)} = \chi_+^{(h')} \sin\tfrac{1}{2}\theta\, e^{-i\phi/2} + \chi_-^{(h')} \cos\tfrac{1}{2}\theta\, e^{-i\phi/2}$$

For the initial nucleon the helicity state already refers to a spin projection on the Oz-axis. Hence

$$\chi_m^{(i)} = \chi_m^{(h)}$$

Since the final spin state, $\chi_m^{(f)}$, is a superposition of helicity states we may write out the amplitude for scattering in the basis of spin states referred to the common z-axis as follows

$$\begin{aligned}(\chi_+^{(f)}, f\chi_m^{(i)}) &= \cos\tfrac{1}{2}\theta\, e^{-i\phi/2}(\chi_+^{(h')}, f\chi_m^{(h)}) - \sin\tfrac{1}{2}\theta\, e^{-i\phi/2}(\chi_-^{(h')}, f\chi_m^{(h)}) \\ &= \cos\tfrac{1}{2}\theta\, e^{-i\phi/2} f_{+m}(\theta, \phi) - \sin\tfrac{1}{2}\theta\, e^{-i\phi/2} f_{-m}(\theta, \phi)\end{aligned}$$

Similarly

$$(\chi_-^{(f)}, f\chi_m^{(i)}) = \sin\tfrac{1}{2}\theta\, e^{i\phi/2} f_{+m}(\theta, \phi) + \cos\tfrac{1}{2}\theta\, e^{i\phi/2} f_{-m}(\theta, \phi)$$

Taking account of parity conservation (4.99) the transition amplitude in the non-relativistic spin representation may be written as a matrix

$$F_{m'm} = g(\theta)\delta_{m'm} + h(\theta)\begin{bmatrix} 0 & e^{-i\phi} \\ -e^{i\phi} & 0 \end{bmatrix}_{m'm}$$

where we have factored out the ϕ-dependence and defined new amplitudes

$$\left.\begin{aligned} g(\theta) &= f_{++}(\theta)\cos\tfrac{1}{2}\theta - f_{-+}(\theta)\sin\tfrac{1}{2}\theta \\ h(\theta) &= -f_{++}(\theta)\sin\tfrac{1}{2}\theta - f_{-+}(\theta)\cos\tfrac{1}{2}\theta \end{aligned}\right\} \quad (4.101)$$

If we introduce the unit vector $\hat{\boldsymbol{n}}$ normal to the scattering plane, with the sign convention that $\hat{\boldsymbol{n}}$ has the direction of

$$\boldsymbol{p}\text{ (initial nucleon)} \wedge \boldsymbol{p}\text{ (final nucleon)}$$

then in this case

$$\hat{\boldsymbol{n}} = (-\sin\phi, \cos\phi, 0)$$

and

$$i\boldsymbol{\sigma}.\hat{\boldsymbol{n}} = \begin{bmatrix} 0 & e^{-i\phi} \\ -e^{i\phi} & 0 \end{bmatrix}$$

so that the matrix F can be written

$$F(\theta, \phi) = g(\theta)1 + ih(\theta)\boldsymbol{\sigma}\cdot\hat{\boldsymbol{n}} \quad (4.102)$$

F is a matrix in the spin space in which all spins are referred to a common z-axis.

The diagonal and off-diagonal elements $g(\theta)$ and $h(\theta)$ are called the no-flip and spin-flip amplitudes respectively. The relatively simple form (4.102) of F can be argued on the grounds of invariance under rotations and space and time reflections. We shall discuss these arguments in chapters 5 and 6.

The partial wave analysis is a resolution of the function $f_{\mu\lambda}(W, \theta, \phi)$ in terms of a set of amplitudes $\mathcal{T}^J_{\mu\lambda}$ in a way which is consistent with rotational and Lorentz invariance. There is one other general constraint which we have not considered: the unitarity of the scattering operator $\mathcal{S}$, which expresses the conservation of probability.

We recall that in a basis of states ϕ_n where the $\mathcal{S}$-matrix elements are

$$\mathcal{S}_{mn} = (\phi_m, \mathcal{S}\phi_n)$$

the unitarity condition takes the form

$$\sum_m \mathcal{S}^*_{mn}\mathcal{S}_{mn'} = \delta_{nn'} \tag{2.15}$$

Let us apply this to the πN system. At energies below the threshold for production of an additional pion, elastic scattering is the only strong interaction process which is allowed. We neglect scattering with the radiation of one or more low energy photons. Then the only non-zero matrix elements of $\mathcal{S}$ are between different states of one π and one N. If we choose CM angular momentum eigenstates Ψ_{WJM} as our basis states, then the scattering operator $\mathcal{S}$ is diagonal

$$(\Psi_{WJ'M'\mu}, \mathcal{S}\Psi_{WJM\lambda}) = \delta_{J'J}\delta_{M'M}\mathcal{S}^J_{\mu\lambda} \tag{4.103}$$

The unitarity condition (2.15) then gives the relations

$$\left.\begin{aligned} |\mathcal{S}^J_{++}|^2 + |\mathcal{S}^J_{+-}|^2 &= 1 \\ \mathcal{S}^{J*}_{++}\mathcal{S}^J_{+-} + \mathcal{S}^{J*}_{+-}\mathcal{S}^J_{++} &= 0 \end{aligned}\right\} \tag{4.104}$$

where use has been made of the relations

$$\mathcal{S}^J_{++} = \mathcal{S}^J_{--}, \quad \mathcal{S}^J_{-+} = \mathcal{S}^J_{+-} \tag{4.105}$$

following from parity conservation.

Equations (4.104) may be further simplified by expressing $\mathcal{S}$ in a basis of parity eigenstates. The transformation law for a πN angular momentum helicity eigenstate under the parity operation P is

$$U(P)\Psi_{JM\lambda} = -(-1)^{J-1/2}\Psi_{JM,-\lambda}$$

as will be shown in chapter 5. Hence for a given JM we may construct eigenstates of the parity operator. The states

$$\Psi^{(\pm)}_{JM} = 2^{-1/2}\{\Psi_{JM+} \mp \Psi_{JM-}\} \tag{4.106}$$

have parity $(-1)^{J\mp 1/2}$. If we calculate the matrix elements of $\mathcal{S}$ in the basis $\Psi^{(\pm)}_{JM}$ it is found that $\mathcal{S}$ is fully diagonal. The two diagonal elements for a given J are

$$\left.\begin{aligned} \mathcal{S}^{J+} &= (\Psi^{(+)}_{JM}, \mathcal{S}\Psi^{(+)}_{JM}) = \mathcal{S}^{J}_{++} - \mathcal{S}^{J}_{+-} \\ \mathcal{S}^{J-} &= (\Psi^{(-)}_{JM}, \mathcal{S}\Psi^{(-)}_{JM}) = \mathcal{S}^{J}_{++} + \mathcal{S}^{J}_{+-} \end{aligned}\right\} \tag{4.107}$$

while

$$(\Psi^{(+)}_{JM}, \mathcal{S}\Psi^{(-)}_{JM}) = (\Psi^{(-)}_{JM}, \mathcal{S}\Psi^{(+)}_{JM}) = 0$$

showing that $\mathcal{S}$ has no matrix elements between states of the same J with opposite parity.

Now that the $\mathcal{S}$-matrix is diagonal, the unitarity condition takes the simple form

$$|\mathcal{S}^{J+}|^2 = |\mathcal{S}^{J-}|^2 = 1$$

We may therefore parametrise the $\mathcal{S}$-matrix by putting

$$\mathcal{S}^{J\pm} = e^{2i\delta_{J\pm}} \tag{4.108}$$

where the *phase shifts* $\delta_{J\pm}(W)$ are real functions of the CM total energy.

Above the threshold for the production of an additional pion, off-diagonal elements appear in the scattering matrix, and from (2.15) the condition on a diagonal matrix element of $\mathcal{S}$ is weakened to

$$|\mathcal{S}_{nn}|^2 = 1 - \sum_{m\neq n} |\mathcal{S}_{mn}|^2$$

or

$$|\mathcal{S}_{nn}| \leqslant 1$$

In the basis $\Psi^{(\pm)}_{JM}$ this reads

$$|\mathcal{S}^{J\pm}| \leqslant 1 \tag{4.109}$$

and the parametrisation (4.108) is replaced by

$$\mathcal{S}^{J\pm} = \eta_{J\pm} e^{2i\delta_{J\pm}} \tag{4.110}$$

where the *inelasticity factors* $\eta_{J\pm}(W)$ are real functions of the CM energy.

It is appropriate here to comment on orbital angular momentum. Orbital angular momentum does not have a natural place in the helicity formulation of the partial wave analysis; however it is possible to relate helicity amplitudes to amplitudes between states of definite orbital angular momentum. For a discussion of the general case the reader is referred to the paper of Jacob and Wick (1959), but for the πN system we may make the connection as follows.

A state of the πN system with orbital angular momentum L is an eigenstate of parity with parity eigenvalue $-(-1)^L$, where the additional sign arises from the negative relative intrinsic parity of the π and N. Now a state of definite L contributes to states of total angular momentum $J = L \pm \frac{1}{2}$.

Conversely a πN state of definite J has contributions from L values as follows

$$J+: \quad L = J + \tfrac{1}{2} \quad \text{with parity} -(-1)^{J+1/2} = (-1)^{J-1/2}$$

$$J-: \quad L = J - \tfrac{1}{2} \quad \text{with parity} -(-1)^{J-1/2} = (-1)^{J+1/2}$$

These two possibilities correspond to the angular momentum parity eigenstates designated $\Psi^{(+)}_{JM}$ and $\Psi^{(-)}_{JM}$ respectively in (4.106). It may be seen that the superscript (±) indicates that the orbital angular momentum is $L = J \pm \frac{1}{2}$. These states are sometimes written $\Psi_{J\pm,M}$ and this is the notation used for the diagonal matrix elements $\mathcal{S}^{J\pm}$ and for the phase shifts and inelasticity factors.

We now give the relation between the transition matrix elements and the phase shifts. The relation between the $\mathcal{S}$ and $\mathcal{T}$ operators is

$$\mathcal{S} = 1 + \mathrm{i}\mathcal{T}$$

and thus for elastic scattering the relation between the matrix elements in the helicity basis is

$$\mathcal{S}^J_{\mu\lambda} = \delta_{\mu\lambda} + \mathrm{i}\mathcal{T}^J_{\mu\lambda}$$

Hence using (4.107) and (4.108) we have

$$\mathcal{T}^J_{++}(W) = \frac{1}{\mathrm{i}}\{\mathcal{S}^J_{++} - 1\}$$

$$= \frac{1}{\mathrm{i}}\{\tfrac{1}{2}(\mathrm{e}^{2\mathrm{i}\delta_{J-}} + \mathrm{e}^{2\mathrm{i}\delta_{J+}}) - 1\}$$

and so

$$\mathcal{T}^J_{++}(W) = a_{J-} + a_{J+} \tag{4.111}$$

where we have defined new partial wave scattering amplitudes

$$a_{J\pm} = \frac{1}{2\mathrm{i}}(\mathrm{e}^{2\mathrm{i}\delta_{J\pm}} - 1) \tag{4.112}$$

Similarly we find

$$\mathcal{T}^J_{+-}(W) = \frac{1}{i}\mathcal{S}^J_{+-}$$

$$= a_{J-} - a_{J+} \qquad (4.113)$$

The scattering amplitude $f_{\mu\lambda}(W, \theta, \phi)$ may be expressed in terms of the $a_{J\pm}$ by means of (4.98), (4.111) and (4.113). We specialise to the case $\phi = 0$ and introduce explicit formulae for the $d^J_{\mu\lambda}(\theta)$ function as follows

$$d^J_{++}(\theta) = \frac{2}{2J+1}\cos\tfrac{1}{2}\theta\,(P'_{J+1/2} - P'_{J-1/2})$$

$$d^J_{-+}(\theta) = \frac{2}{2J+1}\sin\tfrac{1}{2}\theta\,(P'_{J+1/2} + P'_{J-1/2})$$

where

$$P'_L = \frac{\mathrm{d}}{\mathrm{d}(\cos\theta)}\{P_L(\cos\theta)\}$$

denotes the derivative of the Legendre polynomial (Jacob and Wick, 1959).

We find

$$f_{++}(\theta) = \frac{1}{p}\sum_J \cos\tfrac{1}{2}\theta\,(P'_{J+1/2} - P'_{J-1/2})(a_{J-} + a_{J+})$$

$$f_{+-}(\theta) = \frac{1}{p}\sum_J \sin\tfrac{1}{2}\theta\,(P'_{J+1/2} + P'_{J-1/2})(a_{J-} - a_{J+})$$

4.9 Two particle decays

4.9.1 *General formalism*

The helicity formalism is very convenient for the analysis of decays

$$A \to a + b + c + \dots$$

in which the particles carry spin. We shall confine our attention to the case of a two body decay

$$A \to a + b$$

where the masses and spins are respectively $m_A, J; m_a, s_a$; and m_b, s_b.

It is shown in appendix A that the decay width of A is related to the $\mathcal{S}$-matrix element between plane wave helicity eigenstates in the CM, i.e. the rest frame of A, as follows

$$\Gamma = \sum_{\lambda_a\lambda_b} \frac{1}{2m_A}\int \mathrm{d}\Omega\,|f_{\lambda_a\lambda_b M}(\theta, \phi)|^2$$

where

$$f_{\lambda_a\lambda_b M}(\theta,\phi) = (\Phi^{ab}_{W=m_A,\theta\phi\lambda_a\lambda_b}, \mathcal{S}\phi^A_{0M}) \tag{4.114}$$

M is the projection of the spin of A on a conventionally chosen z-axis with respect to which θ and ϕ are measured.

The initial state is an eigenstate of J^2 and J_z with eigenvalues $J(J+1)$ and M, so angular momentum conservation requires that the same be true of the final state $\mathcal{S}\phi^A_{0M}$. Hence the amplitude f to find the decay products a and b in the plane wave state $\Phi^{ab}_{m_A\theta\phi\lambda_a\lambda_b}$ is simply the coefficient of $\Psi^{ab}_{m_A JM\lambda_a\lambda_b}$ in the partial wave expansion (4.87) of $\Phi^{ab}_{m_A\theta\phi\lambda_a\lambda_b}$ multiplied by a decay amplitude.

Proceeding formally we substitute (4.87) in (4.114) and find

$$f_{\lambda_a\lambda_b M}(\theta,\phi) = C_J \mathfrak{D}^{J*}_{M,\lambda_a-\lambda_b}(\phi,\theta,0)\, a_{\lambda_a\lambda_b} \tag{4.115}$$

where we have used angular momentum conservation to write

$$(\Psi^{ab}_{m_A J'M'\lambda_a\lambda_b}, \mathcal{S}\Phi^A_{0M}) = \delta_{J'J}\delta_{M'M} a_{\lambda_a\lambda_b}$$

The label J is omitted from the decay amplitude $a_{\lambda_a\lambda_b}$ since it is fixed throughout. Equation (4.115) gives the angular dependence of the amplitude for decay into definite helicity states. From it, the results of any possible observation of angular distributions or polarisations may be calculated in terms of the $a_{\lambda_a\lambda_b}$, which are to be obtained by comparison with experiment or from a detailed dynamical theory of the decay.

The decay width is obtained by summing over all final spin states and integrating over all angles. We have

$$\Gamma = \sum_{\lambda_a\lambda_b} \frac{2J+1}{4\pi} \int d\Omega\, |\mathfrak{D}^J_{M,\lambda_a-\lambda_b}(\phi,\theta,0)|^2 |a_{\lambda_a\lambda_b}|^2$$

where some of the factors have been absorbed into $a_{\lambda_a\lambda_b}$. The angular integrations may be performed immediately using the properties of the $\mathfrak{D}$-functions, to give

$$\Gamma = \sum_{\lambda_a\lambda_b} |a_{\lambda_a\lambda_b}|^2$$

It is sometimes convenient to work with amplitudes $A_{\lambda_a\lambda_b}$ normalised according to

$$\sum_{\lambda_a\lambda_b} |A_{\lambda_a\lambda_b}|^2 = 1$$

The differential decay distribution into definite spin states is

$$W_{\lambda_a\lambda_b M}(\theta)\, d\Omega = \frac{2J+1}{4\pi} [d^J_{M,\lambda_a-\lambda_b}(\theta)]^2 |A_{\lambda_a\lambda_b}|^2 d\Omega \tag{4.116}$$

It is normalised by

$$\sum_{\lambda_a \lambda_b} \int d\Omega W_{\lambda_a \lambda_b M}(\theta) = 1$$

This formalism will be used among other things to discuss the implication of parity non-conservation in hyperon decays in chapter 5. Here we shall describe briefly some general properties of decay processes.

It will be shown in chapter 5 that if parity is conserved in the two body decay, the amplitudes $a_{\lambda_a \lambda_b}$ (or $A_{\lambda_a \lambda_b}$) satisfy

$$a_{\lambda_a \lambda_b} = \frac{\eta_A}{\eta_a \eta_b} (-1)^{J-s_a-s_b} a_{-\lambda_a, -\lambda_b} \tag{4.117}$$

where η_A, η_a and η_b are the intrinsic parities of the particles.

If we have an unpolarised sample of A particles at rest, then the decay angular distribution is isotropic because there is nothing to define a preferred direction in the A rest frame. Let us check that this result comes out of the formalism.

For an unpolarised sample of A particles we must average over the magnetic quantum number M. This yields the angular distribution into definite helicity states,

$$\overline{W_{\lambda_a \lambda_b}(\theta)} = \frac{1}{2J+1} \sum_M \frac{2J+1}{4\pi} [d^J_{M, \lambda_a - \lambda_b}(\theta)]^2 |A_{\lambda_a \lambda_b}|^2$$

On rewriting one of the d^J as

$$d^J_{M, \lambda_a - \lambda_b}(\theta) = d^J_{\lambda_a - \lambda_b, M}(-\theta)$$

we can use their multiplication property (3.83) to do the M sum.

We find

$$\overline{W_{\lambda_a \lambda_b}(\theta)} = \frac{1}{4\pi} \sum_M d^J_{\lambda_a - \lambda_b, M}(-\theta) d^J_{M, \lambda_a - \lambda_b}(\theta) |A_{\lambda_a \lambda_b}|^2$$

$$= \frac{1}{4\pi} d^J_{\lambda_a - \lambda_b, \lambda_a - \lambda_b}(0) |A_{\lambda_a \lambda_b}|^2$$

hence

$$\overline{W_{\lambda_a \lambda_b}(\theta)} = \frac{1}{4\pi} |A_{\lambda_a \lambda_b}|^2$$

so $|A_{\lambda_a \lambda_b}|^2$ is the relative rate for decay of an unpolarised sample into the helicity state (λ_a, λ_b).

It is important to be able to count the number of independent amplitudes for a decay process. This is relatively straightforward in the present formalism. There are *a priori* $(2s_a+1)(2s_b+1)$ amplitudes $a_{\lambda_a\lambda_b}$ required to describe the decay $A \to a+b$, if both a and b are massive particles.

This number may be reduced for various reasons, which we now list.

A particle of spin J cannot decay into a state of the a–b system in which the component of total angular momentum along any direction is greater than J. So $\lambda_a - \lambda_b$, which is the angular momentum component along the direction of the relative momentum in the CM, cannot exceed J. If the spins of A, a and b are such that $s_a + s_b > J$ then those amplitudes $a_{\lambda_a\lambda_b}$ for which

$$|\lambda_a - \lambda_b| > J$$

are zero. Formally this condition follows from the properties of the $\mathfrak{D}^J_{MM'}$-function in (4.115): it is not defined (zero) for M or M' greater than J.

An interesting case is radiative meson decay into a spinless meson and a photon, for example,

$$X \to \gamma + \pi$$

This is forbidden if X has spin zero, as we now show.

In the decay amplitude

$$f_{\lambda M}(\theta) = C_J d^J_{M\lambda}(\theta) a_\lambda$$

we have $\lambda = \pm 1$, not 0, for a photon, and this is incompatible with $J=0$. In physical terms the component of angular momentum along the decay direction is ± 1 in the final state and 0 in the initial state.

It is clear that we have a general rule: a transition from a $J=0$ state to a $J=0$ state with emission of a photon is strictly forbidden.

Thus the non-observation of the radiative K-meson decay

$$K \to \pi + \gamma$$

was early evidence that the K-meson had spin zero.

If parity is conserved in the decay, (4.117) leads to relations between different amplitudes and further restricts the number of independent ones.

If either of the final particles is massless then there are at most two helicity states $\lambda = \pm s$.

Finally if the two final particles are identical there are restrictions from the Fermi or Bose statistics.

Time reversal invariance does not restrict the number of amplitudes, but may lead to relations between the decay amplitudes and a–b scattering amplitude (final state theorem: see chapter 6).

Let us consider some examples, using the notation $J \to s_a + s_b$.

(i) $\frac{1}{2} \to \frac{1}{2} + 0$: Two amplitudes reduced to one by parity (P) conservation. The virtual Yukawa process $N \to N + \pi$ which conserves P illustrates this: there is one πN coupling constant. $\Lambda \to N\pi$ is a parity violating decay, hence there are two amplitudes. We shall consider this example in chapter 5.

(ii) $\frac{3}{2} \to \frac{1}{2} + 0$: Again there are two amplitudes in general, reduced to one by P-conservation. Hence two amplitudes are required to describe the weak processes $\Omega^- \to \Lambda K^-$ or $\Omega^- \to \Xi^- \pi^0, \Xi^0 \pi^-$.

(iii) $0 \to \frac{1}{2} + \frac{1}{2}$: two amplitudes reduced to one by P-conservation; e.g. $K_2^0 \to e^+ + e^-$.

(iv) $1 \to 1$ (photon) $+ 0$: since the photon has two helicity states there are two amplitudes, reduced to one by P-conservation; e.g. $\omega^0 \to \gamma\pi^0$.

(v) $\frac{1}{2} \to \frac{1}{2} + 1$ (photon) e.g. $\Sigma^0 \to \Lambda^0\gamma$: there are two amplitudes (Λ^0 and γ helicities must be both positive or both negative) reduced to one by P-conservation.

The reader may verify that in any particular case the same number of amplitudes is obtained if the LS-coupling scheme is used.

Thus in $\frac{3}{2} \to \frac{1}{2} + 1$, there are six helicity amplitudes $a_{\lambda_a \lambda_b}$ reduced to three by parity conservation. In LS-coupling the total spin can be $\frac{3}{2}$ or $\frac{1}{2}$. The possible L-values must be such that L and S can add vectorially to give $J = \frac{3}{2}$. So for $S = \frac{3}{2}, L = 0, 1, 2$ or 3, and for $S = \frac{1}{2}, L = 1$ or 2, and again we have six amplitudes a_{LS}.

4.9.2 *The Adair analysis*

To illustrate the use of the formalism developed above we shall end by describing, in the helicity language, Adair's method (Adair, 1955) for the determination of hyperon spins.

Using the Λ^0s produced in the reaction

$$p + \pi^- \to \Lambda^0 + K^0 \qquad (4.118)$$

we seek to determine the Λ^0 spin from the angular distribution in the subsequent decay

$$\Lambda^0 \to N + \pi$$

The spins of the other particles are assumed to be known.

The essence of the method is to select scattering events kinematically so as to restrict the population of the spin states of the Λ^0, and then to see what effect this has on the decay distribution.

We take the z-axis in the CM along the direction of the incident proton. Then for Λ^0 produced along the direction of the incident proton $\theta \approx 0$, there can be no component of orbital angular momentum along the direction of the Λ^0 and hence the helicities of the initial p and final Λ^0 must be equal,

$$\theta \approx 0: \mu_\Lambda = \mu_p$$

(We use μ for helicity in this section.) Similarly for Λ^0 produced in the backward direction $\theta \approx \pi$, the components of angular momentum of Λ^0 and p along the $+z$-axis must be equal, and since the momenta are oppositely directed, the helicities must satisfy

$$\theta \approx \pi: \mu_\Lambda = -\mu_p$$

Proceeding formally we recall from §4.6 that the amplitude for the reaction (4.118) may be written

$$f_{\mu_\Lambda \mu_p}(\theta) = \frac{1}{2p} \sum (2j+1) d^j_{\mu_p \mu_\Lambda}(\theta) \mathcal{T}^j_{\mu_\Lambda \mu_p}(W)$$

Now for all j

$$d^j_{\mu\mu'}(\theta = 0) = \delta_{\mu\mu'}$$

and hence

$$f_{\mu_\Lambda \mu_p}(\theta = 0) = 0 \quad \text{unless} \quad \mu_\Lambda = \mu_p$$

Similarly

$$d_{\mu\mu'}(\theta = \pi) = \text{const.}\, \delta_{\mu, -\mu'}$$

and hence

$$f_{\mu_\Lambda \mu_p}(\theta = \pi) = 0 \quad \text{unless} \quad \mu_\Lambda = -\mu_p$$

In the case of Λ^0 production on a sample of unpolarised protons, the states $\mu_p = \pm \frac{1}{2}$ will be equally likely, hence for Λ^0s produced in the forward or backward direction the helicity states $\mu_\Lambda = \pm \frac{1}{2}$ are equally likely.

Now let us turn to the decay process. We require the decay distribution of a sample of Λ^0s of spin J (necessarily half-integral) into spin $\frac{1}{2}$ and spin 0, in which the substates $M = \mu_\Lambda = \pm \frac{1}{2}$ are equally populated and the other substates are not populated. The decay is described most easily in the CM of the Λ^0. This is reached from the production CM by a Lorentz transformation along the line of flight. This, as we have seen, does not change the helicity of the particle, so we are justified in equating μ_Λ with M in the decay formalism.

The angular distribution of the decay proton in the Λ^0 CM is given by

$$W(\theta) = \sum_{\lambda=\pm 1/2} \frac{1}{2} \sum_{M=\pm 1/2} \frac{2J+1}{4\pi} [d^J_{M,\lambda}(\theta)]^2 |A_\lambda|^2$$

where the helicity of the decay proton λ is summed (not observed).

Now from the properties of the d^J-functions

$$\sum_{M=\pm 1/2} [d^J_{M,\lambda}(\theta)]^2 = [d^J_{+1/2,\lambda}(\theta)]^2 + [d^J_{-1/2,\lambda}(\theta)]^2$$

is seen to have the same value for $\lambda = \pm\frac{1}{2}$ (the two terms exchange places). Hence we can take out the $|A_\lambda|^2$ and then use

$$\sum_\lambda |A_\lambda|^2 = 1$$

to write

$$W(\theta) = \sum_{M=\pm 1/2} \frac{2J+1}{4} [d^J_{M,\lambda}(\theta)]^2$$

where $\lambda = +\frac{1}{2}$ or $-\frac{1}{2}$, and the trivial integration over ϕ has been done. It only remains to calculate $W(\theta)$ for various J.

For $J = \frac{1}{2}$ the detailed formulation is unnecessary, for then *all* the helicity states $M = \pm\frac{1}{2}$ are equally populated, and hence as noted above the decay distribution can only be isotropic.

For higher values of J, substitution of explicit expressions for the $d^J_{M\lambda}$-functions yields the following results:

$$J = \tfrac{1}{2}: \quad W(\theta) = \tfrac{1}{2}$$

$$J = \tfrac{3}{2}: \quad W(\theta) = \tfrac{1}{4}(1 + 3\cos^2\theta)$$

$$J = \tfrac{5}{2}: \quad W(\theta) = \tfrac{3}{8}(1 - 2\cos^2\theta + 5\cos^4\theta)$$

$$J = \tfrac{7}{2}: \quad W(\theta) = \tfrac{1}{32}(9 + 45\cos^2\theta - 165\cos^4\theta + 175\cos^6\theta)$$

Note that for a spin J the highest power of $\cos\theta$ is $2J-1$. The $\theta \to \pi - \theta$ symmetry of the decay distribution follows from the fact that two substates with opposite M-values are equally populated.

The observed distribution can be tested against these possibilities.

This method was applied by Eisler *et al.* (1958) to the Λ^0 decay, and the value $J = \frac{1}{2}$ was indicated. It should be noted that no assumption about parity conservation was involved, and thus the analysis remains valid even though it is now known that parity is violated in the weak decay.

A defect of the method is that it is wasteful of data because only forward and backward production events can be used. An improved method for spin determinations (Lee and Yang, 1958) enables all the decay events to be used. For a given J, they give certain test functions of $\cos\theta$, whose averages against the data must satisfy stated inequalities for the spin to be J. Further tests have been formulated by Byers and

Fenster (1963). These and other methods are reviewed by Koch (1964) and Tripp (1965).

A discussion of cascade decays

$$\Xi \to \Lambda + \pi$$

$$\Lambda \to \mathrm{N} + \pi$$

using the helicity formalism has been given by Ueda and Okubo (1963).

4.10 Further Lorentz transformation properties of helicity states†

We end our discussion of the helicity formalism by considering the effect of an arbitrary Lorentz transformation on helicity states and helicity matrix elements. This is of direct practical importance in that it is often necessary to transform quantities from, for example, the laboratory (LAB) frame of reference to the centre of momentum (CM) frame in which the theoretical analysis is simpler.

4.10.1 *The Wigner rotation*

The main result to be obtained in this section can be described as follows. Consider a spin-$\frac{1}{2}$ particle which in a frame of reference Σ is in the state $\phi_{p,\lambda=+1/2}$, i.e. fully polarised in its direction of motion. From a second frame of reference Σ' moving relative to the first, the particle will have momentum $p' = \Lambda p$ where Λ is the Lorentz transformation from Σ to Σ'. In general however the particle will not be in an eigenstate of helicity for Σ'; it will be in a superposition of helicity states

$$C_+\phi_{p',+1/2} + C_-\phi_{p',-1/2}$$

where the $C_\pm$ depend on both p and Λ. The situation is described by saying that the spin undergoes a rotation called the Wigner rotation, when seen from a moving frame of reference. In certain circumstances this rotation may be zero, as for example when the Lorentz transformation from Σ to Σ' is along the direction of motion of the particle. In this case as we saw in §4.4, the helicity is unchanged.

We consider the effect of a Lorentz transformation Λ on a helicity state $\phi_{p\lambda}$ of a particle of mass $m \neq 0$ and spin s. The transformed state $U(\Lambda)\phi_{p\lambda}$ obtained with the aid of the unitary operator $U(\Lambda)$ corresponding to Λ, can be regarded as a new state relative to the original

† The material in this section is not necessary for understanding the later chapters.

reference frame denoted by Σ (active point of view), or as the description of the same system from a new reference frame Σ', where the coordinates in Σ and Σ' are related by the transformation Λ (passive point of view).

We have used the active interpretation up to now, but the second point of view will be more appropriate when considering changes from for example the LAB to the CM frame.

The same formulae appropriately interpreted hold in both cases.

We proved above (see (4.35)) that the state $U(\Lambda)\phi_{p\lambda}$ has momentum p', where

$$p'_\mu = \sum_\nu \Lambda_{\mu\nu} p_\nu$$

and

$$p_\mu = ((p^2+m^2)^{1/2}, p), \quad p'_\mu = ((p'^2+m^2)^{1/2}, p')$$

The corresponding relation between the three-momenta will be written

$$p' = \Lambda p \tag{4.119}$$

The proof was for the spin zero case. When a helicity label is present the result remains true, but the helicity of the transformed state need not be the same as that of the original state. Nevertheless it must be possible to expand the transformed state in terms of the states $\phi_{p'\lambda'}$ $(-s \leqslant \lambda' \leqslant s)$, which form a complete set of states for the particle belonging to momentum p'. Thus we have

$$U(\Lambda)\phi_{p\lambda} = \sum_{\lambda'} M_{\lambda'\lambda}\phi_{p'\lambda'} \tag{4.120}$$

where the expansion coefficients have been written as a matrix for future convenience.

We now introduce the defining equations of the helicity state (4.55) written more briefly as

$$\phi_{p\lambda} = U(H_p)\phi_{0\lambda} \tag{4.121}$$

where H_p denotes the Lorentz transformation consisting of a boost along the z-axis followed by a rotation

$$H_p = R(\phi, \theta, 0)\mathcal{Z}_p \tag{4.122}$$

The corresponding operators satisfy

$$U(H_p) = U(R(\phi, \theta, 0))\, U(\mathcal{Z}_p)$$

Similarly for each state of momentum p'

$$\phi_{p'\lambda'} = U(H_{p'})\phi_{0\lambda'} \tag{4.123}$$

Inserting these expressions into (4.120) we have

$$U(\Lambda)U(H_p)\phi_{o\lambda} = \sum_{\lambda'} M_{\lambda'\lambda}U(H_{p'})\phi_{o\lambda'}$$

On multiplying through by $U^{-1}(H_{p'}) = U(H_{p'}^{-1})$ and making use of the group property of unitary operators (2.42) we obtain

$$U(H_{p'}^{-1}\Lambda H_p)\phi_{o\lambda} = \sum_{\lambda'} \phi_{o\lambda'}M_{\lambda'\lambda} \tag{4.124}$$

According to this equation the transformation $H_{p'}^{-1}\Lambda H_p$ connects two rest-states of the particle. It must therefore represent a pure rotation which we denote by R_W

$$R_W = H_{p'}^{-1}\Lambda H_p \tag{4.125}$$

R_W is called the Wigner rotation. It is important that $\Lambda, \boldsymbol{p}$ and $\boldsymbol{p}'$ in this expression are related by (4.119). The axis and angular magnitude of the rotation R_W depend on $\boldsymbol{p}$ and Λ. R_W will be calculated for a special case in the next section.

Returning to the analysis we note that on the left-hand side of (4.124) we have a pure rotation operator acting on a member of a standard set of angular momentum eigenstates $\phi_{o\lambda}$ (recall (4.48)), hence (4.124) may be written as

$$U(R_W)\phi_{o\lambda} = \sum_{\lambda'} \phi_{o\lambda'}\mathfrak{D}^s_{\lambda'\lambda}(R_W)$$

where $\mathfrak{D}^s_{\lambda'\lambda}$ is a rotation matrix.

These results may be combined to give the general transformation law of a helicity state

$$\begin{aligned}
U(\Lambda)\phi_p &= U(\Lambda)U(H_p)\phi_{o\lambda} \\
&= U(\Lambda H_p)\phi_{o\lambda} \\
&= U(H_{p'}H_{p'}^{-1}\Lambda H_p)\phi_{o\lambda} \\
&= U(H_{p'})U(H_{p'}^{-1}\Lambda H_p)\phi_{o\lambda} \\
&= U(H_{p'})\{\textstyle\sum_{\lambda'}\phi_{o\lambda'}\mathfrak{D}^s_{\lambda'\lambda}(R_W)\} \\
&= \textstyle\sum_{\lambda'} U(H_{p'})\phi_{o\lambda'}\mathfrak{D}^s_{\lambda'\lambda}(R_W)
\end{aligned}$$

(we used the group property repeatedly and introduced $1 = H_{p'}H_{p'}^{-1}$ which changes nothing, at an appropriate point).

So finally we have the desired transformation law,

$$U(\Lambda)\phi_{p\lambda} = \sum_{\lambda'} \phi_{p'\lambda'}\mathfrak{D}^s_{\lambda'\lambda}(R_W) \tag{4.126}$$

where

$$R_W = H_{p'}^{-1}\Lambda H_p \tag{4.125}$$

and

$$p' = \Lambda p$$

This law which was first derived by Wigner (1939) is particularly simple. The effect of a Lorentz transformation Λ on the momentum variable is clear, and the effect on the helicity indices is described by the now familiar $\mathfrak{D}$-function. R_W is independent of the mass of the particle and of its spin: it is a kinematical effect depending only on the multiplication properties of the Lorentz transformations. This will become clear when we consider a special case and derive an expression for R_W.

4.10.2 *The Wigner rotation in a special case*

Let $\boldsymbol{p}$ lie in the xz-plane making an angle θ with the z-axis

$$\boldsymbol{p} = (p \sin\theta, 0, p\cos\theta)$$

and let Λ be a pure Lorentz transformation in the z-direction so that $\boldsymbol{p}'$ given by (4.119) also lies in the xz-plane,

$$\boldsymbol{p}' = (p' \sin\theta', 0, p' \cos\theta')$$

Thus the y-coordinates remain unchanged throughout and the Wigner rotation must be a rotation about the y-axis. We denote the rotation angle by ω.

In writing down the Lorentz transformation H_p of (4.122) (with $\phi = 0$) it is convenient to parametrise the boost along the z-axis by the hyperbolic parameter κ (sometimes called the rapidity).

We define κ by

$$1 : \cosh\kappa : \sinh\kappa = m : (p^2 + m^2)^{1/2} : |\boldsymbol{p}|$$

Then we have

$$H_p = R(0, \theta, 0)\mathfrak{Z}(\kappa)$$

where

$$\mathfrak{Z}(\kappa) = \begin{array}{c} \begin{matrix} t & x & z \end{matrix} \\ \begin{bmatrix} \cosh\kappa & 0 & \sinh\kappa \\ 0 & 1 & 0 \\ \sinh\kappa & 0 & \cosh\kappa \end{bmatrix} \begin{matrix} t \\ x \\ z \end{matrix} \end{array} \tag{4.127}$$

and

$$R(0, \theta, 0) = \begin{bmatrix} 1 & 0 & 0 \\ 0 & \cos\theta & \sin\theta \\ 0 & -\sin\theta & \cos\theta \end{bmatrix} \tag{4.128}$$

Here we have omitted the y-row and column of the Lorentz transformation matrices since as noted above the y-coordinate is unchanged.

The matrix $H_{p'}$ corresponding to the transformed momentum p' is given by

$$H_{p'} = R(0, \theta', 0)\mathcal{Z}(\kappa') \tag{4.129}$$

where $\mathcal{Z}(\kappa')$ and $R(0, \theta', 0)$ are given by similar matrices with κ' and θ' replacing κ and θ. κ' is defined by

$$1 : \cosh \kappa' : \sinh \kappa' = m : (p'^2 + m^2)^{1/2} : |p'|$$

The Lorentz transformation Λ is

$$\Lambda = (1 - u^2)^{-1/2} \begin{bmatrix} 1 & 0 & u \\ 0 & (1-u^2)^{1/2} & 0 \\ u & 0 & 1 \end{bmatrix} \tag{4.130}$$

The Wigner rotation R_ω is defined by

$$H_{p'} R_\omega = \Lambda H_p \tag{4.131}$$

On writing

$$R_\omega = \begin{bmatrix} 1 & 0 & 0 \\ 0 & \cos\omega & \sin\omega \\ 0 & -\sin\omega & \cos\omega \end{bmatrix} \tag{4.132}$$

and substituting (4.127)–(4.130) and (4.132) in (4.131) we obtain enough equations to solve for κ' and θ' (the parameters defining p') and ω in terms of κ and θ and the velocity u of the Lorentz transformation Λ.

If we denote the velocity of the particle in the state $\phi_{p\lambda}$ by v_p

$$v_p = \frac{|p|}{E_p} = \tanh \kappa$$

then the expression for ω can be written

$$\tan \omega = \frac{u \sin\theta\, (1 - v_p^2)^{1/2}}{v_p + u\cos\theta} \tag{4.133}$$

Equation (4.126) written out for this case becomes

$$U(\Lambda)\phi_{p\lambda} = \sum_{\lambda'=-s}^{+s} \phi_{p'\lambda'} d^s_{\lambda'\lambda}(\omega)$$

We may say that in the transformed state the spin has been rotated by an angle ω relative to the direction of the momentum p' so that while Λ boosts p towards the z-axis ($\theta' < \theta$), the spin lags behind by an angle ω. The situation is illustrated in fig. 4.4.

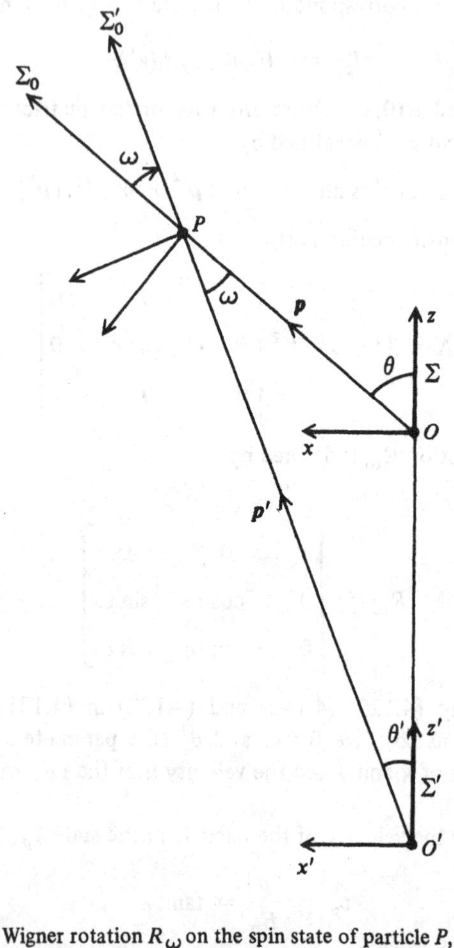

Fig. 4.4. The Wigner rotation R_ω on the spin state of particle P, corresponding to the Lorentz transformation from Σ to Σ' ($\omega < \theta - \theta'$).

The following special cases of (4.133) may be noted. If $\theta = 0$, $\tan \omega = 0$ and hence $\omega = 0$: this corresponds to a Lorentz transformation along the direction of motion of the particle and, as we have already seen, this leaves the helicity unchanged, so there is no spin transformation. This is no longer true if the Lorentz transformation reverses the direction of motion: this corresponds to the denominator of (4.133) passing through zero.

An important case is when the particle is ultra-relativistic $|\,p\,| \gg m$, so that $v_p \approx 1$ and (4.133) gives $\omega \approx 0$. Thus for an ultra-relativistic particle the helicity is more nearly invariant under a Lorentz transformation. This is consistent with the behaviour of a zero-mass particle for which the helicity is truly a Lorentz invariant quantity.

Finally we examine the non-relativistic limit. This corresponds to $u \ll 1$ and $v_p \ll 1$.

Then

$$\tan \omega \approx \frac{u \sin \theta}{v_p + u \cos \theta} \tag{4.134}$$

We may think of a classical non-relativistic spinning object in translational motion with momentum $\boldsymbol{p}$ in a frame Σ. From the point of view of a second frame Σ' the axis of spin has the same absolute orientation while the momentum becomes $\boldsymbol{p}' = \boldsymbol{p} + m\boldsymbol{u}$. If the spin axis is initially parallel to $\boldsymbol{p}$, it makes an angle with $\boldsymbol{p}'$ equal to the angle between $\boldsymbol{p}'$ and $\boldsymbol{p}$, which in the notation used earlier is $(\theta - \theta')$. It is not difficult to show that the right-hand side of (4.134) is equal to $\tan(\theta - \theta')$.

So the effect of using the Lorentz transformation in place of the Galilean transformation is to cause ω to be less than the value $\theta - \theta'$. This deviation of ω from its non-relativistic value is related to the Thomas precession (Thomas, 1926, 1927) familiar in atomic physics. The common origin of these effects lies in the fact that the resultant of two pure Lorentz transformations along different directions is not itself a pure Lorentz transformation, but a Lorentz transformation combined with a rotation.

4.10.3 *Transformation of $\mathcal{S}$-matrix elements*

From the general transformation law of helicity states (4.126) we may derive the transformation law of helicity matrix elements of any operator A whose Lorentz transformation properties are known. We confine ourselves to the important special case of the scattering operator $\mathcal{S}$ which is invariant under Lorentz transformations. Thus

$$\mathcal{S} = U^{-1}(\Lambda)\, \mathcal{S}\, U(\Lambda)$$

It follows that the transition operator $\mathcal{S}$ defined by

$$\mathcal{S} = 1 + \mathrm{i}\mathcal{T}$$

is also an invariant operator,

$$\mathcal{T} = U^{-1}(\Lambda)\, \mathcal{T}\, U(\Lambda) \tag{4.135}$$

Let us further consider πN scattering and use the notation of §4.8.

The scattering amplitude $f_{\mu\lambda}(\theta)$ is the matrix element of $\mathfrak{T}$ between πN helicity states $\phi_{p\lambda}$ and $\phi_{q\lambda}$ where θ is the angle between $\boldsymbol{p}$ and $\boldsymbol{q}$.

On taking the $\phi_{q\mu}$–$\phi_{p\lambda}$ matrix element of (4.135), re-arranging, and using the transformation law (4.126) we have

$$\begin{aligned}(\phi_{q\mu}, \mathfrak{T}\phi_{p\lambda}) &= (\phi_{q\mu}, U^{-1}(\Lambda)\,\mathfrak{T}U(\Lambda)\,\phi_{p\lambda}) \\ &= (U(\Lambda)\,\phi_{q\mu}, \mathfrak{T}U(\Lambda)\,\phi_{p\lambda}) \\ &= \sum_{\mu'\lambda'} \mathfrak{D}^{(1/2)*}_{\mu'\mu}(R_2)(\phi_{q'\mu'}, \mathfrak{T}\phi_{p'\lambda'})\mathfrak{D}^{(1/2)}_{\lambda'\lambda}(R_1)\end{aligned} \tag{4.136}$$

where R_2 and R_1 denote the Wigner rotations corresponding to the momenta $\boldsymbol{q}$ and $\boldsymbol{p}$, and Lorentz transformation Λ.

If $\phi_{p\lambda}$ is an πN state in the CM, $\phi_{p'\lambda'}$ denotes a state of relative momentum $\boldsymbol{p}'$ and non-zero total momentum, e.g. in the LAB frame, so (4.136) is the relation between the transition matrix elements in the CM and the LAB. In some circumstances it is not necessary to know the Wigner rotations. For example if the spin average of the squared matrix elements are taken, the $\mathfrak{D}$-matrices drop out upon using the standard property

$$\sum_{\lambda} |\mathfrak{D}^{s}_{\lambda'\lambda}(R_1)|^2 = 1$$

so that we have

$$\tfrac{1}{2}\sum_{\lambda\mu} |(\phi_{q\mu}, \mathfrak{T}\phi_{p\lambda})|^2 = \tfrac{1}{2}\sum_{\lambda'\mu'} |(\phi_{q'\mu'}, \mathfrak{T}\phi_{p'\lambda'})|^2 \tag{4.137}$$

which expresses the equality of the spin averaged transition probabilities in the two reference frames.

CHAPTER 5

PARITY

We now turn to the symmetry of space inversion which, although applicable to classical systems, only gains its full significance in the study of systems described by quantum mechanics. In the passive picture space inversion involves changing of the spatial reference frame from a right-handed to a left-handed one. Hence invariance under space inversion is equivalent to the indistinguishability of left and right.

Unlike the principles of rotational and Lorentz invariance discussed in preceding chapters, space inversion is known not to be a universal symmetry of Nature on the microscopic scale. The discovery of the violation of space inversion invariance in weak interactions in 1956 led to a critical re-examination of all the accepted symmetry principles from both the experimental and theoretical points of view. This critical attitude has continued ever since. In contrast to previous chapters where the task was to set up the detailed formalism which builds in the accepted principles of rotational and Lorentz invariance, we shall be concerned here with tests of space inversion invariance as well as with consequences of assuming the symmetry to be true. These considerations are of course intimately related.

It is interesting to note that considerations of left–right symmetry arise in chemical and biological processes. For example, certain complex molecules exist in two forms: isomers having the same formula, but whose structures are related to each other by the operation of space inversion. Solutions of these substances which contain an excess of one type of molecule are optically active, and the two isomers can thus be distinguished as dextro- or laevo-rotary (D or L). The amino acids of living matter are of this type and it is found that only L amino acids occur in living systems, while in laboratory syntheses D and L types are produced in equal amounts. This tells us that it is not the laws of chemistry but rather something in Nature which has exerted a preference for one isomer. This preference is closely related to questions of the origin of life.

5.1 Elementary theory of the parity operator

5.1.1 *Parity of wavefunctions*

In 1924 as a result of a study of the iron spectrum, Laporte noted that the energy levels of the iron atom could be classified into two types such that the observed transitions take place only between levels of opposite types. Soon after the discovery of quantum mechanics Wigner (1927) observed that Laporte's rule is a consequence of space inversion invariance in the radiation process.

Let us start by considering a single electron in a central field of force. The Schrödinger wavefunctions $\psi(\mathbf{r})$ describing the energy eigenstates are found to be either even or odd functions of the coordinate $\mathbf{r}$. Indeed for a Hamiltonian of this kind

$$H = -\frac{\hbar^2}{2m}\nabla^2 + V(|\mathbf{r}|)$$

the energy eigenfunctions are of the form

$$\psi_{nlm}(\mathbf{r}) = \mathcal{R}_{nl}(r)\, Y_{lm}(\theta, \phi)$$

Under the transformation of inversion in the origin, $\mathbf{r} \to -\mathbf{r}$, which in spherical polar coordinates is represented by

$$r \to r$$

$$\theta \to \pi - \theta$$

$$\phi \to \phi + \pi$$

we find from the properties of the spherical harmonics (§3.1.2) that

$$Y_{lm}(\pi - \theta, \phi + \pi) = (-1)^l Y_{lm}(\theta, \phi)$$

so that

$$\psi_{nlm}(-\mathbf{r}) = (-1)^l \psi_{nlm}(\mathbf{r}) \qquad (5.1)$$

$(-1)^l$ is called the parity of the state and in this case it is determined by the orbital angular momentum. States for which

$$\psi(-\mathbf{r}) = +\psi(\mathbf{r})$$

are said to have positive or even parity, while those for which

$$\psi(-\mathbf{r}) = -\psi(\mathbf{r})$$

have negative or odd parity.

The parity of the system at any moment cannot be obtained from the spatial distribution of mass or charge since these are given by $|\psi|^2$.

However, for a state of definite parity, $|\psi|^2$ must be symmetric with respect to inversion in the origin.

For a many electron atom the Hamiltonian is more complicated, particularly if account is taken of electron–electron interactions, and exact calculations are not possible. Nevertheless every stationary state can be assigned a definite parity, as Laporte found. In this case we have to rely on more general theory.

5.1.2 *Formal theory of the parity operator*

The hypothesis of space inversion invariance requires that the left-handed coordinate frame Σ' obtained from frame Σ by changing the sign of all three coordinates

$$P: r \rightarrow r' = -r \tag{5.2}$$

is an equally valid frame for expressing the laws of physics.

The state described by the function $\psi(r)$ in Σ is in Σ' assigned the wavefunction $\psi'(r')$ defined by

$$\psi'(r') = \psi(r) \tag{5.3}$$

where r' and r are related by (5.2). So the state considered is assigned the wavefunction

$$\psi'(r') = \psi(-r')$$

by Σ'. We are simply following the general theory of §2.3 here.

In the active interpretation $\psi(-r)$ is a new state of the system in the old frame Σ.

We define the unitary operator U_P by

$$U_P\psi(r) = \psi(-r) \tag{5.4}$$

When U_P is applied to any wavefunction it generates the space inverted wavefunction. The unitary property of U_P is easily verified: for any two functions ϕ and ψ we have

$$(U_P\phi, U_P\psi) = \int d^3r\phi^*(-r)\psi(-r)$$

$$= \int d^3r'\phi^*(r')\psi(r')$$

$$= (\phi, \psi)$$

Since

$$(U_P\phi, U_P\psi) = (\phi, U_P^\dagger U_P\psi)$$

it follows that

$$U_P^\dagger U_P = 1 \tag{5.5}$$

The geometrical operation of space inversion has the special property that when applied twice it is equivalent to the identity transformation, i.e. no change. In symbols

$$P^2 = 1 \tag{5.6}$$

It follows that

$$U_P^2 = 1 \tag{5.7}$$

where 1 denotes the unit operator. Equations (5.5) and (5.7) together imply that U_P is Hermitian and equal to its inverse,

$$U_P^\dagger = U_P = U_P^{-1} \tag{5.8}$$

U_P can therefore represent an observable quantity.

An eigenfunction of U_P satisfies

$$U_P\psi(r) = \eta\psi(r)$$

where the eigenvalue η is called the parity of the state. U_P is sometimes referred to as the parity operator. On applying U_P twice we have

$$U_P^2\psi(r) = \eta^2\psi(r)$$

but by (5.7)

$$U_P^2\psi(r) = \psi(r)$$

so

$$\eta^2 = 1$$

and hence

$$\eta = \pm 1$$

We see immediately that the even and odd functions of r are the eigenfunctions of U_P with eigenvalues $+1$ and -1 respectively.

A system such as an atom or nucleus will admit space reflection invariance if the Hamiltonian is invariant under the parity operation (5.2). In the case of an atom the Hamiltonian is the sum of all the kinetic energy terms and the mutual Coulomb interactions of the electrons and the nucleus. The formal substitution $r \rightarrow -r$ leaves such a Hamiltonian invariant. This is expressed by

$$U_P H U_P^{-1} = H \tag{5.9}$$

The operator on the left may be interpreted as the Hamiltonian for an observer using the inverted coordinate frame Σ'. Equation (5.9) states that the Hamiltonian for the two observers Σ' and Σ is the same, and hence Schrödinger's equation is equally valid for both.

A more particular consequence is as follows. Equation (5.9) can be written

$$[U_P, H] = 0 \tag{5.10}$$

expressing the fact that U_P and H commute. By a general theorem, it follows that energy eigenfunctions can be chosen to be eigenfunctions of U_P, that is to have a definite parity. We stress that this is so, even though it may not be possible to calculate explicitly the energy eigenfunctions. This result is one ingredient of the proof of Laporte's rule.

5.1.3 *Transformation of operators*

The parity transform of an operator A is defined by

$$A^P = U_P A U_P^{-1} \tag{5.11}$$

A^P has the same significance for the inverted frame Σ' as A has for Σ. As we showed in general in chapter 2, A^P has the same expectation value for the state $U_P\psi$ as A has for ψ.

Thus we have

$$\boldsymbol{r} \to \boldsymbol{r}^P = U_P \boldsymbol{r} U_P^{-1}$$

$\boldsymbol{r}^P$ is what we previously denoted by $\boldsymbol{r}'$, and so

$$U_P \boldsymbol{r} U_P^{-1} = -\boldsymbol{r} \tag{5.12}$$

Similarly

$$\boldsymbol{p} \to \boldsymbol{p}^P = U_P \boldsymbol{p} U_P^{-1}$$

and since

$$\boldsymbol{p} = -\mathrm{i}\hbar\boldsymbol{\nabla}$$

it follows that

$$U_P \boldsymbol{p} U_P^{-1} = -\boldsymbol{p} \tag{5.13}$$

As an illustration of these relationships, the parity transform of the plane wave state

$$\phi_{\boldsymbol{k}} = \mathrm{e}^{\mathrm{i}\boldsymbol{k}\cdot\boldsymbol{x}}$$

is

$$U_P\phi_{\boldsymbol{k}} = \mathrm{e}^{\mathrm{i}\boldsymbol{k}\cdot(-\boldsymbol{r})} = \phi_{-\boldsymbol{k}}$$

$\phi_{\boldsymbol{k}}$ is an eigenstate of the momentum operator $\boldsymbol{p}$,

$$\boldsymbol{p}\phi_{\boldsymbol{k}} = \boldsymbol{k}\phi_{\boldsymbol{k}}$$

and $U_P\phi_{\boldsymbol{k}}$ is an eigenstate of the space inverted momentum operator $\boldsymbol{p}^P = -\boldsymbol{p}$ belonging to the same eigenvalue $\boldsymbol{k}$ as the original state

$$\boldsymbol{p}^P U_P \phi_{\boldsymbol{k}} = \boldsymbol{k} U_P \phi_{\boldsymbol{k}}$$

From (5.12) and (5.13), the parity transformation law of other quantities may be determined. For example

$$\boldsymbol{L} = \boldsymbol{r} \wedge \boldsymbol{p} \to U_P \boldsymbol{L} U_P^{-1} = \boldsymbol{L} \tag{5.14}$$

This states that U_P and the orbital angular momentum commute, and is in accord with our earlier observation that angular momentum eigenstates in a central field problem are parity eigenstates too.

We have neglected spin so far. An atomic or nuclear Hamiltonian has spin dependence via a spin–orbit coupling interaction of the form

$$\boldsymbol{L}\cdot\boldsymbol{S}f(|\boldsymbol{r}|) \tag{5.15}$$

so it is necessary to know the parity transformation properties of the spin variables. Since $\boldsymbol{S}$ is an angular momentum, (5.14) suggests that we require

$$U_P \boldsymbol{S} U_P^{-1} = \boldsymbol{S} \tag{5.16}$$

Then the total angular momentum $\boldsymbol{J}$ also satisfies

$$U_P \boldsymbol{J} U_P^{-1} = \boldsymbol{J} \tag{5.17}$$

We can see that a spin–orbit interaction term (5.15) in the Hamiltonian is compatible with parity invariance.

It follows from the foregoing considerations that the helicity operator

$$\mathcal{H} = \frac{\boldsymbol{J}\cdot\boldsymbol{p}}{|\boldsymbol{p}|}$$

satisfies

$$U_P \mathcal{H} U_P^{-1} = -\mathcal{H} \tag{5.18}$$

so that its eigenvalues change sign under space inversion.

The classification of quantities as scalar, vector, etc. according to their behaviour under rotations may be extended to cover behaviour under space inversions. Scalar quantities are called true scalars or pseudo-scalars according as they do not or do change sign under inversions: energy and helicity are respective examples.

Similarly a *true* or *polar vector* is one which, like the space coordinate $\boldsymbol{r}$, changes sign under inversion, while a *pseudo- (or axial) vector* such as angular momentum $\boldsymbol{L}$ is unchanged in sign under inversion.

This classification applies equally well to the operators representing these quantities and to their eigenvalues.

5.1.4 *A remark on the definition of symmetry transformations*

We formulated the theory of the parity operator using the coordinate representation of quantum mechanics. One may avoid committing oneself to this or any other particular representation by making the transformation laws of $\boldsymbol{r}$ and $\boldsymbol{p}$ the starting point of the discussion. This more fundamental approach proceeds by writing down the space inversion transformation

$$\left.\begin{aligned} r \to r^P &= -r \\ p \to p^P &= -p \end{aligned}\right\} \tag{5.19}$$

by analogy with the classical case. Then one notes that the fundamental commutation relations

$$[r_i, p_j] = i\hbar\delta_{ij} \tag{5.20}$$

remain valid for the transformed operators

$$\begin{aligned} [r_i^P, p_j^P] &= (-1)^2 i\hbar\delta_{ij} \\ &= i\hbar\delta_{ij} \end{aligned}$$

A transformation with this property is called *canonical* and it follows from quantum mechanical transformation theory that a unitary operator U_P exists with the properties

$$\left.\begin{aligned} U_P r U_P^{-1} &= r^P \\ U_P p U_P^{-1} &= p^P \end{aligned}\right\} \tag{5.21}$$

Equations (5.19) and (5.21) together correspond to (5.12) and (5.13) written down earlier.

From the present point of view we say that space inversion is a symmetry of the system if in the inverted frame of reference, the commutation relations (5.20) and the Hamiltonian $H(r, p)$ retain the same form.

In the previous treatment we had already satisfied the commutation relations by substituting

$$p \to -i\hbar\nabla$$

in H.

5.1.5 *Parity conservation in reactions*

We have considered parity for stationary states. Now let us consider parity in scattering and reaction processes. If the initial state in some process is an eigenstate of parity and if the interaction is invariant under space inversion, then the final state must have the same parity. Let us prove this. Space inversion invariance is expressed by

$$U_P \mathcal{T} U_P^{-1} = \mathcal{T} \tag{5.22}$$

$\mathcal{T}$ is the transition operator, and is a function of the Hamiltonian, so that (5.22) follows from (5.9). On rearranging (5.22) and taking the ϕ_b–ϕ_a matrix element we have

$$(\phi_b, U_P \mathcal{T} \phi_a) = (\phi_b, \mathcal{T} U_P \phi_a)$$

If ϕ_a and ϕ_b are eigenstates of parity we have

$$\eta_b(\phi_b, \mathcal{T}\phi_a) = \eta_a(\phi_b, \mathcal{T}\phi_a)$$

which implies that the amplitude for a transition from ϕ_a to ϕ_b is zero unless ϕ_a and ϕ_b have the same parity.

Scattering states are not eigenstates of parity. They are more nearly plane wave states and are superpositions of states of different l values and hence superpositions of states of opposite parities. Exceptions occur when the reaction takes place at low enough energy that only the lowest l value can contribute. However in general it is necessary to proceed differently.

Typically we measure the probability W for transitions between a state ϕ_a and a state ϕ_b. W will depend on the momenta and spins of the particle present in the initial state prepared and the final state detected. We represent this by writing $W(\boldsymbol{p}, \boldsymbol{S})$ where $\boldsymbol{p}$ stands for all the momenta (initial and final) and $\boldsymbol{S}$ symbolises all the spins. It might be more appropriate to replace $\boldsymbol{S}$ by a density matrix or polarisation vector for each of the particles with spin, but the result of the argument would be the same.

Now

$$W(\boldsymbol{p}, \boldsymbol{S}) = |(\phi_b, \mathcal{T}\phi_a)|^2$$

where the transition operator $\mathcal{T}$ is to be calculated from H in a detailed theory. Now we obtain the same result if on the right-hand side we replace ϕ_a, ϕ_b and $\mathcal{T}$ by $U_P\phi_a$, $U_P\phi_b$ and $U_P\mathcal{T}U_P^{-1}$ where U_P is the parity operator. This would in fact be true for any unitary operator.

Now the effect of U_P on ϕ_a and ϕ_b is to transform the momenta and angular momenta according to the rules

$$\boldsymbol{p} \to -\boldsymbol{p}$$

$$\boldsymbol{S} \to +\boldsymbol{S}$$

Thus it follows that

$$W(\boldsymbol{p}, \boldsymbol{S}) = W^P(-\boldsymbol{p}, \boldsymbol{S})$$

where

$$W^P(\boldsymbol{p}, \boldsymbol{S}) = |(\phi_b, U_P\mathcal{T}U_P^{-1}\phi_a)|^2$$

is computed from the transformed transition operator (or ultimately from the transformed Hamiltonian $U_PHU_P^{-1}$) in the same way as W was from $\mathcal{T}$ or H.

Now if parity is a symmetry operation

$$U_PHU_P^{-1} = H$$

and hence

$$U_P\mathcal{T}U_P^{-1} = \mathcal{T}$$

It follows that W^P is the same function of its arguments as W, and hence

$$W(\boldsymbol{p}, \boldsymbol{S}) = W(-\boldsymbol{p}, \boldsymbol{S}) \tag{5.23}$$

In order to test (5.23) it is sufficient to determine how W depends on quantities such as $\boldsymbol{p}_1 \wedge \boldsymbol{p}_2 \cdot \boldsymbol{p}_3$ and $\boldsymbol{S} \cdot \boldsymbol{p}$ which change sign under space inversion. Equation (5.23) can be interpreted as the statement that the number of events in which such a quantity is positive is equal to the number of events for which it is negative. So to test parity conservation we must perform an experiment to see how W depends on the pseudoscalars formed from the available momentum and spin vectors. In carrying this out it is assumed that all other spins are summed over (not observed) and all other momenta are integrated over.

5.2 Parity in atomic and nuclear physics

The applications of the parity concept in atomic and nuclear physics will be discussed here because some of the methods and results encountered in a familiar context have a wider validity. In addition sensitive tests of parity conservation are possible in these fields.

5.2.1 *Some consequences of parity conservation*

Let us start by completing the proof of Laporte's rule as given by Wigner.

The probability per unit time for an atom to make a transition from a state ψ_a to a state ψ_b with the emission of electric dipole radiation is proportional to the square of the matrix element

$$(\psi_b, \boldsymbol{d}\psi_a) \tag{5.24}$$

where $\boldsymbol{d}$ denotes the electric dipole moment operator

$$\boldsymbol{d} = \sum_i e\boldsymbol{r}^i \tag{5.25}$$

in which the sum goes over all the electrons. Dipole radiation is characterised as that in which the radiation carries off one unit of angular momentum. If the matrix element (5.24) vanishes, the radiative transition $\psi_a \rightarrow \psi_b$ is called *forbidden*. However it may still be possible for the transition to occur with the emission of magnetic dipole (or higher multipole) radiation, but then one finds that the matrix element is reduced by a factor ka where k is the wavenumber of the radiation and a is a length characteristic of the size of the radiating system. For optical emission by atoms, $ka \sim 10^{-3}$.

We have seen that as a consequence of parity invariance of the Hamiltonian for the atom alone, every energy eigenstate can be assigned a parity; even or odd. We denote the parities of the states ψ_a and ψ_b by η_a and η_b. Formally we have

$$U_P\psi_a = \eta_a\psi_a \tag{5.26a}$$

$$U_P\psi_b = \eta_b\psi_b \tag{5.26b}$$

Now it follows immediately from (5.12) that d is a polar vector operator, for which

$$U_P d U_P^{-1} = -d \tag{5.27a}$$

We rearrange this as

$$d = -U_P^{-1} d U_P \tag{5.27b}$$

and take the ψ_b–ψ_a matrix element. After use of equations (5.26) one finds

$$(\psi_b, d\psi_a) = -\eta_b\eta_a(\psi_b, d\psi_a)$$

We conclude that the transition dipole moment $(\psi_b, d\psi_a)$ vanishes unless $\eta_a\eta_b = -1$, that is unless ψ_a and ψ_b are of opposite parity. This is precisely the content of Laporte's rule: dipole transitions are allowed only between states of opposite parity.

The preceding argument is also valid for nuclei. In this case the coordinates r^i in (5.25) refer to the protons in the nucleus. For the emission of gamma rays with energies of the order of MeV from nuclei ($a \sim 10^{-13}$ cm), $ka \sim 10^{-3}$ and higher multipole radiation is normally less intense than electric dipole radiation. The standard reference on multipole radiation from nuclei is still Blatt and Weisskopf (1952).

If, instead of the transition dipole moment, we enquire about the static dipole moment of a quantum-mechanical system (atom, nucleus, elementary particle) in a stationary state, then we have the following result. If the Hamiltonian for the system is invariant under space inversion, then in any stationary state ψ the electric dipole moment vanishes, provided the state ψ is not parity degenerate. ψ is parity degenerate if there exists a second stationary state ψ' degenerate in energy with ψ such that ψ and ψ' have opposite parity. Thus degeneracy due to spin or to L_z is disregarded. The result is easily proved. If the state ψ is non-degenerate and has parity η, we take the expectation value of both sides of (5.27b) as in the discussion of transitions. One has

$$(\psi, d\psi) = -\eta^2(\psi, d\psi)$$

and since $\eta = \pm 1$,

$$(\psi, d\psi) = 0$$

The need for the proviso about degeneracy can be understood by considering the wave-mechanical treatment of the hydrogen atom in an electric field, $\mathcal{E}$ (the Stark effect). An energy shift linear in $\mathcal{E}$ corresponds to a non-zero electric dipole moment. There is no such shift for the 1s-state: for the 2s- and 2p-states which are degenerate (in the simple theory) one finds a linear energy shift. This can be understood by observing that a wavefunction

$$\Psi = \alpha\psi_{2s} + \beta\psi_{2p}$$

can correspond to an asymmetric charge distribution with a non-vanishing dipole moment. Ψ has just the form that the wavefunction for the system takes when the Stark effect is calculated using degenerate perturbation theory (see e.g. Schiff (1968) chapter 8).

5.2.2 *Experimental tests of parity conservation in atomic physics*

We may test the validity of parity conservation in atoms by looking for transitions which are forbidden according to Laporte's rule. This is a typical test of a symmetry principle or conservation law: we seek experimental evidence for the occurrence of some process or effect which is strictly forbidden if that conservation law is valid. We shall meet many instances of this as we proceed. In all these cases it is desirable to be able to express the result of the experiments quantitatively. To do this we must imagine what would happen if the conservation law were violated. In the present case this requires a (hopefully small) additional interaction between electrons and nucleus in the atom which violates parity conservation. As a result the energy eigenstates of this atom cannot be truly eigenstates of parity which is no longer a good quantum number. However we can write the wavefunction of a stationary state in the form

$$\Psi = \psi_{\eta} + F\psi_{-\eta} \tag{5.28}$$

where ψ_{η} and $\psi_{-\eta}$ are of opposite parity.

Provided the hypothetical interaction is small, the coefficient F is small. It expresses the relative amplitude of the state of 'wrong' parity in the state which is 'mostly' of parity η. Even in the presence of parity violation it remains true that the dipole operator has non-vanishing matrix elements only between wavefunctions of opposite parity. So a transition $\Psi_a \to \Psi_b$ which is forbidden in the limit of parity conservation may take place by electric dipole transition (E1) between say Ψ_a and the 'wrong' parity part of Ψ_b (and vice versa). So the relative amplitude for such a transition is of order F, and the probability of order $|F|^2$. Now the electric dipole selection rule is in itself an approximation: transitions between state of the same parity can for example take place by magnetic dipole (M1) transitions but, as noted, these are reduced in intensity by a factor $(a/\lambda)^2$, where λ is the wavelength of emitted radiation, and a is the dimension of radiating system. The experimental evidence on electric dipole transitions for atoms shows no deviation from Laporte's rule beyond those expected from higher order multipoles. We therefore conclude that if any parity violating admixture is present its magnitude is limited by

$$|F|^2 \lesssim (a/\lambda)^2$$

This gives $|F|^2 \lesssim 10^{-6}$ as an upper limit on parity violation in the electromagnetic interaction responsible for atomic structure.

5.2.3 *Tests of parity conservation in nuclear physics*

The general principles of parity conservation and its testing are the same here as in atomic physics. It is found that nuclear energy levels can be assigned a unique parity which is evidence that parity is conserved to a high degree by the strong interaction,

$$U_P H_{st} U_P^{-1} = H_{st}$$

To test this we examine a transition which is forbidden by space inversion invariance. The first such test was made by Tanner (1957) but we shall use as an illustration the case studied by Wilkinson (1958).

The second excited state of ^{6}Li lies at about 3.6 MeV and has spin parity 0^+. (It forms an isospin triplet with ^{6}He and ^{6}Be.) The break-up

$$^6\text{Li}^* \to {}^4\text{He} + \text{d} \tag{5.29}$$

is energetically allowed, but is forbidden by parity conservation while the radiative decay

$$^6\text{Li}^* \to {}^6\text{Li} + \gamma$$

is found to have a much greater probability. ^{4}He and d have J^P equal to 0^+ and 1^+ respectively, so the final state must have orbital angular momentum $l = 1$, and hence negative parity. If parity is violated, the wavefunction for $^6\text{Li}^*$ will have a small admixture of a negative parity (0^-) component, and (5.29) can occur with relative amplitude F in the notation of (5.28).

One could prepare $^6\text{Li}^*$ and search for the small amount of heavy particle emission (5.29) in competition with the radiative decay, but this presents experimental difficulties. Instead one looks for the reaction

$$^4\text{He} + \text{d} \rightarrow {}^6\text{Li}^* \rightarrow {}^6\text{Li} + \gamma$$

the rate for which is controlled by the relatively small rate (width) $\Gamma_{\alpha d}$ for formation of the excited state. In this way Wilkinson put an experimental limit of 0.2 eV on $\Gamma_{\alpha d}$. Nuclear level widths are typically 1 MeV, so we conclude that

$$|F|^2 \lesssim 10^{-7}$$

for the strong interaction.

A different test is to search for circular polarisation of gamma rays emitted by excited nuclear states. Circular polarisation corresponds to the expectation value $\langle \boldsymbol{S}\cdot\boldsymbol{p} \rangle$ of the photon helicity and, as we have seen (§5.1.5), such a quantity must vanish by space inversion invariance. All other spins must be averaged over for this to hold, so the excited states must be produced by unpolarised beams on unpolarised targets.

In a very sensitive experiment, Lobashov *et al.* (1966) studied the 400 keV γ-rays emitted in the transition from a $\frac{9}{2}^-$ excited state to the $\frac{7}{2}^+$ ground state of ^{175}Lu. This transition is an electric dipole transition, like the atomic optical transitions. Admixtures of opposite parity in these two states leads to γ-emission by magnetic dipole (M1) transitions which take place with no parity change. The amplitude for γ-emission of 400 keV can be written

$$A = A_{\text{E1}} + FA_{\text{M1}}$$

and the intensity to first order in F is

$$|A|^2 \simeq |A_{\text{E1}}|^2(1 + FR)$$

It is the interference between the E1- and M1-radiations which leads to a circular polarisation. So the expected polarisation is RF where

$$R = \frac{2\,\text{Re}\,(A_{\text{E1}}A^*_{\text{M1}})}{|A_{\text{E1}}|^2} \sim \left|\frac{A_{\text{M1}}}{A_{\text{E1}}}\right|$$

The enhancement factor R can be estimated from the theory of multipole radiation and a knowledge of nuclear structure. For the present case $R \simeq 50$. The photon polarisation was detected by scattering in magnetised iron, and was found to be

$$P_\gamma = (4 \pm 1) \times 10^{-5}$$

giving a relative amplitude of parity violating interaction

$$|F| \sim 0.8 \times 10^{-6}$$

Similar experiments have been performed by Boehm and Kankeleit (1965) and Bock and Schopper (1965).

A third kind of test of parity conservation is to look for an angular asymmetry in the emission of γ-rays from polarised nuclei: if parity is conserved the angular distribution measured with respect to the polarisation direction cannot contain odd powers of $\cos\theta$, otherwise we should have a non-zero value for $\langle \boldsymbol{S}_{\text{nucl}} \cdot \boldsymbol{p}_\gamma \rangle$. Grodzins and Genovese (1961) looked for a forward–backward asymmetry of photons from polarised ^{57}Fe. To within experimental error none was found.

A similar type of experiment uses polarised neutrons to produce the polarised nuclear level. Abov *et al.* (1965, 1968) studied

$$n_{\text{thermal}} + {}^{113}\text{Cd} \rightarrow {}^{114}\text{Cd}^*(1^+) \rightarrow {}^{114}\text{Cd}\,(0^+) + \gamma$$

taking advantage of the isolated resonance level which makes cadmium such a good absorber of slow neutrons. The $\cos\theta$ term in the decay results from interference between the dominant M1-transition $1^+ \rightarrow 0^+$ and the irregular E1-transition.

So in the notation used above, the coefficient of the $\cos\theta$ term is of order RF. Abov and co-workers found for the asymmetry

$$\epsilon = -(3.7 \pm 0.9) \times 10^{-4}$$

which leads to a value for F of

$$|F| \simeq 4 \times 10^{-7}$$

The experiments discussed here were prompted by the discovery of violation of parity conservation in weak interactions, which will be discussed below. The current–current theory of weak interactions (Feynman and Gell-Mann, 1958) predicts a parity violating nucleon–nucleon potential of strength 10^{-7} relative to the strong interaction. The size of the effects detected by Abov and Lobashov and co-workers are consistent with a weak interaction origin and there is no evidence of parity violation in the strong nuclear interaction.

The parity violating nature of the weak force is now understood so well that the main value of the experiments described here is a test of nuclear structure theory. A further discussion of this aspect and a critical examination of the experiments referred to may be found in Hamilton (1968).

To sum up, there is good evidence that parity is conserved in both the electromagnetic and strong interactions. We shall proceed to discuss parity in elementary particle physics on the assumption that this remains true, and return at the end of the chapter to note some further tests of invariance.

5.3 Parity in elementary particle physics

When we turn to elementary particle processes in which particles can be created and destroyed, new aspects of the parity concept are revealed. The first of these is the notion of the intrinsic parity of a particle.

This can be described in an idealised situation as follows. Suppose a system of particles in a state A of a definite parity makes a transition to another state B with the emission of a neutral pion, *but with no change in the other kinds of particle*:

$$A \to B + \pi^0$$

e.g.

$$\mathrm{p} + \mathrm{n} \to \mathrm{p} + \mathrm{n} + \pi^0$$

If weak or beta decay processes are excluded it is found that the parity of the final state $(B\pi^0)$ is always opposite to that of A. This observation can be reconciled with the law of parity conservation if we suppose that the π^0 has an internal or intrinsic parity, and that the overall parity of the final state $(B\pi^0)$ is the product of the parity due to motion as defined previously and the intrinsic parity of the pion. Thus

$$\eta_{\text{final}} = \eta_{\text{orbital}}\eta_{\pi^0}$$

where η_{orbital} denotes the orbital parity due to relative motion. By taking η_{π^0} to be -1 we can restore the law of parity conservation

$$\eta_{\text{initial}} = \eta_{\text{final}} \tag{5.30}$$

Formally if the particles are described by wavefunctions the intrinsic parity η_P is defined by generalising the transformation law (5.4) to

$$U_P\psi(r) = \eta_P\psi(-r)$$

For consistency we should introduce intrinsic parity factors for all the other particles present in the states A and B. However, we imposed the condition that the types of other particle present should not change, so that these factors cancel out of (5.30). This is possible only because the π^0 is uncharged. A charged pion could be produced only at the expense of a change of charge elsewhere in the system, e.g.

$$\mathrm{p} + \mathrm{p} \to \mathrm{p} + \mathrm{n} + \pi^+$$

and the intrinsic parities of A (pp) and B (pn) do not cancel. The result is that although all particles may be assigned intrinsic parities, some of these must be fixed by definition. We shall discuss this further when describing how some intrinsic parities have been determined.

The other new feature of space inversion symmetry is that it is not a universally valid symmetry: the weak interaction violates parity conservation. Let us describe an experiment which shows this unambiguously.

Λ^0-hyperons are produced by the strong reaction

$$\pi^- + p \to \Lambda^0 + K^0$$

The Λ^0-hyperon decays predominantly by the modes

$$\Lambda^0 \to p + \pi^-$$
$$\to n + \pi^0$$

The Λ^0 lifetime of 2.5×10^{-10} s indicates that this is a weak process.

In a typical production–decay event in a bubble chamber we know the incident pion direction represented by its momentum $\boldsymbol{p}_{\pi 1}$, and from the p and π^- momenta $\boldsymbol{p}_p$ and $\boldsymbol{p}_{\pi 2}$ of the decay

$$\Lambda^0 \to p + \pi^-$$

we can recover $\boldsymbol{p}_\Lambda$. The axial vector $\boldsymbol{p}_{\pi 1} \wedge \boldsymbol{p}_\Lambda$ defines the positive normal to the production plane (see fig. 5.1).

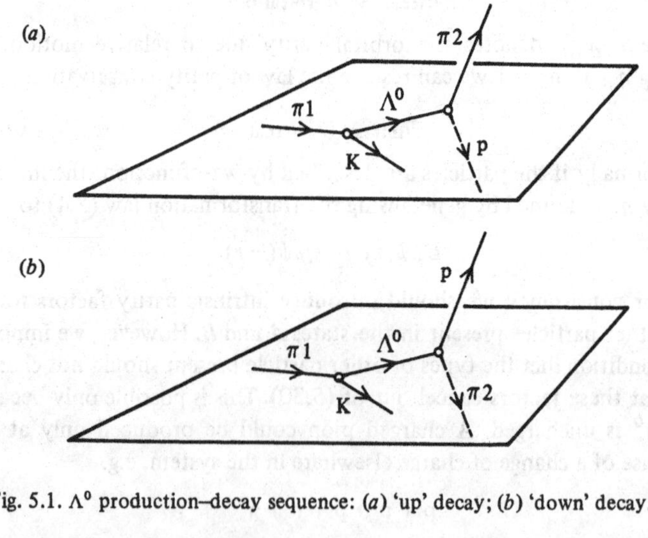

Fig. 5.1. Λ^0 production–decay sequence: (*a*) 'up' decay; (*b*) 'down' decay.

According to the discussion of §5.1.5, if parity is conserved in the process there should be as many decays in which the decay pion comes out on the positive side of the reaction plane (upward) as there are decays on the negative side (downward). For these two possibilities correspond to the pseudo-scalar quantity

$$\boldsymbol{p}_{\pi 1} \wedge \boldsymbol{p}_{\Lambda} \cdot \boldsymbol{p}_{\pi 2}$$

being positive and negative and, as we have seen, parity conservation requires that the mean value of a pseudo-scalar quantity be zero.

In 1957 following the suggestion of Lee and Yang, a group at Berkeley searched for an up–down asymmetry, and found that the decay pion was preferentially emitted upwards

$$\boldsymbol{p}_{\pi 1} \wedge \boldsymbol{p}_{\Lambda} \cdot \boldsymbol{p}_{\pi 2} > 0$$

Crawford *et al.* (1957) observed 353 bubble chamber events and found

$$N_{\mathrm{up}} = 215, \quad N_{\mathrm{down}} = 138$$

These results were later confirmed by other groups.

Thus we have a natural process which enables us to define unambiguously a right-handed coordinate system. To convey to an observer in a distant galaxy which is a right-handed coordinate system it is sufficient to instruct him† to perform the Λ^0 production–decay experiment and observe the three momenta $\boldsymbol{p}_{\pi 1}$, $\boldsymbol{p}_{\Lambda}$ and the preferred direction of decay pions $\boldsymbol{p}_{\pi 2}$. These three vectors taken in order define a right-handed system.

Another way to exhibit the parity violating nature of the observations is to reflect the actual experiment in a mirror. This is shown in fig. 5.2. The operation of space inversion is equivalent to reflection in a plane followed by a rotation through π about the normal to the mirror plane. Rotational invariance is assumed, so a test of mirror reflection symmetry is equivalent to a test of spatial inversion symmetry.

In the actual experiment Λ^0s scattered to the left emit decay pions upwards while right-scattered Λ^0s emit pions downwards (not shown). In the mirror image experiment Λ^0s scattered to the right would emit pions upwards, which is a contradiction with what is observed. So the mirror image of the Λ^0 decay experiment is not a physically realised situation.

In this pictorial description, parity conservation required that the mirror image of an event occurs with the same quantum-mechanical probability as the event itself. In the present case this would imply

$$N_{\mathrm{up}} = N_{\mathrm{down}}$$

† Unless he is an antiobserver using antiprotons to make antihyperons, in which case he will obtain a left-handed system (see §8.3.1).

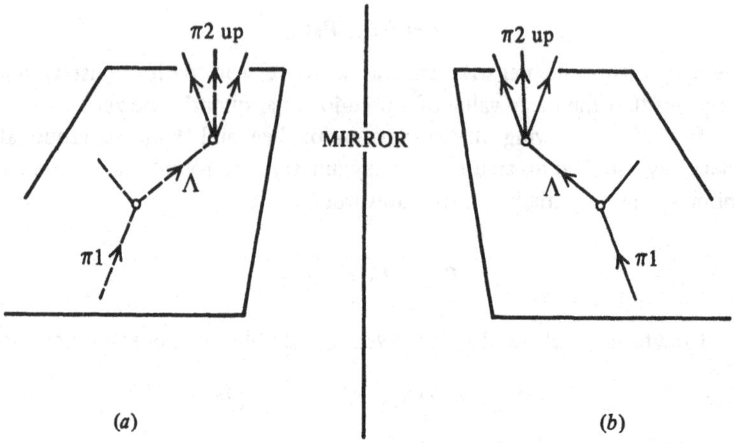

Fig. 5.2. The mirror image (a) of a Λ^0 production–decay experiment (b) is not realised in Nature.

5.4 Space inversion and the helicity description

In this section we shall discuss the transformation properties under space inversion of the helicity states described in the last chapter. This will enable us to derive the consequences of parity conservation for reaction and decay amplitudes. We have already quoted the results without proof in the general discussion of these processes in chapter 4.

The method used is as follows: from the geometrical definition of the parity operation P we can deduce the commutation properties of P with rotations and Lorentz transformations used to define the helicity states of a single particle. The effect of the unitary parity operator on one particle helicity states will then be derived from these commutation rules. The parity transformation properties of plane wave and angular momentum eigenstates of two particle states then follow.

5.4.1 *Parity transformation of one and two particle states*

The space inversion operation P in four-dimensional notation is

$$\begin{pmatrix} x_0 \\ x_1 \\ x_2 \\ x_3 \end{pmatrix} \to \begin{pmatrix} x_0' \\ x_1' \\ x_2' \\ x_3' \end{pmatrix} = \begin{pmatrix} 1 & 0 & 0 & 0 \\ 0 & -1 & 0 & 0 \\ 0 & 0 & -1 & 0 \\ 0 & 0 & 0 & -1 \end{pmatrix} \begin{pmatrix} x_0 \\ x_1 \\ x_2 \\ x_3 \end{pmatrix} \tag{5.31}$$

which makes explicit that P leaves the time coordinate fixed, and shows that P can be regarded as a generalised Lorentz transformation.

We have

$$P^2 = 1 \tag{5.32}$$

Here and in the following P denotes the 4×4 matrix on the right of (5.31). Since P reverses the sign of the space coordinates it does not commute with a spatial displacement.

For a displacement in space–time by a_μ,

$$T_{a_\mu}: x_\mu \to x'_\mu = x_\mu + a_\mu$$

we find

$$PT_{(a_0,\mathbf{a})} = T_{(a_0,-\mathbf{a})}P \tag{5.33}$$

P commutes with rotations, hence

$$PR = RP \tag{5.34}$$

where R denotes a rotation

$$R = \begin{pmatrix} 1 & 0 & 0 & 0 \\ 0 & R_{11} & R_{12} & R_{13} \\ 0 & R_{21} & R_{22} & R_{23} \\ 0 & R_{31} & R_{32} & R_{33} \end{pmatrix}$$

On the other hand, space inversion does not commute with Lorentz transformations: for the case of a pure boost along the z-direction represented by

$$\mathcal{Z}_v = \begin{pmatrix} (1-v^2)^{-1/2} & 0 & 0 & v(1-v^2)^{-1/2} \\ 0 & 1 & 0 & 0 \\ 0 & 0 & 1 & 0 \\ v(1-v^2)^{-1/2} & 0 & 0 & (1-v^2)^{-1/2} \end{pmatrix}$$

we find

$$P\mathcal{Z}_v = \mathcal{Z}_{-v}P \tag{5.35}$$

In general the parity transform $P \Lambda P^{-1}$ of a pure Lorentz transformation with velocity $\mathbf{v}$ is a pure Lorentz transformation with velocity $-\mathbf{v}$.

In a quantum-mechanical system there corresponds to P a unitary operator† $U(P)$ acting on the state vectors of the system. According to

† In what follows we use the notation $U(P)$ rather that U_P to be uniform with the other unitary operators, $U(a)$ etc.

the general theory there corresponds to each of the equations (5.32), (5.33), (5.34) and (5.35) an equation for the corresponding operators

$$U(P)^2 = 1 \tag{5.36}$$

$$U(P)\,U(a_0, \boldsymbol{a}) = U(a_0, -\boldsymbol{a})\,U(P) \tag{5.37}$$

$$U(P)\,U(R) = U(R)\,U(P) \tag{5.38}$$

$$U(P)\,U(\mathcal{Z}_v) = U(\mathcal{Z}_{-v})\,U(P) \tag{5.39}$$

Rather than define $U(P)$ directly by its action on the wavefunction as in §5.1, we choose to define $U(P)$ acting on one particle states so that these equations are satisfied.

First of all we bring (5.37) to a more useful form by considering an infinitesimal displacement, so that as in (4.28) we write with the aid of the four-momentum $P_\mu = (P_0, \boldsymbol{P})$

$$U(a_0, \boldsymbol{a}) = 1 + \mathrm{i}(a_0 P_0 - \boldsymbol{a}\cdot\boldsymbol{P})$$

Substituting in (5.37) we have

$$U(P)\{1 + \mathrm{i}a_0 P_0 - \mathrm{i}\boldsymbol{a}\cdot\boldsymbol{P}\} = \{1 + \mathrm{i}a_0 P_0 + \mathrm{i}\boldsymbol{a}\cdot\boldsymbol{P}\}\,U(P)$$

and on comparing coefficients of the parameters a_0 and $\boldsymbol{a}$ we find

$$P_0 U(P) = U(P) P_0 \tag{5.40}$$

$$\boldsymbol{P} U(P) = -U(P)\boldsymbol{P} \tag{5.41}$$

which state respectively that $U(P)$ commutes with the energy operator (Hamiltonian) and anticommutes with the three-momentum, $\boldsymbol{P}$.

We consider a single particle of mass m and spin s whose helicity states $\phi_{p\lambda}$ are defined by the procedure of §4.3 and §4.4. $\phi_{p\lambda}$ is an eigenstate of P_0 and $\boldsymbol{P}$ with eigenvalues $(p^2 + m^2)^{1/2}$ and $\boldsymbol{p}$. Hence on applying (5.40) and (5.41) in turn to this state, we have

$$\begin{aligned} P_0 U(P)\,\phi_{p\lambda} &= U(P) P_0 \phi_{p\lambda} \\ &= (p^2 + m^2)^{1/2}\,U(P)\,\phi_{p\lambda} \end{aligned}$$

$$\boldsymbol{P} U(P)\,\phi_{p\lambda} = -\boldsymbol{p} U(P)\,\phi_{p\lambda}$$

These equations show that $U(P)\,\phi_{p\lambda}$ is an energy–momentum eigenstate with energy $(p^2 + m^2)^{1/2}$ and three-momentum $-\boldsymbol{p}$. These results are not unexpected: we have shown how they are derived from the space–time transformation properties of the theory.

Consider a set of rest states $\phi_{0\lambda}(-s \leqslant \lambda \leqslant s)$ which were defined to transform in the standard way under rotations in §4.4.

The parity transformed states

$$U(P)\,\phi_{0\lambda}$$

are also rest states; this follows from what we have just done. These states also transform in the standard way under rotation; this follows from (5.38).

$$\begin{aligned} U(R)\,U(P)\,\phi_{0\lambda} &= U(P)\,U(R)\,\phi_{0\lambda} \\ &= U(P)\{\sum_{\lambda'}\phi_{0\lambda'}\mathfrak{D}^{s}_{\lambda'\lambda}(R)\} \\ &= \sum_{\lambda'} U(P)\,\phi_{0\lambda'}\mathfrak{D}^{s}_{\lambda'\lambda}(R) \end{aligned}$$

It may be concluded that the states $U(P)\,\phi_{0\lambda}$ and $\phi_{0\lambda}$ are proportional

$$U(P)\,\phi_{0\lambda} = \eta_P\phi_{0\lambda} \tag{5.42}$$

where η_P is a phase factor. A more careful analysis which we shall not enter into here, shows that the only alternative to (5.42) is that there are two distinct sets of $(2s+1)$ states for the particle at rest. This is not what is observed.

Adopting (5.42), (5.36) then implies that

$$\eta_P^2 = 1$$

and hence

$$\eta_P = \pm 1 \tag{5.43}$$

η_P is the intrinsic parity of the particle.

Next we consider $U(P)$ acting on a state with momentum p directed along the z-axis. By (4.50) and using (5.39) we have

$$\begin{aligned} U(P)\,\phi_{p00\lambda} &= U(P)\,U(\mathcal{Z}_p)\,\phi_{000\lambda} \\ &= U(\mathcal{Z}_{-p})\,U(P)\,\phi_{000\lambda} \end{aligned}$$

and so

$$U(P)\,\phi_{p00\lambda} = \eta_P U(\mathcal{Z}_{-p})\,\phi_{000\lambda} \tag{5.44}$$

Here $\mathcal{Z}_{-p}$ denotes a boost in the negative z-direction, and so the state on the right is apart from a phase factor, the state defined in (4.66). However, rather than casting it in that form, we proceed as follows.

A rotation Y_π through π about the y-axis, takes the $+z$-direction into the $-z$-direction, hence the Lorentz transformations satisfy

$$Y_\pi^{-1}\,\mathcal{Z}_p Y_\pi = \mathcal{Z}_{-p}$$

Correspondingly

$$U(Y_\pi)^{-1}U(\mathcal{Z}_p)\,U(Y_\pi) = U(\mathcal{Z}_{-p}) \tag{5.45}$$

We substitute this into (5.44) and use the known transformation properties under rotations of the rest-states,

$$U(Y_\pi)\phi_{000\lambda} = \sum_{\lambda'} \phi_{000\lambda'} d^s_{\lambda'\lambda}(\pi)$$
$$= (-1)^{s-\lambda}\phi_{000,-\lambda}$$

to obtain

$$U(P)\phi_{p00\lambda} = \eta_P U(Y_{-\pi}) U(\mathcal{Z}_p)(-1)^{s-\lambda}\phi_{000,-\lambda}$$

and so

$$U(P)\phi_{p00\lambda} = \eta_P(-1)^{s-\lambda} U(Y_{-\pi})\phi_{p00,-\lambda} \tag{5.46}$$

This equation states that the effect of the parity operator on a single particle state is the same as that of a certain rotation $Y_{-\pi}$. It does not state that the operators are equal, because it is not true for all states.

Equation (5.46) is the most useful form of the parity transformation law for helicity states. It can be expressed in another way by taking the rotation operator to the other side, so that we have

$$U(Y_\pi)U(P)\phi_{p00\lambda} = \eta_P(-1)^{s-\lambda}\phi_{p00,-\lambda} \tag{5.47}$$

We then observe that the combined operation $Y_\pi P$ is that of reflection in the xz-plane. We denote this by I_y, so we have

$$U(I_y)\phi_{p00\lambda} = \eta_P(-1)^{s-\lambda}\phi_{p00,-\lambda} \tag{5.48}$$

Now a rotation in the xz-plane commutes with reflection in that plane, so we can apply $U(Y_\theta)$ to both sides of (5.48) to obtain

$$U(I_y)U(Y_\theta)\phi_{p00\lambda} = \eta_P(-1)^{s-\lambda}U(Y_\theta)\phi_{p00,-\lambda}$$

or

$$U(I_y)\phi_{p\theta 0\lambda} = \eta_P(-1)^{s-\lambda}\phi_{p\theta 0,-\lambda} \tag{5.49}$$

A similar calculation may be made for the state $\hat{\phi}_{p\pi 0\lambda}$ with p directed along $-z$, defined by (4.66). The result is

$$U(P)\hat{\phi}_{p\pi 0\lambda} = \eta_P(-1)^{s+\lambda}U(Y_\pi)\hat{\phi}_{p\pi 0,-\lambda} \tag{5.50}$$

which can also be put in the form

$$U(I_y)\hat{\phi}_{p\pi 0\lambda} = \eta_P(-1)^{s+\lambda}\hat{\phi}_{p\pi 0,-\lambda} \tag{5.51}$$

and finally applying $U(Y_\theta)$

$$U(I_y)U(Y_\theta)\hat{\phi}_{p\pi 0\lambda} = \eta_P(-1)^{s+\lambda}U(Y_\theta)\hat{\phi}_{p\pi 0,-\lambda} \tag{5.52}$$

For a two particle state

$$\phi^{ab}_{W\theta 0\lambda_a\lambda_b} = U(Y_\theta)\{\phi^a_{p00\lambda_a}\hat{\phi}^b_{p\pi 0\lambda_b}\}$$

Equations (5.49) and (5.52) together give

$$U(I_y)\phi^{ab}_{W\theta 0\lambda_a\lambda_b} = \eta_a\eta_b(-1)^{s_a-\lambda_a+s_b+\lambda_b}\phi^{ab}_{W\theta 0,-\lambda_a,-\lambda_b} \tag{5.53}$$

where η_a and η_b are the intrinsic parities of a and b.

For two particle reactions and decays the reaction plane can be chosen to be the xz-plane, and the transformation law (5.53) for states with momenta in this plane enables us to derive the consequences of parity conservation for such processes.

For completeness we quote the parity transformation law of a two particle angular momentum eigenstate:

$$U(P)\Psi^{ab}_{WJM\lambda_a\lambda_b} = \eta_a\eta_b(-1)^{J-s_a-s_b}\Psi^{ab}_{WJM,-\lambda_a,-\lambda_b} \quad (5.54)$$

The proof which starts from (4.89) is given in Jacob and Wick (1959).

Equation (5.54) shows that the angular momentum eigenstates are not eigenstates of parity, except in the case of spinless particles. However with the aid of this equation it is easy to construct parity eigenstates. Let us consider two examples.

The case of the πN system was already referred to in §4.8.

The product of the πN parities is known to be odd

$$\eta_\pi\eta_N = -1$$

so that we have

$$U(P)\Psi^{\pi N}_{JM\lambda} = -(-1)^{J-1/2}\Psi^{\pi N}_{JM,-\lambda}$$

Hence the normalised states

$$\Psi^{(\pm)}_{JM} = 2^{-1/2}\{\Psi^{\pi N}_{JM,+1/2} \mp \Psi^{\pi N}_{JM,-1/2}\}$$

have parity $(-1)^{J\mp 1/2}$.

For two spin-$\frac{1}{2}$ particles, e.g. pn where the intrinsic parity product is

$$\eta_p\eta_n = +1$$

we have

$$U(P)\Psi^{pn}_{JM\lambda_a\lambda_b} = (-1)^{J-1}\Psi^{pn}_{JM,-\lambda_a,-\lambda_b}$$

So the states

$$2^{-1/2}\{\Psi_{JM++} \pm \Psi_{JM--}\}$$

have parity $\mp(-1)^J$, and the states

$$2^{-1/2}\{\Psi_{JM+-} \pm \Psi_{JM-+}\}$$

have parity $\mp(-1)^J$.

For the case of two identical particles e.g. two protons, there are further restrictions from statistics.

5.4.2 *Consequences of parity conservation for reactions and decays*

The CM scattering amplitude for a two particle reaction

$$a + b \rightarrow c + d$$

is the matrix element of the $\mathcal{T}$-operator between CM plane wave states

$$f_{\lambda_c\lambda_d,\lambda_a\lambda_b}(W\theta,\phi=0)\sim(\Phi^{cd}_{W\theta 0\lambda_c\lambda_d},\mathcal{T}\Phi^{ab}_{W00\lambda_a\lambda_b})$$

where

$$\Phi^{ab}_{W00\lambda_a\lambda_b}\sim\phi^{a}_{p00\lambda_a}\hat{\phi}^{b}_{p\pi 0\lambda_b} \tag{5.55}$$

$$\Phi^{cd}_{W\theta 0\lambda_c\lambda_d}\sim U(Y_\theta)\{\phi^{c}_{p00\lambda_c}\hat{\phi}^{d}_{p\pi 0\lambda_d}\} \tag{5.56}$$

The approximate equality sign indicates the omission of kinematical factors which were dealt with in §4.7, and do not affect the argument here.

Invariance under space inversion is expressed by

$$U(P)^{-1}\,\mathcal{T}U(P)=\mathcal{T}$$

Since rotational invariance of $\mathcal{T}$ is assumed we can substitute $U(Y_\pi)^{-1}\,\mathcal{T}U(Y_\pi)$ for $\mathcal{T}$ on the left and obtain

$$U(I_y)^{-1}\,\mathcal{T}U(I_y)=\mathcal{T}$$

On taking the matrix element of this equation between the states (5.55) and (5.56) and using the transformation laws (5.53), (5.48) and (5.51) we obtain

$$\frac{\eta_a\eta_b(-1)^{s_a-\lambda_a+s_b+\lambda_b}}{\eta_c\eta_d(-1)^{s_c-\lambda_c+s_d+\lambda_d}}(\Phi^{cd}_{W\theta 0,-\lambda_c,-\lambda_d},\mathcal{T}\Phi^{ab}_{W00,-\lambda_a,-\lambda_b})$$

$$=(\Phi^{cd}_{W\theta 0\lambda_c\lambda_d},\mathcal{T}\Phi^{ab}_{W00\lambda_a\lambda_b})$$

Although the η and $(-1)^{s\pm\lambda}$ are real (in fact, ± 1), we shall manipulate them as if they were complex phase factors for which $\eta^*=\eta^{-1}$, so that when they are associated with complex conjugate wavefunctions they are taken to the denominator. Expressing this result in terms of the helicity scattering amplitude we have

$$f_{\lambda_c\lambda_d,\lambda_a\lambda_b}(W,\theta,0)=\frac{\eta_a\eta_b(-1)^{s_a-\lambda_a+s_b+\lambda_b}}{\eta_c\eta_d(-1)^{s_c-\lambda_c+s_d+\lambda_d}}f_{-\lambda_c,-\lambda_d,-\lambda_a,-\lambda_b}(W,\theta,0) \tag{5.57}$$

This is the consequence of parity conservation for the reaction amplitude.

For πN elastic scattering $s_a=s_c=\frac{1}{2}$, $s_b=s_d=0$, we obtain the result quoted in §4.8

$$\left.\begin{aligned} f_{--}(\theta)&=f_{++}(\theta)\\ f_{+-}(\theta)&=-f_{-+}(\theta)\end{aligned}\right\} \tag{4.99}$$

For completeness, we give the condition of parity conservation on the scattering amplitude in the angular momentum representation. With the aid of (5.54) it may be shown that

$$\frac{\eta_a\eta_b(-1)^{J-s_a-s_b}}{\eta_c\eta_d(-1)^{J-s_c-s_d}}\mathcal{T}^J_{-\lambda_c,-\lambda_d,-\lambda_a,-\lambda_b}(W) = \mathcal{T}^J_{\lambda_c\lambda_d,\lambda_a\lambda_b}(W) \tag{5.58}$$

In order to see the implications of parity conservation in reactions such as

$$\mathrm{K}+\mathrm{N}\rightarrow\pi+\Lambda$$

it is useful to consider the general case of a spin-0–spin-½ reaction.

We take

$$s_a = s_c = \tfrac{1}{2}, \quad s_b = s_d = 0$$

the indices λ_b and λ_d are suppressed and the values $\pm\frac{1}{2}$ for λ_a and λ_c are denoted by $\pm$. Equation (5.57) gives

$$f_{\mu\lambda}(W,\theta) = \eta_r(-1)^{\mu-\lambda}f_{-\mu,-\lambda}(W,\theta) \tag{5.59}$$

where we have defined the reaction parity

$$\eta_r = \eta_a\eta_b/\eta_c\eta_d \tag{5.60}$$

Restoring the ϕ-dependence, (5.59) becomes

$$f_{\mu\lambda}(W,\theta,\phi) = \eta_r(-1)^{\mu-\lambda}e^{2i\lambda\phi}f_{-\mu,-\lambda}(W,\theta,\phi)$$

So as a matrix in the helicity basis, $f_{\mu\lambda}$ is

$$f(W,\theta,\phi) = \begin{bmatrix} f_{++} & -\eta_r e^{-i\phi}f_{-+} \\ f_{-+} & \eta_r e^{-i\phi}f_{++} \end{bmatrix}$$

To compare with the non-relativistic approach we make the transformation of the final helicity states to the same z-axis of quantisation as the initial spin-½ particle. This was described in §4.8. With the more general expression here, we obtain

$$F(W,\theta,\phi) = \begin{bmatrix} \cos\frac{1}{2}\theta f_{++} - \sin\frac{1}{2}\theta f_{-+} & -\eta_r\sin\frac{1}{2}\theta e^{-i\phi}f_{++} - \eta_r\cos\frac{1}{2}\theta e^{-i\phi}f_{-+} \\ \sin\frac{1}{2}\theta e^{i\phi}f_{++} + \cos\frac{1}{2}\theta e^{i\phi}f_{-+} & \eta_r\cos\frac{1}{2}\theta f_{++} - \eta_r\sin\frac{1}{2}\theta f_{-+} \end{bmatrix} \tag{5.61}$$

The helicity amplitudes on the right are evaluated with $\phi = 0$, i.e. the ϕ-dependence is explicit. The case $\eta_r = +1$ corresponds to the πN scattering treated before where we showed that F can be written

$$F = g(\theta)1 + i\boldsymbol{\sigma}\cdot\hat{\boldsymbol{n}}h(\theta) \tag{5.62}$$

and $\hat{\boldsymbol{n}}$ is the normal to the scattering plane.

In the other case $\eta_r = -1$ we can write

$$F(W, \theta, \phi) = j(\theta)\boldsymbol{\sigma}\cdot\hat{\boldsymbol{p}}_i + k(\theta)\boldsymbol{\sigma}\cdot\hat{\boldsymbol{p}}_f \tag{5.63}$$

where $\hat{\boldsymbol{p}}_i$ and $\hat{\boldsymbol{p}}_f$ are unit vectors along the incident and final spin-$\frac{1}{2}$ particles so

$$\boldsymbol{p}_i = (0, 0, 1)$$

$$\boldsymbol{p}_f = (\sin\theta\cos\phi, \sin\theta\sin\phi, \cos\theta)$$

and thus

$$\boldsymbol{\sigma}\cdot\hat{\boldsymbol{p}}_i = \begin{pmatrix} 1 & 0 \\ 0 & -1 \end{pmatrix}$$

$$\boldsymbol{\sigma}\cdot\hat{\boldsymbol{p}}_f = \begin{pmatrix} \cos\theta & \sin\theta\, e^{-i\phi} \\ \sin\theta\, e^{i\phi} & -\cos\theta \end{pmatrix}$$

Equation (5.63) agrees with (5.61) if we define the amplitudes $j(\theta)$ and $k(\theta)$ by

$$k(\theta)\sin\theta = \sin\tfrac{1}{2}\theta f_{++} + \cos\tfrac{1}{2}\theta f_{-+}$$

$$j(\theta) + k(\theta)\cos\theta = \cos\tfrac{1}{2}\theta f_{++} - \sin\tfrac{1}{2}\theta f_{-+}$$

To summarise, if the overall intrinsic parity does not change, $\eta_r = +1$, and the scattering amplitude is of the form

$$F = g(\theta)1 + ih(\theta)\frac{\boldsymbol{\sigma}\cdot\boldsymbol{p}_i \wedge \boldsymbol{p}_f}{|\boldsymbol{p}_i \wedge \boldsymbol{p}_f|} \tag{5.64}$$

while if the overall intrinsic parity does change, $\eta_r = -1$, and we have

$$F = j(\theta)\boldsymbol{\sigma}\cdot\hat{\boldsymbol{p}}_i + k(\theta)\boldsymbol{\sigma}\cdot\hat{\boldsymbol{p}}_f \tag{5.65}$$

These forms can be derived by elementary invariance arguments as follows. F is a 2×2 matrix in spin space and may be formed from the unit operator in spin space 1, and the vector of Pauli spin matrices $\boldsymbol{\sigma}$. Rotational invariance requires that F be a scalar. The available vectors to form a scalar with $\boldsymbol{\sigma}$ are constructed from the CM momenta and can be taken as the unit vectors $\hat{\boldsymbol{p}}_i$, $\hat{\boldsymbol{p}}_f$ and $\hat{\boldsymbol{n}}$, where

$$\hat{n} = \frac{\boldsymbol{p}_i \wedge \boldsymbol{p}_f}{|\boldsymbol{p}_i \wedge \boldsymbol{p}_f|}$$

If the intrinsic parity product in the final state is the same as that in the initial state, F must be a true scalar under reflections. So since $\hat{\boldsymbol{p}}_i$ and $\hat{\boldsymbol{p}}_f$ change sign on reflection while $\hat{\boldsymbol{n}}$ does not, we arrive at the allowed form (5.64). On the other hand if the overall intrinsic parity changes, F must be a pseudo-scalar under reflections and can only contain the terms $\boldsymbol{\sigma}\cdot\boldsymbol{p}_i$ and $\boldsymbol{\sigma}\cdot\boldsymbol{p}_f$ and we have (5.65).

These elementary arguments are not easily generalised to higher spins. In such cases and when massless particles are involved, the helicity description is preferable.

Finally we turn to parity conserving decays. The amplitude for the decay

$$A \to a + b$$

at rest is

$$f_{\lambda_a\lambda_b M}(\theta) = (\Phi^{ab}_{m_A\theta 0\lambda_a\lambda_b}, \mathcal{S}\phi^A_{0M})$$

where we have specialised (4.114) to the xz-plane, $\phi = 0$. Proceeding as before we take the matrix element of

$$U(I_y)^{-1}\,\mathcal{S}\,U(I_y) = \mathcal{S}$$

between the A-state and the ab-state. For the rest-state of A, we have

$$\begin{aligned} U(I_y)\,\phi_{0M} &= U(Y_\pi)\,U(P)\,\phi_{0M} \\ &= \eta_A U(Y_\pi)\,\phi_{0M} \\ &= \eta_A(-1)^{J-M}\,\phi_{0,-M} \end{aligned}$$

while for the ab-state we use (5.53) to obtain

$$\frac{\eta_A(-1)^{J-M}}{\eta_a\eta_b(-1)^{s_a-\lambda_a+s_b+\lambda_b}}\, f_{-\lambda_a,-\lambda_b,-M}(\theta) = f_{\lambda_a\lambda_b M}(\theta) \quad (5.66)$$

The explicit form of the angular dependence of $f(\theta)$,

$$f_{\lambda_a\lambda_b M}(\theta) = C_J d^J_{M,\lambda_a-\lambda_b}(\theta)\,a_{\lambda_a\lambda_b} \quad (4.115a)$$

may be introduced into (5.66) to give

$$\frac{\eta_A(-1)^{J-M}}{\eta_a\eta_b(-1)^{s_a-\lambda_a+s_b+\lambda_a}}\, d^J_{-M,-\lambda_a+\lambda_b}(\theta)\,a_{-\lambda_a-\lambda_b} = d^J_{M,\lambda_a-\lambda_b}(\theta)\,a_{\lambda_a\lambda_b}$$

and on using the d-function identity

$$d^J_{MM'}(\theta) = (-1)^{M'-M} d^J_{-M,-M'}(\theta)$$

on the right, the d^J-functions cancel to give

$$\frac{\eta_A}{\eta_a \eta_b}(-1)^{J-s_a-s_b} a_{-\lambda_a,-\lambda_b} = a_{\lambda_a \lambda_b} \tag{4.117a}$$

which is the condition of parity conservation for the decay amplitude quoted already.

5.4.3 *Parity and massless particles*

We discussed the special features of the Lorentz invariant description of massless particles in §4.5. We argued there that it is consistent with Lorentz invariance for a massless particle of spin s to exist only in one helicity state $\lambda = s$ or $-s$. In other words the projection of the spin on the direction of motion is a Lorentz invariant quantity when $m = 0$.

Now let us require the parity operator to be applicable to the states of a massless particle. We cannot start with a rest-state of such a particle; instead we consider for example the state in which the particle is moving along the $+z$-axis. Also we work with the operation of reflection $I_y = Y_\pi P$.

Consider then

$$U(I_y)\phi_{p00\lambda}$$

I_y leaves the xz-plane invariant and hence this state still has momentum along $+Oz$. The helicity operator $\mathcal{H}$ is a pseudo-scalar under inversion, so the state has helicity $-\lambda$. Formally, (5.18) and the rotational invariance of $\mathcal{H}$ (under Y_π) together imply

$$U(I_y)\mathcal{H} = -\mathcal{H}U(I_y)$$

So by a now familiar argument,

$$\begin{aligned}\mathcal{H}U(I_y)\phi_{p00\lambda} &= -U(I_y)\mathcal{H}\phi_{p00\lambda} \\ &= -\lambda U(I_y)\phi_{p00\lambda}\end{aligned}$$

This proves an assertion made in §4.5: the requirement that the parity operator be applicable to the states of a massless particle implies the existence of both helicity states $\pm\lambda$ $(= \pm s)$.

We set

$$U(I_y)\phi_{p00\lambda} = \eta_P \phi_{p00,-\lambda} \tag{5.67}$$

where

$$\eta_P = \pm 1$$

η_P is independent of p, because I_y commutes with Lorentz transformations in the z-direction which change p on both sides. Also I_y commutes with Y_θ, hence as for massive particles we can generalise to a state of motion in the xz-plane

$$U(I_y)\phi_{p00\lambda} = \eta_P\phi_{p\theta 0,-\lambda} \qquad (5.68)$$

We note that this equation agrees with (5.49) for a massive particle in case $\lambda = s$. η_P is called the intrinsic parity.

The full significance of these considerations only becomes apparent when we consider particles in interaction. In order to construct a parity conserving interaction involving a massless particle it is necessary that the particle should exist in both helicity states, $\lambda = \pm s$.

The two known massless particles, the photon and neutrino, exemplify the possibilities. We have already noted that the $\lambda = \pm 1$ helicity states of the photon of spin-1 exist and correspond to quanta of the right and left circularly polarised plane waves,

$$A_\pm = \mp 2^{-1/2}(e_x \pm ie_y)\,\mathrm{e}^{ipz} \qquad (4.63a)$$

A more detailed exploration of the connection requires a discussion of the quantisation of the radiation field, and is beyond the scope of this book.

We can see from (4.63a) that since I_y changes the sign of the e_y but not e_x, the I_y transform of $A_\pm$ is $-A_\mp$. Comparing this classical argument with the corresponding rule (5.67) for quantum states we deduce that the intrinsic parity of the photon is -1. This result depends on the overall sign on the right of (4.63a), however these signs were required by earlier conventions. For example, a rotation Y_π must take the photon state $\phi_{p00,+1}$ into $\phi_{p\pi 0,+1}$ (no ˆ).

Correspondingly for the classical wave

$$A_+ \xrightarrow{Y_\pi} -2^{-1/2}(-e_x + ie_y)\,\mathrm{e}^{-ipz}$$
$$= 2^{-1/2}(e_x - ie_y)\,\mathrm{e}^{-ipz}$$

which has spin component -1 along the $+z$-axis, and hence positive helicity.

The neutrino of spin $\frac{1}{2}$, is found to exist only in one helicity state, $\lambda = -\frac{1}{2}$. Correspondingly the interactions involving the neutrino violate space inversion invariance. This does not fully explain parity violation in Nature, because the weak processes which do not involve leptons also violate the parity conservation law. Furthermore, the situation for the neutrino is complicated by the fact that neutrinos carry a generalised leptonic charge (lepton number), so the above argument applies only to a neutrino of definite lepton number. One finds in Nature an antineutrino with the opposite helicity; this is as we shall see in accord with further general arguments (the *CPT* theorem).

5.5 Determination of intrinsic parities

5.5.1 *General remarks*

A variety of methods has been used to determine the intrinsic parities of particles and resonances. Some of these are described below in discussion of concrete examples involving the strong and electromagnetic interactions which, as we have seen, conserve parity. If one attempts to determine intrinsic parities using parity violating reactions it is impossible to obtain an unambiguous result.

When introducing the concept of intrinsic parity in §5.3 we considered a process

$$A \rightarrow B + \pi^0$$

such that the particle (π^0) was produced without a change in the number and kind of other particles present. For some particles this condition can never be satisfied owing to other conservation laws. For example, protons can never be produced singly, since a reaction in which two protons interact to produce three protons and nothing else violates the conservation law of electric charge which is believed to be absolutely obeyed in Nature. In fact, this reaction is independently forbidden by angular momentum conservation: the total angular momentum J in the initial state can only assume integral values while since orbital angular momentum is necessarily integral, J in the final state is half-integral. This latter rule also forbids the production of one neutron in a proton–proton interaction which would conserve electric charge.

Considerations of this sort show that an intrinsic parity cannot be assigned to all particles. However, if we formally introduce an intrinsic parity factor η for every particle, some of these may be defined arbitrarily to be positive (say), and then others are fixed by experiment relative to these definitions. This is illustrated in the remainder of this section.

5.5.2 *Parities of the neutral pion and photon*

It would be natural to start with the π^0 which has an intrinsic parity defined independently of any conventions. However, a reaction such as

$$\mathrm{p} + \mathrm{p} \rightarrow \mathrm{p} + \mathrm{p} + \pi^0$$

is not convenient to use because of the three particle final state. An analysis of decay mode

$$\pi^0 \to \gamma + \gamma \to (e^+e^-) + (e^+e^-)$$

permits an unambiguous determination of the π^0-parity, and shows it to be negative.

Another parity which is absolutely defined is that of the photon, and in §5.4.3 we showed on theoretical grounds that

$$\eta_\gamma = -1$$

5.5.3 *Parity of the charged pion*

The intrinsic parity of the negative pion has been determined by an analysis of its capture at rest in a liquid deuterium target. The following reactions are seen with the relative frequencies indicated

$$\pi^- + d \to n + n \qquad 70 \text{ per cent} \qquad (5.69a)$$

$$\text{(at rest)} \to n + n + \gamma \qquad 30 \text{ per cent} \qquad (5.69b)$$

$$\to n + n + \pi^0 \qquad \sim 0 \text{ per cent} \qquad (5.69c)$$

The last is not observed although it is energetically allowed: the $\pi^- - \pi^0$ mass difference exceeds the binding energy of the deuteron by about 2.4 MeV.

The negative pion is slowed down by ionisation until it is captured into an atomic orbit around the deuteron. An analysis of the relative rates shows that the pion makes radiative transitions to the 1s-state before being captured by the deuteron. Since the deuteron has spin 1, we conclude that capture takes place from an initial state of total angular momentum $J = 1$.

Now the final state contains two identical spin-$\frac{1}{2}$ particles, and Fermi statistics requires states of total spin 0 and 1 to be odd and even respectively under exchange of the space coordinates, and hence to have orbital angular momentum odd and even respectively. Hence the final state with $J = 1$ can only be a triplet p-state: $^3P_1$, in the spectroscopic notation $^{2S+1}L_J$. Since this state has odd spatial parity while the deuteron has even parity, reaction (5.69a) can occur with conservation of parity only if the intrinsic parity of the π^- is negative,

$$\eta_{\pi^-} = -1$$

Proceeding more formally we may assign intrinsic parities to all the particles. Then the parity of the initial state is

$$\eta_{\pi^-}\eta_p\eta_n\eta_{orb} = \eta_{\pi^-}\eta_p\eta_n$$

Here the orbital parity of the deuteron is + 1, because deuteron ground state is mainly $l = 0$ with a small admixture of $l = 2$.

The parity of the final state is

$$\eta_n^2\eta_{orb} = -\eta_n^2 = -1$$

Equating the two expressions we have

$$\eta_{\pi^-}\eta_p\eta_n = -1$$

So, we may say that the $pn\pi^-$ relative parity is negative.

It is usual to take

$$\eta_p = +1 \quad \text{(definition)} \tag{5.70}$$

and

$$\eta_n = +1 \quad \text{(definition)} \tag{5.71}$$

This choice is natural in view of isospin symmetry in which proton and neutron are regarded as different charge states of the same particle. We then find that the parity of the π^- is negative.

It is now significant and not a matter of definition, that the parities of neutral and negative pions are the same.

5.5.4 *Intrinsic parities of strange particles*

It was observed early in the study of strange particles that they cannot be produced singly but only in pairs (associated production). This tells us that we are not going to be able to assign intrinsic parities to strange particles in an absolute way. In precise terms we have the law of conservation of strangeness S (or equivalently, hypercharge) which is obeyed by the strong interaction, so that a single particle carrying one unit of S cannot be produced from a collision between pions and/or nucleons all of which have zero strangeness. Although strangeness conservation is not absolute in weak decays such as $\Sigma, \Lambda \to \pi N$, $\Xi \to \pi\Lambda$, or $K \to \mu\nu$ this does not permit an independent determination of strange particle parities because the weak interaction is not invariant under space inversion. We have already illustrated this in §5.3 for a non-leptonic process, the hyperon decay. We shall discuss parity violation in leptonic decays in §5.6.

Strange particle reactions involve a different part of the strong interaction Hamiltonian from that responsible for non-strange interactions between pions and nucleons, so we may ask whether this new interaction conserves parity. Indirect evidence that it does comes from low energy

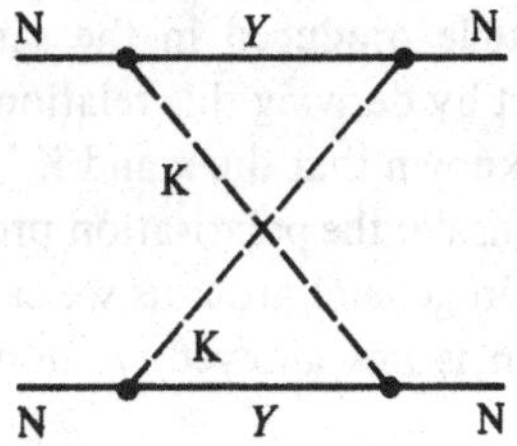

Fig. 5.3. Two kaon exchange contribution to the nucleon–nucleon interaction.

nuclear physics. The interaction between nucleons may be expected to receive contributions from strange particles via second order Yukawa processes with the exchange of two K-mesons, illustrated in fig. 5.3.

So any parity violation in the strange particle interaction would show up in the nucleon–nucleon interaction. Now the high degree of parity conservation in low energy nuclear physics already noted shows that any parity violating term in the strange particle strong interaction is small, so we proceed on the assumption that parity is conserved in these processes.

The determination of strange particle parities has been described by Dalitz (1962).

The parity of one strange particle is a matter of definition, so we put

$$\eta_\Lambda = +1 \quad \text{(definition)} \tag{5.72}$$

and then $\eta_{K^\pm}$, $\eta_{\Sigma^\pm}$ etc. are to be determined by experiment. Extensive experimentation on the production and decay of hypernuclei led to the assignment

$$\eta_K = -1 \tag{5.73}$$

See, for example, Block, Lendinari and Monari (1962).

5.5.5 *Experiments with polarised targets to determine strange particle parities*

It is possible to determine the KNΛ and KNΣ relative parities by a study of the reactions

$$\pi + p \to K + Y$$

$$\bar{K} + p \to \pi + Y$$

(Y denotes Λ or Σ) with a polarised proton target. The determination rests on a relation between (*a*) the left–right asymmetry in the angular distribution in the reaction with a polarised target and (*b*) the polarisation of the spin-$\frac{1}{2}$ particle produced in the same reaction on an unpolarised target. We start by deriving this relation.

It is assumed to be known that the π and K have spin 0 and that the Λ and Σ have spin $\frac{1}{2}$. Consider the polarisation produced in a reaction on an unpolarised target. On general grounds we can argue that a non-zero longitudinal polarisation is not allowed. A non-vanishing expectation value

$$\langle \boldsymbol{\sigma}\cdot\boldsymbol{p}_{\mathrm{f}}\rangle \neq 0$$

where $\boldsymbol{p}_{\mathrm{f}}$ is the final CM momentum would correspond to a non-zero expectation of a pseudo-scalar quantity, which is forbidden by space inversion invariance. On the other hand a net polarisation perpendicular to the reaction plane corresponding to

$$\langle \boldsymbol{\sigma}\cdot\boldsymbol{p}_{\mathrm{i}}\wedge\boldsymbol{p}_{\mathrm{f}}\rangle \neq 0$$

where $\boldsymbol{p}_{\mathrm{i}}$ is the initial CM momentum, is not ruled out as it is a true scalar. We consider $\pi^-\mathrm{p}\to\Sigma^+\mathrm{K}^-$ for definiteness. Proceeding formally we may regard an unpolarised proton target as a mixture of equal numbers with positive and negative helicity. If the amplitude for the reaction is denoted by $f_{\mu\lambda}(W,\theta,\phi)$, as in the discussion of §5.4.2, then for the case of initial spin up the final Σ spin state at the energy and angle considered is simply

$$\chi_\mu = f_{\mu+}(W,\theta,\phi) = \begin{pmatrix} f_{++} \\ f_{-+} \end{pmatrix} \tag{5.74}$$

The final polarisation state is defined by the expectation value of the Pauli spin operator ((B.15) of appendix B),

$$\boldsymbol{P}_+ = \frac{(\chi,\boldsymbol{\sigma}\chi)}{(\chi,\chi)}$$

hence with the expression (5.74) we have

$$(\chi,\sigma_x\chi) = (f^*_{++}f^*_{-+})\begin{pmatrix} 0 & 1 \\ 1 & 0 \end{pmatrix}\begin{pmatrix} f_{++} \\ f_{-+} \end{pmatrix}$$

$$= f^*_{++}f_{-+} + f^*_{-+}f_{++}$$

and hence

$$P_{x'+} = \frac{2\,\mathrm{Re}\,(f_{++}f^*_{-+})}{|f_{++}|^2+|f_{-+}|^2}$$

Similarly we find

$$P_{y'+} = -\frac{2\,\mathrm{Im}\,(f_{++}f_{-+}^*)}{|f_{++}|^2+|f_{-+}|^2}$$

$$P_{z'+} = \frac{|f_{++}|^2-|f_{-+}|^2}{|f_{++}|^2+|f_{-+}|^2}$$

For that fraction of protons with negative helicity the final Σ^- spin state is

$$\chi = \begin{pmatrix} f_{+-} \\ f_{--} \end{pmatrix}$$

and the polarisation, denoted in this case by P_- has components

$$P_{x'-} = \frac{2\,\mathrm{Re}\,(f_{+-}^*f_{--})}{|f_{+-}|^2+|f_{--}|^2}$$

$$P_{y'-} = \frac{2\,\mathrm{Im}\,(f_{+-}^*f_{--})}{|f_{+-}|^2+|f_{--}|^2}$$

$$P_{z'-} = \frac{|f_{+-}|^2-|f_{--}|^2}{|f_{+-}|^2+|f_{--}|^2}$$

Since the target is an incoherent mixture of spin up and spin down, the observed result is obtained by taking one half of the sum of P_+ and P_-. We make use of parity conservation in the reaction; as expressed by

$$f_{+-}(W,\theta,\phi) = -\eta_r e^{-i\phi} f_{-+}(W,\theta,\phi) \qquad (5.75a)$$

$$f_{--}(W,\theta,\phi) = \eta_r e^{-i\phi} f_{++}(W,\theta,\phi) \qquad (5.75b)$$

where

$$\eta_r = \eta_a\eta_b/\eta_c\eta_d$$

is the reaction parity, as discussed in §5.4.2. On noting that the reaction cross-section off an unpolarised target is

$$\frac{\overline{d\sigma}}{d\Omega} = \tfrac{1}{2}\sum_{\mu\lambda}|f_{\mu\lambda}|^2 = |f_{++}|^2+|f_{-+}|^2$$

we find for the final Σ polarisation

$$P_{x'} = 0$$

$$\frac{\overline{d\sigma}}{d\Omega}P_{y'} = -2\,\mathrm{Im}\,(f_{++}f_{-+}^*)$$

$$P_{z'} = 0$$

So the final polarisation is normal to the scattering plane (as expected from general arguments) and of magnitude

$$P_{y'} = -\frac{2\,\mathrm{Im}\,(f_{++}f_{-+}^*)}{|f_{++}|^2+|f_{-+}|^2} \tag{5.76}$$

which is independent of the reaction parity η_r.

Next we consider the angular distribution in the same reaction on a polarised target. The calculation here is most conveniently done with the aid of the density matrix†. The relation between the initial and final density matrices is

$$\rho^f = f\rho^i f^\dagger \tag{B.23}$$

where f denotes the scattering amplitude $f_{\mu\lambda}(W, \theta, \phi)$ written as a matrix. ρ^i and ρ^f may be expressed in terms of the polarisation vectors of the initial and final spin-$\frac{1}{2}$ particles by

$$\rho^i = \tfrac{1}{2}(1+\boldsymbol{P}^i\cdot\boldsymbol{\sigma}) \tag{5.77a}$$

$$\rho^f = \tfrac{1}{2}(1+\boldsymbol{P}^f\cdot\boldsymbol{\sigma}) \tag{5.77b}$$

As noted in appendix B, $\boldsymbol{P}^f$ is a vector whose components refer to the helicity frame of the final Σ particle, i.e. the frame reached from the initial frame by the standard helicity rotation $R(\phi, \theta, 0)$. We substitute (5.77) into (B.23) and use the conditions (5.75) of parity conservation. After some algebra one can identify the components of the final particle polarisation

$$\boldsymbol{P}^f = \tfrac{1}{2}\,\mathrm{Tr}\,(\boldsymbol{\sigma}\rho^f)$$

and the differential cross-section from the initial state with polarisation $\boldsymbol{P}^i$. One finds

$$\frac{d\sigma}{d\Omega} = \{|f_{++}|^2+|f_{-+}|^2\} \mp 2\eta_r P^i_y\,\mathrm{Im}\,(f_{++}f_{-+}^*)$$

$$\frac{d\sigma}{d\Omega}P^f_{x'} = \pm\eta_r P^i_x\{|f_{++}|^2-|f_{-+}|^2\}+2P^i_z\,\mathrm{Re}\,(f_{++}f_{-+}^*)$$

$$\frac{d\sigma}{d\Omega}P^f_{y'} = -2\,\mathrm{Im}\,(f_{++}f_{-+}^*) \pm 2\eta_r P^i_y\{|f_{++}|^2+|f_{-+}|^2\}$$

$$\frac{d\sigma}{d\Omega}P^f_{z'} = \mp 2\eta_r P^i_x\,\mathrm{Re}\,(f_{++}f_{-+}^*)+P^i_z\{|f_{++}|^2-|f_{-+}|^2\}$$

where the upper and lower signs refer to scattering at the azimuthal angles $\phi = 0$ and π respectively, i.e. to left and right scattering as indicated in fig. 5.4. The helicity non-flip and flip amplitudes f_{++} and f_{-+} on

† See appendix B for a brief introduction to the density matrix.

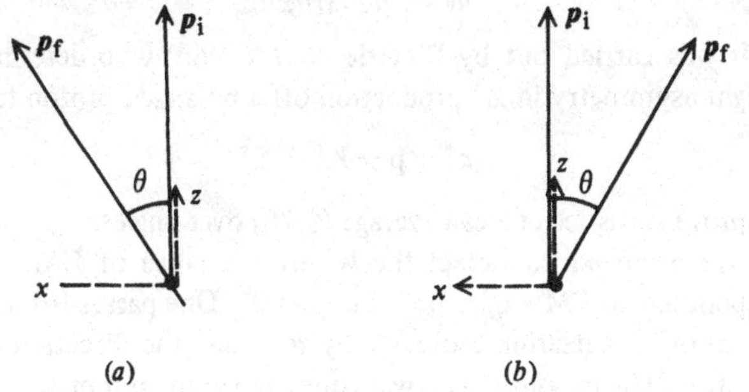

Fig. 5.4. (a) left scattering, $\phi = 0$; (b) right scattering, $\phi = \pi$.

the right are evaluated at $\phi = 0$. Note that as part of the helicity convention the polarisation vectors are referred to different axes in the two cases.

This set of equations describes the cross-sections and polarisations for the spin-0 reaction off a spin-$\frac{1}{2}$ target of arbitrary polarisation. The coefficients of P_z^i and P_x^i in $(d\sigma/d\Omega)P_{x'}^f$ divided by $d\sigma/d\Omega$ are called the (centre of mass) A and R parameters. Together with $d\sigma/d\Omega$ they serve to determine the two complex amplitudes f_{++} and f_{-+} to within an unobservable overall phase. The particular result of interest here is that it is the transverse component P_y^i of initial polarisation which can influence the differential cross-section. If we define the left-right asymmetry function for the process by

$$\epsilon_{LR}(W,\theta) = \frac{\left.\dfrac{d\sigma}{d\Omega}\right)_{\phi=0} - \left.\dfrac{d\sigma}{d\Omega}\right)_{\phi=\pi}}{\left.\dfrac{d\sigma}{d\Omega}\right)_{\phi=0} + \left.\dfrac{d\sigma}{d\Omega}\right)_{\phi=\pi}}$$

then we have

$$\epsilon_{LR}(W,\theta) = -2\eta_r \frac{\mathrm{Im}\,(f_{++}f_{-+}^*)}{|f_{++}|^2 + |f_{-+}|^2} P_y^i \tag{5.78}$$

Comparing (5.76) and (5.78) we see that over all energies and angles the final Σ polarisation produced in the reaction on an unpolarised proton target is proportional to the left–right asymmetry in the Σ differential cross-section for the same reaction off a polarised p target. This relationship was originally discovered by Wolfenstein and Ashkin (1952) in the context of nucleon–nucleus scattering.

It was pointed out by Bilenky (1958) that in general the proportionality factor contains the reaction parity, so that we have a method for determining the intrinsic parity product

$$\eta_r = \eta_p \eta_\pi / \eta_\Sigma \eta_K$$

This was carried out by Dieterle *et al.* (1968) who determined the left–right asymmetry in Σ^+ production off a polarised proton target:

$$\pi^+ + p \rightarrow K^+ + \Sigma^+$$

To improve statistics one can average (5.78) over angles.

It was arranged to detect the K^+ over a range of LAB momenta corresponding to CM angles $45° < \theta < 100°$. One passes from left scattering to right scattering geometry by reversing the direction of target polarisation. The measured ϵ_{exp} was compared with asymmetry predicted with the aid of (5.78) and results for the polarisation produced in previous experiments with unpolarised targets. ϵ_{exp} was within 1.1 standard deviations of that predicted by (5.78) for η_r even, but 2.7 standard deviations from the value predicted for odd η_r. One may therefore conclude that η_r is even, and hence

$$\eta_{K^+} \eta_{\Sigma^+} = -1 \tag{5.79}$$

5.5.6 *The decay $\Sigma^0 \rightarrow \Lambda^0 + \gamma$ and the $\Sigma\Lambda$ relative parity*

The parity conserving electromagnetic decay

$$\Sigma^0 \rightarrow \Lambda^0 + \gamma \tag{5.80}$$

permits the determination of the Σ^0 parity relative to the convention $\eta_\Lambda = +1$. The correlation between the spin of the γ and the spin of the Λ^0 produced from the decay of a polarised Σ^0 is sensitive to the sign of η_Σ. This is a difficult experiment. To see that nothing simpler such as the measurement of the polarisation of just the Λ^0 will suffice, we first calculate the Λ^0 polarisation from a polarised Σ when the γ is not observed.

The CM amplitude for the process (5.80) is

$$f_{\lambda\gamma,M}(\theta) = \frac{1}{2\pi} d^{1/2}_{M,\lambda-\gamma}(\theta) A_{\lambda\gamma}$$

where M denotes the spin projection of the Σ and γ and λ are the helicities of the photon and Λ-hyperon. Parity conservation in the decay requires that

$$A_{\lambda\gamma} = -\frac{\eta_\Sigma}{\eta_\Lambda \eta_\gamma} A_{-\lambda,-\gamma}$$

The final spin component along the direction of relative motion,

$\lambda-\gamma$, can only be $\pm\frac{1}{2}$ and since $\gamma=\pm 1$ for the photon, the only possible values for the pair (λ, γ) are $(+\frac{1}{2}, +1)$ and $(-\frac{1}{2}, -1)$. We may therefore label A by λ alone, and further indicate the values $\pm\frac{1}{2}$ by $\pm$. Parity conservation now reads

$$A_+ = +\eta_\Sigma A_-$$

and there is just one amplitude to describe the decay.

The polarisation of the Λ^0 produced from the decay of a sample of Σ^0s all with spin up, $M=+\frac{1}{2}$, is given by the expectation value of σ in the final state,

$$P_\Lambda = (f, \boldsymbol{\sigma} f)$$
$$= \sum_\gamma \sum_{\lambda'\lambda} f^*_{\lambda'\gamma,+1/2}(\boldsymbol{\sigma})_{\lambda'\lambda} f_{\lambda\gamma,+1/2}$$

where we have summed over the photon helicity which is not observed.

Remembering that $\boldsymbol{P}_\Lambda$ is referred to the helicity frame of the final Λ, one finds that it is directed along the momentum of the Λ^0 and has magnitude

$$|P_\Lambda| = -|A_+|^2 \cos\theta$$

If $M=-\frac{1}{2}$, $\boldsymbol{P}_\Lambda$ has the same magnitude and opposite direction. From this we may find the polarisation of the Λ in the decay of a sample of Σs of polarisation $\boldsymbol{P}_\Sigma$. We take the z-axis of the Σ rest-frame along $\boldsymbol{P}_\Sigma$, then the sample consists of fractions $\frac{1}{2}(1 \pm |P_\Sigma|)$ with spin components $M=\pm\frac{1}{2}$. Hence the polarisation of the resulting Λ^0s is given in magnitude by

$$|P_\Lambda| = \tfrac{1}{2}(1+|P_\Sigma|)(-|A_+|^2\cos\theta) + \tfrac{1}{2}(1-|P_\Sigma|)(+|A_+|^2\cos\theta)$$
$$= -|P_\Sigma||A_+|^2\cos\theta$$

There is no dependence on the relative parity here, as only $|A_+|^2$ and $|A_-|^2$ entered the calculation.

Following Gatto (1957), we next consider the polarisation of a Λ^0 produced in coincidence with a *linearly* polarised photon from the decay of a polarised Σ.

With the conventions for photon states established in §4.5 a photon travelling along the z-direction with polarisation vector along the y-axis,

$$\boldsymbol{e}_y = i2^{-1/2}(\boldsymbol{e}_+ + \boldsymbol{e}_-)$$

is described by a superposition of helicity states,

$$i2^{-1/2}(\phi^\gamma_{p00,+1} + \phi^\gamma_{p00,-1})$$

It follows that

$$F_\lambda = i2^{-1/2}\{f_{\lambda,+1,M}(\theta) + f_{\lambda,-1,M}(\theta)\}$$

is the amplitude for the decay of a Σ^0 with spin component M into a Λ^0 with helicity λ, and a photon with polarisation vector along the y-axis in its helicity frame. This is opposite to the direction of the y-axis in the decay frame of the Σ^0 because of the conventions for the 'second' particle, see fig. 4.2. We then obtain the polarisation of the final Λ^0,

$$\boldsymbol{P}^\Lambda = \langle \boldsymbol{\sigma} \rangle = \sum_{\lambda'\lambda} F^*_{\lambda'} \boldsymbol{\sigma}_{\lambda'\lambda} F_\lambda$$

Assuming the Σ^0 fully polarised ($M = +\frac{1}{2}$) one finds

$$P^\Lambda_{x'} = -\tfrac{1}{2}\eta_\Sigma |A_+|^2 \sin\theta$$

$$P^\Lambda_{y'} = 0$$

$$P^\Lambda_{z'} = -\tfrac{1}{2}|A_+|^2 \cos\theta$$

If we interpret the polarisation vectors of the Σ^0, Λ^0 and photon in the same frame, we may say that in the decay, the hyperon polarisation vector is rotated through 180° about $\boldsymbol{e}^\gamma$ if $\eta_\Sigma = -1$ and about $\boldsymbol{p}_\Lambda \wedge \boldsymbol{e}^\gamma$ through 180° if $\eta_\Sigma = +1$.

The Λ^0 polarisation may be determined by the angular asymmetry in its subsequent decay. One way to determine the photon polarisation is to allow the photon to convert into an e^+–e^- pair: the plane of the pair is correlated with the plane of polarisation. Byers and Burkhardt (1961) showed that if the photon undergoes internal conversion

$$\Sigma^0 \to \Lambda^0 + \gamma_{\text{virtual}} \to \Lambda^0 + e^+ + e^- \qquad (5.81)$$

the normal to the plane of the Dalitz pair is correlated with $\boldsymbol{\epsilon}^\gamma$. No successful measurement of η_Σ by this type of experiment has been reported. However Feinberg (1958) and Feldman and Fulton (1958) showed that the effective mass distribution of the e^+–e^- pair in (5.81) is sensitive to the Σ–Λ relative parity.

Courant *et al.* (1963) and Alff *et al.* (1965) used this method on Σ^0 produced by

$$K^- + p \to \Sigma^0 + \pi^0$$

with stopped K^-. Their results were only consistent even with parity. Thus

$$\eta_{\Sigma^0} = +1 \qquad (5.82)$$

Assuming that the other charge states of the Σ have positive parity the result of the preceding section (5.79) gives

$$\eta_{K^+} = -1$$

Thus both the K and its antiparticle $\bar{K}$ have odd parity.

5.5.7 *Parity of* Ξ

The reaction

$$\Xi^- + p \rightarrow \Lambda^0 + \Lambda^0 \tag{5.83}$$

conserves strangeness and is a strong interaction. The presence of two Λ^0s means that the intrinsic parity of the Ξ^- is defined relative to that of the proton, independently of the convention $\eta_{\Lambda^0} = +1$.

Treiman (1959) pointed out that a study of (5.83) in which the Ξ^- is captured at rest, can serve to determine the Ξ^- parity.

No measurement of Ξ^- by this method has been reported. An alternative method for the determination of η_Ξ would be to apply the 'polarisation-asymmetry relation' described in §5.5.5 to the reaction

$$K^- + p \rightarrow K^+ + \Xi^-$$

on polarised protons.

p, n, Λ and Σ have been found to have even parity (subject to the conventions established) and the success of the *SU*(3) symmetry scheme, which required the cascade particles to have the same parity as the other six baryons, is strong evidence for even Ξ parity.

5.5.8 *Parity of* Ω^-

Proposals have been made for determining the Ω^- parity (Bilenky and Ryndin, 1965). The existence of the strong reaction

$$K^- + p \rightarrow \Omega^- + K^+ + K^0$$

shows that η_Ω is defined relative to the other particle parities without further conventions.

Since the Ω^- can only be produced in three or more particle final states a determination would require very large statistics.

5.6 Parity violation in weak interactions

5.6.1 *The* τ–θ *puzzle*

The original motivation for the experiments which led to the discovery of parity violation came from the 'τ–θ puzzle'. In the years preceding 1957 the θ- and τ-mesons had been identified by their decay modes

$$\theta^+ \to \pi^+ + \pi^0$$

$$\tau^+ \to \pi^+ + \pi^+ + \pi^0$$

and careful studies had shown that the masses and lifetimes of the two mesons were equal within experimental errors. It was natural to conclude that these were simply two different decay modes of the same meson. However, Dalitz made a spin parity analysis of the decays (assuming parity conservation) and concluded that the θ and τ must have opposite parity.

Consider first the θ decay. The θ spin was then unknown. Suppose it to be J; then since the pions are spinless and both of negative parity, the final orbital angular momentum is J and the final parity is $(-1)^2(-1)^J$. Hence the spin parity assignment of the θ^+ can only be $0^+, 1^-, 2^+$, etc.

The τ decay is more complicated because of the three particles in the final state. The total angular momentum in the final state is the vector sum of the relative orbital angular momentum $\boldsymbol{L}_1$ of, say, the two π^+, and the orbital angular momentum $\boldsymbol{L}_2$ of the π^- relative to the CM of the two π^+:

$$\boldsymbol{L}_{\text{tot}} = \boldsymbol{L}_1 + \boldsymbol{L}_2$$

If the τ spin is assumed to be zero the eigenvalue of $\boldsymbol{L}_{\text{tot}}$ must be zero which is possible only if the eigenvalues corresponding to $\boldsymbol{L}_1$ and $\boldsymbol{L}_2$ are equal: $\ell_1 = \ell_2$. Now the orbital parity of a three particle state is generally given in terms of the orbital angular momenta by $(-1)^{\ell_1+\ell_2}$. So the final state parity is $(-1)^3(-1)^{\ell_1+\ell_2}$, which is negative for τ spin zero. So if parity is conserved and the θ and τ have spin zero, then they cannot be the same particle. We shall not discuss the possibility of higher τ spin. An analysis of the energy distribution in the 3π-state showed that the spin parity assignment of the τ is most probably 0^-. This then is the puzzle: two particles have all the same properties except that they are of opposite intrinsic parity.

5.6.2 *The Wu experiment*

It is clear that if parity is not conserved in the decay of τ and θ then the above analysis is invalid and there is no problem. Thus, after several attempted explanations, Lee and Yang (1956*a*) in a celebrated paper made a systematic analysis of the evidence for parity conservation in elementary particle processes. They found that while there was plenty of evidence for the validity of parity conservation in electromagnetic and strong interactions (from atomic and nuclear physics as we have noted in §5.2), there was no experimental evidence whatsoever for parity conservation in β-decay or the weak decays of the mesons and hyperons then known.

Following on this analysis Wu, Ambler, Hayward, Hoppes and Hudson (1957) showed conclusively that parity is violated in β-decay. As we have seen, parity conservation forbids the existence of non-vanishing expectation value $\langle \boldsymbol{S}\cdot\boldsymbol{p}\rangle$ where $\boldsymbol{S}$ is a spin and $\boldsymbol{p}$ a momentum, so a possible test is to examine the angular distribution of electrons from the β-decay of polarised nuclei. Wu and coworkers used a sample of polarised ^{60}Co nuclei

$$^{60}\text{Co} \rightarrow {}^{60}\text{Ni} + \text{e}^- + \bar{\nu}$$

and found the angular distribution of the electrons to be given by

$$W(\theta) = 1 - \alpha P \cos\theta \quad (\text{with } \alpha > 0.7)$$

where θ is the angle of emission of the electron relative to the polarisation vector P of the cobalt sample. It was found that $\alpha \cong v/c$, where v is the electron velocity in agreement with β-decay theory. Thus electrons are preferentially emitted opposite to the spin direction.

As with the Λ^0 decay experiment we can reflect the experiment in a mirror, see fig. 5.5. The correlation between the current flow in the solenoid polarising the ^{60}Co sample and the direction of preferential electron emission when reflected in the mirror corresponds to a situation not realised in Nature.

It was later shown by Frauenfelder *et al.* (1957) that the electrons emitted in β-decay are longitudinally polarised, i.e. $\langle \boldsymbol{\sigma}\cdot\boldsymbol{p}_{\text{e}}\rangle \neq 0$, which is again evidence of parity violation.

5.6.3 *Pion decay and muon decay*

The form of the weak interaction was elucidated with the aid of the wealth of information on nuclear β-decays. We shall not discuss this aspect of weak interactions here as good reviews are available elsewhere. Instead we shall consider some elementary aspects of the π–μ–e sequence of decays, in order to make clear the special properties of the neutrino and their connection with the symmetries of leptonic weak decays. We proceed on the assumption that the neutrino mass is identically zero.

The principal decay mode of the π^- is

$$\pi^- \rightarrow \mu^- + \nu \qquad (5.84)$$

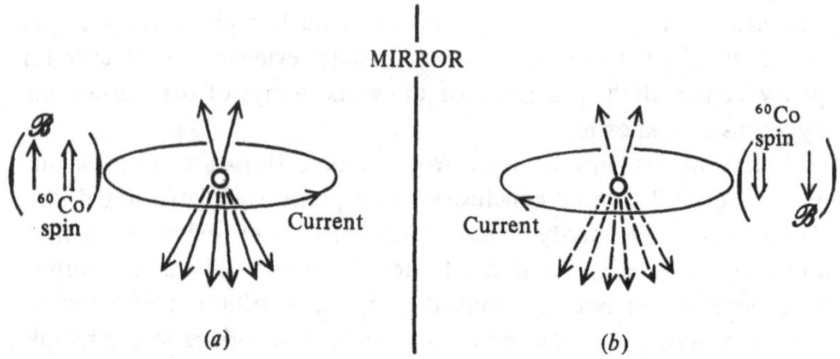

Fig. 5.5. Wu experiment. (*a*) Observed electron distribution relative to the sense of the (positive) current in the polarising solenoid. (*b*) Mirror image experiment not realised. The directions of the magnetic field $\mathscr{B}$ and the nuclear spin orientation are shown.

The μ^- subsequently decays by

$$\mu^- \rightarrow e^- + \nu + \nu'$$

Two neutrinos are emitted which may or may not be identical. We shall see that they are in fact distinct.

The properties of the muon in muonic atoms show that it has spin $\frac{1}{2}$ and, since the pion has spin 0, (5.84) tells us that the neutrino has half-integer spin. In fact the μ and ν must have antiparallel spins, or more precisely equal helicities, and so $|\lambda_\nu| = \frac{1}{2}$, see fig. 5.6.

Several experiments have been carried out to measure the muon helicity in the case of

$$\pi^- \rightarrow \mu^- + \nu$$

(Alikanov *et al.*, 1960, Backenstoss *et al.*, 1961, and Bardon *et al.*, 1961). In the last-mentioned experiment pions decay in flight and the left–right asymmetry $\epsilon_{LR}(\theta)$ in the rescattering of the decay muons off a lead target is measured. As we have seen $\epsilon_{LR}(\theta)$ gives a measure of the transverse polarisation of the muons via the spin–orbit interaction of the μ^- in the Coulomb field of the nuclei. When converted to the rest-frame of the pion by the Wigner rotation the results show that the μ^- is always emitted with positive helicity. This of course shows that parity is violated in the decay. Indeed it is *maximally violated* since not only is the mean value of the μ^- helicity non-zero, it is equal to its maximum possible value $+\frac{1}{2}$. This result can also be correlated with the theoretical possibility which we have alluded to in discussing massless particles: we

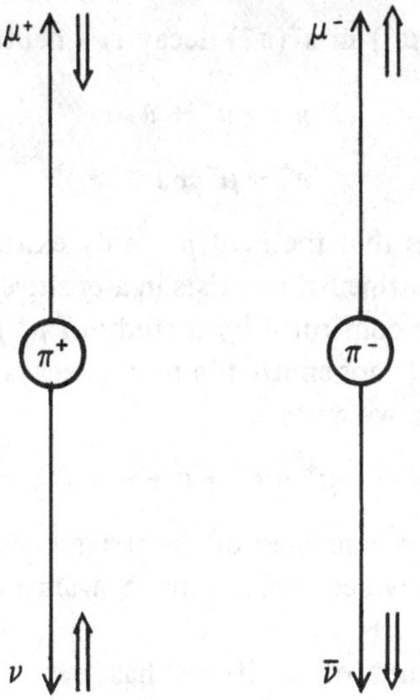

Fig. 5.6. $\pi^{\pm}$ decay.

make the hypothesis that the neutrino which couples to the μ^- only exists in the positive helicity state, i.e. as a right-handed neutrino. We have seen in §4.5 and §5.4.3 that this is consistent with Lorentz invariance but not with parity conservation and it provides a natural mechanism for maximal parity violation.

In the case of the π^+ decay

$$\pi^+ \to \mu^+ + \nu$$

there is evidence that the μ^+ is always emitted with *negative* helicity, again indicated maximal parity violation, but the neutrino coupled to the μ^+ has negative helicity, i.e. is a left-handed neutrino.

We now appear to have the two helicity states $\pm\frac{1}{2}$ of the neutrino, but one might then ask why do not both these neutrino states couple in to each of the processes

$$\pi^{\pm} \to \mu^{\pm} + \nu?$$

The answer is provided by the law of *lepton conservation* which assigns non-zero lepton number L only to electrons, muons and neutrinos, and zero lepton number to other particles, and requires that L be conserved in all interactions. The particles μ^- and e^- have $L = +1$. Processes such as pair production $\gamma \to e^+e^-$ then require that the *antiparticles* μ^+ and e^+ have $L = -1$. The pion decays then imply that the massless particles emitted with the $\mu^+(\mu^-)$ in $\pi^+(\pi^-)$ decay is a neutrino (antineutrino). So we write

$$\pi^- \to \mu^- + \bar{\nu}$$

$$\pi^+ \to \mu^+ + \nu$$

Thus our hypothesis is that the neutrino only exists in a negative helicity state, while the antineutrino exists in a positive helicity state only. These assignments are confirmed by a study of $\mu^\pm$ decay. To conserve lepton number the neutrinos emitted in muon decays must be a neutrino and an antineutrino, so we write

$$\mu^+ \to e^+ + \nu + \bar{\nu} \qquad (5.85)$$

Let us test some consequences of the assignment of lepton numbers and neutrino helicities which we have made, against observations on the muon decay.

The energy spectrum of positrons has been measured (see e.g. Bardon *et al.*, 1965) and is strongly peaked at the upper end of the range (the maximum kinematically allowed energy is $E_{max} = (m_\mu^2 + m_e^2)/2m_\mu$). The emission of a positron of maximum energy corresponds to the configuration shown in fig. 5.7 in which the two massless particles go off in the same direction with equal momenta. If the two neutrinos were identical this configuration would be suppressed by the Pauli exclusion principle, and the positron spectrum would go to zero at E_{max}. We are assuming that the neutrino of spin $\frac{1}{2}$ obeys Fermi statistics. Thus we conclude that the neutrinos are distinct: a neutrino and antineutrino, with negative and positive helicity respectively. They are even more different: an electron neutrino ν_e and muon antineutrino $\bar{\nu}_\mu$.

As we have seen, the pion decay affords a natural source of polarised muons, and these can be stopped in some materials (e.g. carbon) without depolarisation. The resulting angular distribution of positrons is peaked in the direction opposite to the line of flight of the stopping μ^+. Since

$$\langle \boldsymbol{\sigma}_\mu \cdot \boldsymbol{p}_e \rangle \neq 0$$

this asymmetry confirms parity violation in muon decay (Garwin, Lederman and Weinrich, 1957; Friedman and Telegdi, 1957). Now we know that the stopping μ^+ have negative helicity, so the positrons are preferentially emitted in the direction of the μ^+ spin. This favoured configuration is shown in fig. 5.7(*a*). The two neutrinos carry off no net angular momentum along the line of momenta, and thus to conserve angular momentum the positron helicity must be positive.

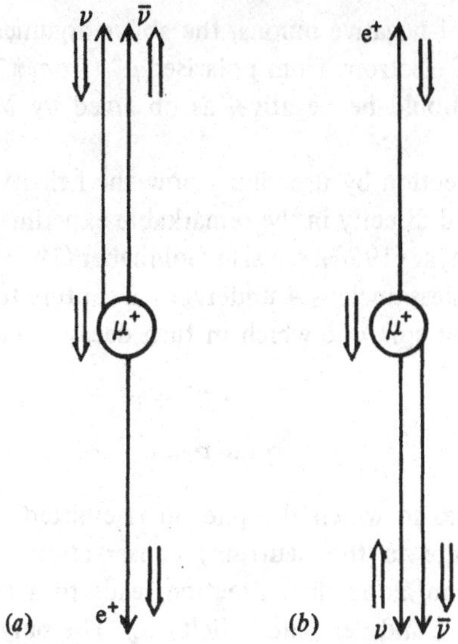

Fig. 5.7. μ^+ decay: two configurations in which the positron has maximum energy and is emitted (*a*) parallel and (*b*) antiparallel to the muon spin.

It is possible to argue why configuration (*a*) is favoured over configuration (*b*) in which the positron is emitted opposite to the spin, with negative helicity. The maximum energy electrons are highly relativistic. If we could neglect their mass entirely, they would behave like neutrinos. The positron is an antilepton and like the antineutrinos would prefer the positive helicity state. Taken together with angular momentum conservation, this shows that configuration (*a*) is preferred over configuration (*b*).

Now the positron does exist in both helicity states. A more exact argument using a Lorentz invariant interaction involving electron and neutrino fields shows that the amplitudes for emission of positrons in the positive and negative helicity states are in the ratio

$$\left(\frac{E+p}{2E}\right)^{1/2} : \left(\frac{E-p}{2E}\right)^{1/2}$$

and when $E \gg m_e$, $E \approx |\boldsymbol{p}|$ and we again come to the conclusion stated above.

This is true whatever the polarisation state of the muons and therefore predicts that the positrons from decay of *unpolarised* muons should have positive helicity. The experiments of Macq, Crowe and Haddock (1957) and Culligan *et al.* (1957) confirm this.

For the case of negative muons, the above arguments lead to backward emission of electrons from polarised μ^- (from π^- decay) and that the e^- helicity should be negative, as observed by Macq, Crowe and Haddock (1957).

We end this section by describing how the helicity of the neutrino has been measured directly in the remarkable experiment of Goldhaber, Grodzins and Sunyar (1958); see also Goldhaber (1958).

Suppose a spinless nucleus A undergoes K capture to an excited state of a nucleus B^* of spin one which in turn decays to its ground state B of spin 0

$$e^- + A \to B^* + \nu$$

$$B^* \to B + \gamma$$

For those cases in which the photon is emitted along the line of B^* recoil (opposite to the neutrino) conservation of component of angular momentum along that direction leads to a correlation of the photon helicity λ_γ and neutrino helicity λ_ν. The only contribution to the initial angular momentum is from the electron spin and so we have

$$\pm\tfrac{1}{2} = \lambda_\gamma - \lambda_\nu$$

which requires $\lambda_\nu = 2\lambda_\gamma$ because $\lambda_\gamma = 0$ is forbidden. Thus those photons emitted along the B^* recoil direction should be 100 per cent polarised. Although the neutrino is not observed, the desired events can be selected by the following ingenious method. One may show that if the sequence $A \to B^* \to B$ is such that the recoil momentum of the B^* (i.e. the energy of the neutrino) is nearly equal to the energy difference ΔMc^2 between B^* and B, then those photons emitted opposite to the neutrino direction have exactly the right energy to undergo resonant rescattering off a secondary target of B nuclei,

$$\gamma + B \to B^* \to B + \gamma$$

Goldhaber, Grodzins and Sunyar used ^{152}Eu ($J^P = 0^-$) which undergoes K capture to an excited state ^{152}Sm (1^-) which in turn decays to its 0^+ ground state.

The neutrino energy is 840 keV and the ΔMc^2 is 960 keV, so the polarisation was not quite 100 per cent. However the resonant rescattering was clearly seen and the photon helicity was detected by selective absorption in magnetised iron. The photon, and hence the neutrino, was found to have negative helicity.

5.6.4 *Phenomenology of hyperon decays*

In this section we shall describe the phenomenological analysis of hyperon decays. For definiteness we consider the decay

$$\Lambda^0 \to p + \pi^-$$

The formalism also applied to Σ decays

$$\Sigma^\pm \to n + \pi^\pm, \quad \Sigma^+ \to p + \pi^0$$

In all these cases we have a spin-$\frac{1}{2}$ particle decaying into a spin-$\frac{1}{2}$ and a spin-0 particle. The same formalism applied to both stages of the cascade decay,

$$\Xi^- \to \Lambda^0 + \pi^- \to p + \pi^- + \pi^-$$

In Λ^0 decay the $p\pi^-$ system in the final state can have orbital angular momentum 0 or 1. If parity was conserved only one of these possibilities would be allowed. However as we have seen parity is violated in the Λ^0 decay and the decay process is described by an s- and a p-wave of amplitudes a_s and a_p.

We shall use the helicity formalism for decays of §4.9.1. The amplitude for decay of a spin-$\frac{1}{2}$ hyperon at rest with the emission of the proton in the direction $\Omega = (\theta, \phi)$ is

$$f_{\lambda M}(\theta, \phi) = (2\pi)^{-1/2} \mathfrak{D}^{1/2}_{M\lambda}(\phi, \theta, 0)\, a_\lambda \tag{5.86}$$

where M and λ denote the spin projection of the Λ^0 and the helicity of the proton. We denote the two decay amplitudes by a_+ and a_-. Parity conservation would imply $a_- = -a_+$. The total decay rate is

$$\Gamma = |a_+|^2 + |a_-|^2$$

In the original analysis of this decay, Lee and Yang (1957) discussed three types of experiment.

The first is:

The angular distribution of the decay proton from a polarised hyperon at rest. This is given by $W_+(\theta, \phi)\, d\Omega$ where

$$\begin{aligned} W_+(\theta, \phi) &= \Gamma^{-1} \sum_\lambda |f_{\lambda,+1/2}(\theta, \phi)|^2 \\ &= (2\pi\Gamma)^{-1} \sum_\lambda |a_\lambda|^2 [d^{1/2}_{+1/2,\lambda}(\theta)]^2 \\ &= (2\pi\Gamma)^{-1} \{|a_+|^2 \cos^2 \tfrac{1}{2}\theta + |a_-|^2 \sin^2 \tfrac{1}{2}\theta\} \end{aligned}$$

After rearrangement this can be written:

$$W_+(\theta, \phi) = (4\pi)^{-1}(1 + \alpha \cos \theta)$$

where

$$\alpha = \frac{|a_+|^2 - |a_-|^2}{|a_+|^2 + |a_-|^2} \tag{5.87}$$

which would be zero if parity was conserved. We have taken the hyperon spin to be up. If $M = -\frac{1}{2}$, we find instead

$$W_-(\theta, \phi) = (4\pi)^{-1}(1 - \alpha \cos \theta)$$

The hyperon is produced in a strong interaction and may only have a polarisation perpendicular to the production plane. We therefore take the z-axis in the Λ^0 rest-frame along the normal to the production plane. Then a sample of Λ^0 with polarisation P_Λ corresponds to fractions $\frac{1}{2}(1 + P_\Lambda)$ with $M = +\frac{1}{2}$ and $\frac{1}{2}(1 - P_\Lambda)$ with $M = -\frac{1}{2}$. The resultant angular distribution for such a sample is

$$\begin{aligned} W(\theta, \phi) &= \tfrac{1}{2}(1 + P_\Lambda)\, W_+ + \tfrac{1}{2}(1 - P_\Lambda)\, W_- \\ &= (4\pi)^{-1}(1 + \alpha P_\Lambda \cos \theta) \end{aligned} \tag{5.88}$$

Here P_Λ is measured in the hyperon rest-frame.

Such a sample is obtained experimentally by selecting those Λ^0s produced in a definite plane. However P_Λ is not known independently so a different method is required to determine α.

Consider then:

The longitudinal polarisation of the nucleon emitted in the decay of unpolarised hyperons at rest. The expectation of the nucleon helicity from a sample of decays of unpolarised Λ^0s is

$$\begin{aligned} P_N &= (2\Gamma)^{-1} \sum_M \int d\Omega \{\tfrac{1}{2}|f_{+1/2,M}|^2 - \tfrac{1}{2}|f_{-1/2,M}|^2\} \\ &= (2\Gamma)^{-1} \sum_M \int d\Omega \sum_\lambda \lambda |f_{\lambda M}|^2 \\ &= (2\Gamma)^{-1} \sum_M \sum_\lambda \lambda |a_\lambda|^2 (2\pi)^{-1} \int d\Omega\, |d^{1/2}_{M\lambda}(\theta)|^2 \\ &= \Gamma^{-1} \sum_\lambda \lambda |a_\lambda|^2 \end{aligned}$$

where we used

$$\begin{aligned} \sum_{M'} (d^J_{MM'}(\theta))^2 &= \sum_{M'} d^J_{MM'}(-\theta) d^J_{M'M}(\theta) \\ &= d^J_{MM}(0) = 1 \end{aligned}$$

Thus

$$P_N = \frac{1}{2}\frac{|a_+|^2 - |a_-|^2}{|a_+|^2 + |a_-|^2} = \frac{1}{2}\alpha$$

Thus the asymmetry parameter α is also the degree of longitudinal polarisation (in units of $\frac{1}{2}\hbar$). A non-zero helicity is of course direct evidence of parity violation as we observed before.

We can obtain an unpolarised sample experimentally by considering all Λ^0 decays without reference to their production plane. P_N, and hence α, may be determined by allowing the decay protons to rescatter off nuclei, e.g. ^{12}C. The relation between the azimuthal asymmetry and the transverse nucleon polarisation must be known from independent experiments. We note that in the LAB or more correctly the proton–^{12}C CM frame, the longitudinal polarisation is rotated as a result of the kinematical transformation from the decay frame. The angle of rotation ω is given by (4.133)

$$\tan\omega = \frac{u(1-v^2)^{1/2}\sin\theta}{v+u\cos\theta}$$

where v is the velocity of the decay nucleon in the Λ^0 decay frame and u is the velocity of the Λ^0 in the LAB. α has been determined by this method in several experiments. For example, Cronin and Overseth (1963) studied 1156 Λ^0 decays and found

$$\alpha = +0.62 \pm 0.07$$

The current average is

$$\alpha = +0.647 \pm 0.013$$

(Particle Data Group, 1974). Thus the proton prefers positive helicity. Once α is known the experiment of type (*a*) can be used to determine P_Λ in the production process.

The two complex decay parameters $a_\pm$ involve three real numbers if we neglect an overall phase factor.

The decay rate Γ and α furnish two numbers. The determination of the remaining decay parameter requires:

Measurement of the transverse polarisation of the nucleon emitted in a given direction in the decay of a polarised hyperon. The polarisation vector of the proton produced in the decay of a Λ^0 with polarisation P_Λ is most conveniently calculated using density matrix methods.

As before we take the production normal as the z-axis in the Λ^0 decay frame. Then the sample of Λ^0 has density matrix

$$\rho^i = \tfrac{1}{2}(1 + P_\Lambda \boldsymbol{\sigma}_z)$$

The density matrix for the final proton is given by

$$\rho^{\mathrm{f}}_{\lambda'\lambda} = \sum_{MM'} f_{\lambda'M'}\rho^{\mathrm{i}}_{M'M} f^*_{\lambda M} \tag{5.89}$$

where $f_{\lambda M}(\theta, \phi)$ is given by (5.86).

The polarisation vector of the proton is

$$\mathscr{I} \boldsymbol{P}_{\mathrm{N}} = \mathrm{Tr}\,(\rho^{\mathrm{f}}\boldsymbol{\sigma})$$

where $\mathscr{I}$ is the intensity for the decay

$$\mathscr{I} = \mathrm{Tr}\,(\rho^{\mathrm{f}})$$

Because we are using the helicity formalism $\boldsymbol{P}_{\mathrm{N}}$ is a vector referred to the rest-frame (x', y', z') of the nucleon.

From the form of $f_{\lambda M}(\theta, \phi)$ and the fact that ρ^{i} is diagonal it can be seen that ρ^{f} is independent of ϕ. This is an instance of a general result that there is no azimuthal dependence in the decay when the initial state is an incoherent superposition of states of different M, i.e. no off-diagonal terms in ρ^{i}.

We therefore specialise to $\phi = 0$. The geometry is shown in fig. 5.8. On substituting for f and ρ^{i} we find

$$4\pi\rho^{\mathrm{f}} = \begin{bmatrix} (1 + P_\Lambda \cos\theta)\,|a_+|^2 & -\frac{1}{2}P_\Lambda \sin\theta\, a_+ a_-^* \\ -\frac{1}{2}P_\Lambda \sin\theta\, a_+^* a_- & (1 - P_\Lambda \cos\theta)\,|a_-|^2 \end{bmatrix}$$

The decay intensity is

$$\mathscr{I}(\theta, \phi) = \mathrm{Tr}\,(\rho^{\mathrm{f}}) = (4\pi)^{-1}\Gamma(1 + P_\Lambda \alpha \cos\theta)$$

which checks with (5.88).

The components of $\boldsymbol{P}_{\mathrm{N}}$ are given by

$$\mathscr{I} P_{\mathrm{N}x'} = -(4\pi)^{-1} P_\Lambda \sin\theta \cdot 2\,\mathrm{Re}\,(a_+ a_-^*)$$

$$\mathscr{I} P_{\mathrm{N}y'} = (4\pi)^{-1} P_\Lambda \sin\theta \cdot 2\,\mathrm{Im}\,(a_+ a_-^*)$$

$$\mathscr{I} P_{\mathrm{N}z'} = (4\pi)^{-1}\{(|a_+|^2 - |a_-|^2) + P_\Lambda(|a_+|^2 + |a_-|^2)\cos\theta\}$$

We define two further decay parameters

$$\beta = \frac{2\,\mathrm{Im}\,(a_+ a_-^*)}{|a_+|^2 + |a_-|^2} \tag{5.90}$$

and

$$\gamma = \frac{2\,\mathrm{Re}\,(a_+ a_-^*)}{|a_+|^2 + |a_-|^2} \tag{5.91}$$

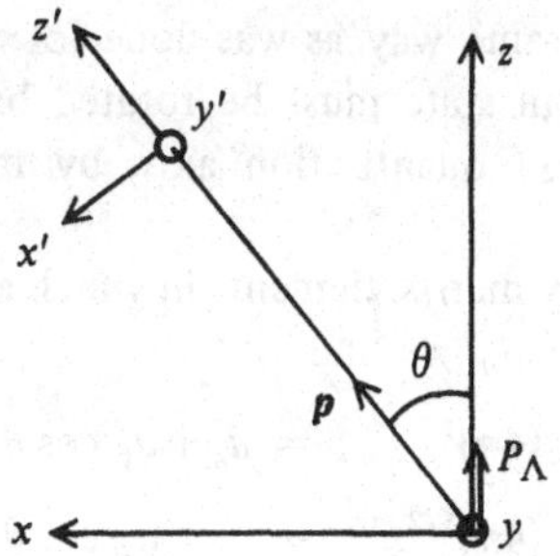

Fig. 5.8. Geometry of the Λ^0 decay. $x'y'z'$ is the helicity frame of the decay nucleon.

So that finally

$$P_{Nx'} = -\frac{\gamma P_\Lambda \sin\theta}{1+\alpha P_\Lambda \cos\theta}$$

$$P_{Ny'} = \frac{\beta P_\Lambda \sin\theta}{1+\alpha P_\Lambda \cos\theta}$$

$$P_{Nz'} = \frac{\alpha + P_\Lambda \cos\theta}{1+\alpha P_\Lambda \cos\theta}$$

The three parameters α, β and γ satisfy

$$\alpha^2+\beta^2+\gamma^2 = 1$$

A third independent parameter (with Γ and α) is the phase angle ϕ, defined by

$$\beta = (1-\alpha^2)^{1/2} \sin\phi$$

$$\gamma = (1-\alpha^2)^{1/2} \cos\phi$$

We refer the reader to Cronin and Overseth (1963) for a description of the measurement of β and γ. The current average value of ϕ given by the Particle Data Group (1974) for $\Lambda^0 \to p\pi^-$ is

$$\phi = -(6.5 \pm 3.5)^\circ$$

The corresponding value of β is small and possibly zero. It is a measure of the phase difference between a_+ and a_-. We shall see in §6.4.2 that time reversal invariance has implications for the relative phase of the two decay amplitudes.

We end this section by noting the connection between the helicity decay amplitudes $a_\pm$ and the s- and p-wave amplitudes a_s and a_p which are commonly met with in the literature. The relations between them are obtained in the same way as was done for πN scattering in §4.8. The final nucleon spin state must be rotated back to the z-direction which is the Λ^0 spin quantisation axis, by means of the rotation $R(\phi, \theta, 0)^{-1}$.

Denoting the decay matrix elements in which all spins are referred to a common z-axis by $F_{m'm}$ we find

$$\left.\begin{aligned}(4\pi)^{1/2}F_{++} &= a_s + a_p \cos\theta \\ (4\pi)^{1/2}F_{--} &= a_s - a_p \cos\theta \\ (4\pi)^{1/2}F_{+-} &= a_p \sin\theta e^{-i\phi} \\ (4\pi)^{1/2}F_{-+} &= a_p \sin\theta e^{i\phi}\end{aligned}\right\} \tag{5.92}$$

where

$$a_s = 2^{-1/2}(a_+ + a_-)$$

$$a_p = 2^{-1/2}(a_+ - a_-)$$

are the amplitudes for decay into a final state of definite parity (and hence definite ℓ).

Equation (5.92) is equivalent to the vectorial form

$$\Gamma = (4\pi)^{-1/2}(a_s + a_p \boldsymbol{\sigma}\cdot\hat{\boldsymbol{p}})$$

where $\hat{\boldsymbol{p}}$ is a unit vector in the direction of the nucleon momentum in the Λ^0 rest-frame. This form can be argued on the grounds of rotational invariance.

In terms of a_s and a_p the decay parameters are given by

$$\alpha = \frac{2\,\mathrm{Re}\,(a_s^* a_p)}{|a_s|^2 + |a_p|^2} \tag{5.93}$$

$$\beta = \frac{2\,\mathrm{Im}\,(a_s^* a_p)}{|a_s|^2 + |a_p|^2} \tag{5.94}$$

$$\gamma = \frac{|a_s|^2 - |a_p|^2}{|a_s|^2 + |a_p|^2} \tag{5.95}$$

The $\Sigma^\pm$ decays can be analysed with the same formalism.

5.7 Tests of parity conservation

Some tests of conservation of parity in atomic and nuclear physics have already been described in §5.2.2 and §5.2.3. In this section we shall discuss further tests which have been made.

Finally we summarise the general situation.

5.7.1 *Strong interactions*

The rationale of the experimental tests was explained in §5.1.5. We search for a pseudo-scalar quantity with a non-vanishing expectation value. For example,

$\langle \boldsymbol{\sigma}_a \cdot \boldsymbol{p}_a \rangle$: non-vanishing longitudinal polarisation

$\langle \boldsymbol{p}_a \wedge \boldsymbol{p}_b \cdot \boldsymbol{p}_c \rangle$

$\langle \boldsymbol{\sigma}_a \cdot \boldsymbol{p}_b \rangle$: angular distribution of b relative to spin of a.

There are not enough independent momenta in a two particle reaction to test a quantity such as $\langle \boldsymbol{p}_a \wedge \boldsymbol{p}_b \cdot \boldsymbol{p}_c \rangle$ so we must go to a three or more particle final state, or a double scattering process.

As an instance of the first of these possibilities Pais (1959) showed that in $\bar{\mathrm{p}}\mathrm{p}$ annihilation in flight

$$\bar{\mathrm{p}} + \mathrm{p} \to m_a + m_b + x$$

parity conservation predicts

$$W(\phi_b) = W(-\phi_b)$$

where ϕ_b is the azimuthal angle of meson b relative to the plane of meson a and p. Dobrzynski *et al.* (1966) studied several thousand annihilations at 1.2 GeV/c. Their results were consistent with a parity violating amplitude of relative amount

$$|F| = (0.1 \pm 1.0) \times 10^{-2}$$

Since both pions and kaons are produced we are subjecting both the non-strange and strange particle parts of the strong interaction Hamiltonian to the test.

A direct test of parity conservation of the strong interaction is possible in the double scattering of protons off unpolarised nuclei. In this experiment a beam of protons of momentum $\boldsymbol{p}_\mathrm{i}$ is incident on the first target. Scattered protons of momentum $\boldsymbol{p}'$ impinge on a second target to give second scattered protons of momentum $\boldsymbol{p}'_\mathrm{f}$. Parity conservation requires that the angular distribution for the process must be independent of any pseudo-scalar quantity such as $\boldsymbol{p}_\mathrm{i} \wedge \boldsymbol{p}' \cdot \boldsymbol{p}_\mathrm{f}$. This would show as an up–down asymmetry in the second scattering relative to the plane of the first scattering. Chamberlain *et al.* (1954) found no evidence for such effects and deduced that

$$|F|^2 < 10^{-4}$$

A more accurate test of the magnitude of any parity violating contribution to the nucleon–nucleon interaction was made by Jones *et al.* (1958) who searched for longitudinal polarisation of neutrons produced by proton interactions in beryllium. As usual any longitudinal polarisation must be converted to a transverse polarisation (in this case by causing spin precession in a magnetic field) which can be detected by left–right asymmetry in a second scattering. They found

$$|F|^2 \lesssim 4 \times 10^{-6}$$

As a test of parity conservation in strong interactions of strange particles we may search for longitudinal hyperon polarisation in

$$K^+ + p \rightarrow \pi^- + \Sigma^+$$

It is amusing to note that the parity violating decay of the Σ enables its polarisation state to be determined. If we use §5.6.4 but we choose the direction of the Σ^+ momentum as quantisation axis rather than the normal to the reaction plane, then the angular distribution in the Σ^+ CM is

$$W(\theta') = (4\pi)^{-1}(1 + \alpha P_\Sigma \cos\theta')$$

where θ' is measured with respect to the quantisation axis. P_Σ is now the longitudinal polarisation and $\alpha \neq 0$ because parity is violated in the decay. Leitner *et al.* (1959) found no significant asymmetry, consistent with parity conservation in the production process. Lander, Powell and White (1959) made a similar study of $K^-p \rightarrow \pi^0\Lambda^0$.

Although the limits on the size of the parity violating strong interaction obtained in these experiments are much less stringent than in the experiments of Abov and Lobashov, they remain significant tests because the energies involved are higher. It could be that a parity violating interaction only shows up at high energies. We should expect this if, for instance, it is a much shorter range interaction than the parity conserving strong interaction. For this reason it is important to make tests of a conservation law over a wide range of energies.

A different type of test of space inversion invariance involves static electric dipole moments. We saw above in §5.2.1 that if parity is conserved the static electric dipole moment of any eigenstate of the Hamiltonian must vanish. However, it was pointed out by Landau (1957) that if time reversal invariance holds, then the static electric dipole moment must still vanish even if parity is not conserved.

We defer further discussion to the section on tests of time reversal invariance (§6.6).

5.7.2 *Summary*

In summarising the status of parity conservation, we may consider in turn the various classes of interaction.

Strong interaction. The experiments on angular asymmetry (Abov *et al.*, 1965, 1968) and circular polarisation (Lobashov *et al.*, 1966) in nuclear γ transitions show that any parity violating part of the strong interaction has at most an amplitude F_{st} of order

$$F_{st} \sim 10^{-7}$$

Electromagnetic interaction. Since the electronic structure of atoms is determined almost purely by the electromagnetic interaction, the non-occurrence of 'forbidden' optical transitions in atoms is direct evidence for parity conservation in the electromagnetic interaction. The limit placed on the parity violating amplitude was seen to be

$$|F_{em}| \sim 10^{-3}$$

The stringent limit on the parity violating amplitude in the strong interaction is also evidence for parity conservation in the electromagnetic interaction. Any parity violating part of the electromagnetic interaction could by virtual processes lead to wrong parity admixtures in the nuclear levels of order of the fine structure constant, $\alpha = e^2/\hbar c$. So the value of F_{st} above leads to a limit on F_{em} of order 10^{-4}.

Weak interactions. These may be divided into (i) purely hadronic: hyperon decays and $K \to 2\pi$ and 3π; (ii) hadron–lepton: classical beta decay and leptonic decays of other hadrons, e.g. $\pi \to \mu\nu$; (iii) purely leptonic: muon decay. In all these cases we have seen unambiguous evidence that space inversion invariance is violated.

See also the postscript on Recent Developments (p. 366) added in the 1980 reprint.

CHAPTER 6

TIME REVERSAL

The symmetry of time reversal which would be more appropriately called motion reversal, is of a different nature from the other symmetries considered in this book. The discussion here therefore differs on several important points of principle from, for example, that on space inversion in the preceding chapter.

We shall start by considering time reversal in Newtonian mechanics, and then in wave mechanics, before proceeding to the helicity formulation and applications to elementary particles.

6.1 Time reversal in classical mechanics

Consider the motion of a classical point particle in a field of force which depends only on the position of the particle (no velocity dependence). The motion is governed by Newton's second law, and we have

$$m \frac{\mathrm{d}^2}{\mathrm{d}t^2} \boldsymbol{r}(t) = \boldsymbol{F}(\boldsymbol{r}(t)) \tag{6.1}$$

where m is the mass of the particle.

To a particular motion $\boldsymbol{r}(t)$, we define the *time reversed motion* $\boldsymbol{r}^T(t)$ as that motion in which the position of the particle at time t is the same as the position of the particle in the original motion at time $-t$, that is

$$\boldsymbol{r}^T(t) = \boldsymbol{r}(-t) \tag{6.2}$$

This is illustrated in fig. 6.1.

Let us now show that the time reversed motion is a possible particle motion in the given force field $\boldsymbol{F}$, that is, that $\boldsymbol{r}^T(t)$ satisfies (6.1). This is immediately seen to be true on replacing t by $-t$ in (6.1) and using (6.2). The important points are that only the second derivative of $\boldsymbol{r}$ with respect to t occurs in (6.1) and that we have assumed a force law which depends only on position and not, for example, on the velocity of the particle.

Thus under these conditions the time reversed motion of any

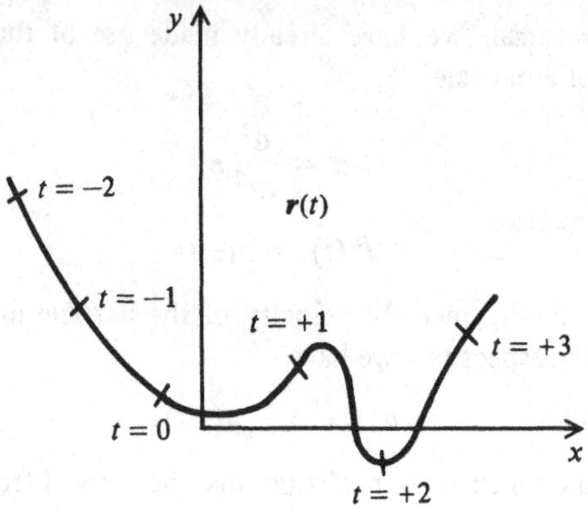

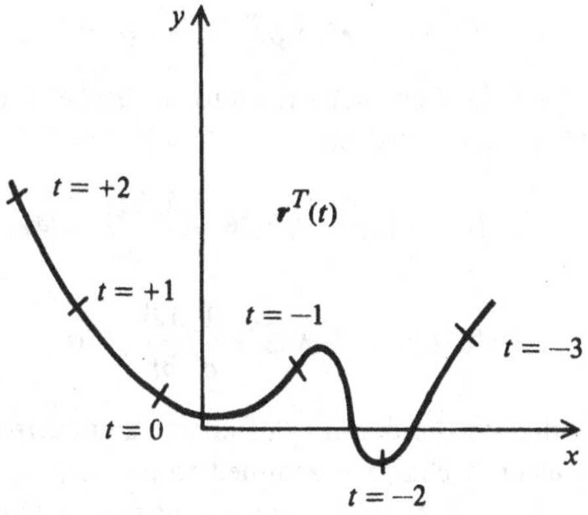

Fig. 6.1. A motion and its time reverse.

possible motion is itself a possible motion and we say that the equation of motion (6.1) is time reversal invariant. In this approach we see that no metaphysical notion of reversal of the direction of flow of time is involved. We are led to consider time reversed processes (see §6.2.4), but not reversal of time itself. Although motion reversal and motion reversal invariance would be better names, we shall adhere to the accepted, if imprecise, usage.

Let us next consider how various physical quantities transform under time reversal. We have already made use of the fact that the acceleration of a particle

$$a = \frac{d^2}{dt^2}r$$

satisfies the equation

$$a^T(t) = a(-t) \tag{6.3}$$

On the other hand, since the velocity of the particle involves the first derivative with respect to t we have

$$v^T(t) = -v(-t) \tag{6.4}$$

All dynamical quantities in mechanics may be derived from the position r and velocity v, so their time reversal properties follow from (6.1) and (6.4).

Thus momentum

$$p = mv$$

obeys

$$p^T(t) = -p(-t) \tag{6.5}$$

and angular momentum obeys

$$L^T(t) = r^T \wedge p^T = -L(-t) \tag{6.6}$$

We may extend these considerations to classical electromagnetism as described by Maxwell's equations,

$$\left.\begin{aligned} \nabla\cdot\mathfrak{E} = 4\pi\rho, \quad \nabla\wedge\mathfrak{B} - \frac{1}{c}\frac{\partial\mathfrak{E}}{\partial t} = 4\pi j \\ \nabla\cdot\mathfrak{B} = 0, \quad \nabla\wedge\mathfrak{E} + \frac{1}{c}\frac{\partial\mathfrak{B}}{\partial t} = 0 \end{aligned}\right\} \tag{6.7}$$

where ρ and j denote the densities of charge and current respectively. If the sign of electric charge is assumed to be unchanged, then under the substitution $t \to -t$, ρ will be unchanged in sign while j will change sign because it involves the velocities of the elementary charges,

$$\rho^T(r, t) = \rho(r, -t), \quad j^T(r, t) = -j(r, -t) \tag{6.8}$$

On making the substitution $t \to -t$ in Maxwell's equations, we see that the time reversed electric and magnetic fields defined by

$$\mathfrak{E}^T(r, t) = \mathfrak{E}(r, -t), \quad \mathfrak{B}^T(r, t) = -\mathfrak{B}(r, -t) \tag{6.9}$$

satisfy Maxwell's equations with ρ^T and j^T as the sources, and are therefore physically realisable fields.

This behaviour of $\mathfrak{E}$ and $\mathfrak{B}$ under time reversal could also have been deduced from consideration of the motion of a charged particle in prescribed electric and magnetic fields. In this case (6.1) is replaced by

$$m \frac{\mathrm{d}^2}{\mathrm{d}t^2} r = \mathrm{e} \left\{ \mathfrak{E} + \frac{1}{c} \frac{\mathrm{d}r}{\mathrm{d}t} \wedge \mathfrak{B} \right\} \tag{6.10}$$

where e is the charge on the particle. Because the velocity is present in the Lorentz force on the right, this equation is only invariant under the substitution $t \rightarrow -t$, if at the same time we replace $\mathfrak{B}$ by $-\mathfrak{B}$. Thus $r^T(t)$ is a possible motion in the fields $\mathfrak{E}$, $-\mathfrak{B}$ if $r(t)$ is a possible motion in the fields $\mathfrak{E}, \mathfrak{B}$.

6.2 Time reversal in non-relativistic quantum mechanics

6.2.1 *Spinless particles*

Let us try to find the time reverse of a wavefunction for a simple system described by a Hamiltonian

$$H = -\frac{\hbar^2}{2m} \nabla^2 + V(r)$$

which is the quantum-mechanical analogue of that discussed above. The time development of a state is described by Schrödinger's equation

$$\mathrm{i}\hbar \frac{\partial}{\partial t} \psi(r, t) = H\psi(r, t) \tag{6.11}$$

On replacing t by $-t$ we obtain

$$-\mathrm{i}\hbar \frac{\partial}{\partial t} \psi(r, -t) = H\psi(r, -t) \tag{6.12}$$

which differs from the original by a sign. So $\psi(r, -t)$ is not a solution of Schrödinger's equation (6.11). However, if we take the complex conjugate of (6.12) we have

$$\mathrm{i}\hbar \frac{\partial}{\partial t} \psi^*(r, -t) = H^*\psi^*(r, -t) \tag{6.13}$$

Now

$$H^* = H$$

and (6.13) becomes

$$\mathrm{i}\hbar \frac{\partial}{\partial t} \psi^*(r, -t) = H\psi^*(r, -t) \tag{6.14}$$

This suggests that we define the time reversed wavefunction by

$$\psi^T(r,t) = \psi^*(r,-t) \tag{6.15}$$

With this definition we have shown that the Schrödinger equation is invariant under time reversal.

For example if ψ describes a particle propagating along the $+z$-axis with momentum k and energy $\omega = k^2/2m$, then

$$\psi(r,t) = \mathrm{e}^{ikz - i\omega t}$$

and the time reversed state is

$$\psi^T(r,t) = \mathrm{e}^{-ikz - i\omega t}$$

that is, a particle propagating along $-z$ with the same energy.

We may check that this definition of time reversal agrees with the classical one in the correspondence principle limit. Denoting by a subscript the time at which the wavefunctions are evaluated, the expectation value of r at time t in the state ψ^T is equal to that at $-t$ in the state ψ:

$$(\psi^T, r\psi^T)_t = \int \psi^{T*}(r,t)\, r\, \psi^T(r,t)\,\mathrm{d}^3r$$

$$= \int \psi(r,-t)\, r\, \psi^*(r,-t)\,\mathrm{d}^3r$$

and thus

$$(\psi^T, r\psi^T)_t = (\psi, r\psi)_{-t} \tag{6.16}$$

This is to be compared with (6.2).

On the other hand, for the momentum expectation value one finds after an integration by parts

$$(\psi^T, p\,\psi^T)_t = -(\psi, p\,\psi)_{-t} \tag{6.17}$$

So the expectation value of the momentum changes sign (compare (6.5)).

The time reversal properties of other observables now follow. Thus for the orbital angular momentum

$$L = r \wedge p$$

we have

$$(\psi^T, L\,\psi^T)_t = -(\psi, L\,\psi)_{-t} \tag{6.18}$$

Finally we note that the transformation preserves the normalisation

$$(\psi^T, \psi^T)_t = (\psi, \psi)_{-t}$$

as is essential. The time subscript could have been omitted since normalisations are preserved in time.

6.2.2 *Spin half particles*

In the non-relativistic (Pauli) theory a spin half particle is described by a two component wavefunction ψ_σ, $\sigma = \pm\frac{1}{2}$, which can be written as a column vector

$$\begin{pmatrix} \psi_{+1/2}(r, t) \\ \psi_{-1/2}(r, t) \end{pmatrix}$$

According to (6.18) the orbital angular momentum changes sign under time reversal and we require the same should be true for spin angular momentum: thus a spin up state becomes a spin down state. This means that (6.15) must be generalised to include a transformation on the spin coordinate.

We put

$$\psi_\sigma^T(r, t) = \sum_{\sigma'} M_{\sigma\sigma'}\, \psi_{\sigma'}^*(r, -t) \tag{6.19}$$

and seek to determine the 2×2 matrix $M_{\sigma\sigma'}$, by requiring that

$$(\psi^T, \boldsymbol{S}\,\psi^T)_t = -(\psi, \boldsymbol{S}\,\psi)_{-t} \tag{6.20}$$

in analogy to (6.18). $M_{\sigma\sigma'}$ must be unitary to conserve probability. $\boldsymbol{S}$ is the spin operator with components given in terms of the Pauli matrices by

$$\boldsymbol{S} = \tfrac{1}{2}\boldsymbol{\sigma}$$

The bracket notation now implies sums over the spin coordinates σ. We have

$$\begin{aligned}(\psi^T, \boldsymbol{S}\,\psi^T)_t &= \sum_{\sigma\sigma'} \int \mathrm{d}^3r\, \psi_{\sigma'}^{T*}(r, t)\boldsymbol{S}_{\sigma'\sigma}\, \psi_\sigma^T(r, t) \\ &= \sum_{\substack{\sigma\sigma' \\ \tau\tau'}} \int \mathrm{d}^3r M_{\sigma'\tau'}^*\, \psi_{\tau'}(r, -t)\boldsymbol{S}_{\sigma'\sigma} M_{\sigma\tau}\, \psi_\tau^*(r, -t) \\ &= \sum_{\tau\tau'} \int \mathrm{d}^3r\, \psi_\tau^*(r, -t)(\widetilde{M}\widetilde{\boldsymbol{S}}M^*)_{\tau\tau'}\, \psi_{\tau'}(r, -t)\end{aligned}$$

where ~ denotes the matrix transpose.

If we choose M so that

$$\widetilde{M}\widetilde{\boldsymbol{S}}M^* = -\boldsymbol{S} \tag{6.21}$$

then (6.20) will be satisfied.

Using the Hermiticity of S

$$\tilde{S}^* = S$$

and the unitarity of M

$$M^{-1} = \tilde{M}^*$$

(6.21) becomes

$$M^{-1}SM = -S^*$$

and hence

$$M^{-1}\boldsymbol{\sigma} M = -\boldsymbol{\sigma}^* \tag{6.22}$$

Since σ_x and σ_z are real while σ_y is purely imaginary, (6.22) requires that M shall commute with σ_y and anticommute with σ_x and σ_z. Any multiple of σ_y itself has these properties, so we put

$$M = \alpha\sigma_y$$

which is unitary provided $|\alpha| = 1$. A common convention is to put $\alpha = -\mathrm{i}$, so that

$$M = \begin{pmatrix} 0 & -1 \\ +1 & 0 \end{pmatrix}$$

The time reverse of a state of a spin half particle is therefore

$$\psi_\sigma^T(r, t) = \sum_\tau (-\mathrm{i}\sigma_y)_{\sigma\tau}\, \psi_\tau^*(r, -t) \tag{6.23}$$

It may be noted that the argument up to (6.22) is valid for a particle of arbitrary spin: the specialisation to spin half came with the determination of M.

6.2.3 *Formal properties of the time reversal operator*

Before proceeding further it is convenient to define a time reversal operator, which is that operator which changes a state into its time reverse:

$$\psi^T(r, t) = O_T\, \psi(r, t) \tag{6.24}$$

For a spinless particle we have shown that

$$O_T\, \psi(r, t) = \psi^*(r, -t) \tag{6.25}$$

and for a spin half particle

$$O_T\, \psi(r, t) = -\mathrm{i}\sigma_y\, \psi^*(r, -t) \tag{6.26}$$

where spin indices have been suppressed.

All the operators we have dealt with so far have the property that complex numbers can be commuted past them, that is

$$A(\alpha\psi) = \alpha(A\psi)$$

Because of the complex conjugation involved O_T does not have this property and care must be exercised even with the order of numerical factors. O_T applied to a complex multiple of a wavefunction obeys the rule

$$O_T(\alpha\psi) = \alpha^* O_T\psi \tag{6.27}$$

This is often expressed more generally by

$$O_T\{c_1\psi_1 + c_2\psi_2\} = c_1^* O_T\psi_1 + c_2^* O_T\psi_2 \tag{6.28}$$

and an operator with this property is called *antilinear*.

We now consider the properties of O_T. For both the spin-0 and spin half cases we showed that a state and its time reverse satisfy

$$(\psi^T, \psi^T) = (\psi, \psi)$$

and hence in terms of O

$$(O_T\psi, O_T\psi) = (\psi, \psi) \tag{6.29}$$

More generally we have for the overlap integral of two states

$$(O_T\chi, O_T\phi)_t = (\phi, \chi)_{-t} = (\chi, \phi)^*_{-t} \tag{6.30}$$

An operator with this property is called *antiunitary*. Equation (6.30) can be derived from (6.29) by substituting $\psi = \phi + \chi$ and $\psi = \phi + i\chi$ in turn and making use of (6.27).

The time reverse A^T of an operator A may be defined by

$$A^T = O_T A O_T^{-1} \tag{6.31}$$

O_T^{-1} is the antiunitary operator with the properties

$$O_T^{-1} O_T = O_T O_T^{-1} = 1$$

A^T has the following property: its expectation value with respect to a time reversed state $O_T\psi$ at time t is equal to the complex conjugate of the expectation value of A in the original state ψ at time $-t$. Thus

$$(\psi^T, A^T\psi^T)_t = (\psi, A\psi)^*_{-t}$$

If A is a Hermitian observable, the right-hand side is real, so we have

$$(\psi^T, A^T\psi^T)_t = (\psi, A\psi)_{-t} \tag{6.32}$$

Comparing this formula with equations (6.16) to (6.20) we find for the time reverses of the common observables

$$r^T = O_T r O_T^{-1} = +r$$

$$p^T = -p$$

$$L^T = -L$$

$$S^T = -S$$

the helicity

$$\mathcal{H} = J\cdot P/|P|$$

obeys

$$\mathcal{H}^T = O_T \mathcal{H} O_T^{-1} = +\mathcal{H} \tag{6.33}$$

Next we consider the fact that the operation of time reversal $t \to -t$ when performed twice restores the original state. Thus in the classical case taking the time reverse of the trajectory $r^T(t)$ brings us back to the original $r(t)$. This follows from (6.2).

Similarly for a spinless particle in quantum mechanics we see from (6.15) that the time reverse of $\psi^T(r, t)$ is $\psi(r, t)$. Formally

$$O_T^2 \psi(r, t) = O_T\{\psi^*(r, -t)\} = \psi(r, t)$$

Hence

$$O_T^2 = 1 \quad \text{(spinless particles)} \tag{6.34}$$

However, in the case of a spin half particle we have

$$\begin{aligned} O_T^2 \psi(r, t) &= O_T\{-i\sigma_y \psi^*(r, -t)\} \\ &= (-i\sigma_y)^*(-i\sigma_y)\psi(r, t) \\ &= -\psi(r, t) \end{aligned}$$

because $-i\sigma_y$ is real and

$$(-i\sigma_y)^2 = -1$$

Thus we have

$$O_T^2 = -1 \quad \text{(spin half particles)} \tag{6.35}$$

The distinction between (6.34) and (6.35) is an absolute one. This is due to the antilinear property of O_T. One may try to redefine O_T by multiplying it by a complex number ω which must have unit modulus, to preserve the antiunitary property. However, the sign of the square of the operator is not changed, for one has

$$\begin{aligned} (O_T')^2 &= (\omega O_T)^2 = \omega O_T \omega O_T = \omega\omega^* O_T^2 \\ &= O_T^2 \end{aligned}$$

because $|\omega| = 1$. Obviously the antiunitary property of O_T was essential here. In the case of the *unitary* parity operator we usually *choose* to have

$$U(P)^2 = 1$$

but we could equally well define

$$U(P)' = \omega U(P)$$

leading to

$$\{U(P)'\}^2 = \omega^2 U(P)^2 = \omega^2$$

and this would not change any physics.

In the case of time reversal O_T^2 is necessarily equal to ± 1: which value is taken depends on the system considered. We shall see later that for the states of a particle of spin s

$$O(T)^2 = (-1)^{2s}$$

On the other hand for a state of N particles of spin half

$$O_T^2 = (-1)^N$$

The considerations of this section and the two preceding ones are due to Wigner (1932).

6.2.4 *Time reversal invariance in scattering processes and reactions*

Our discussion of time reversal in scattering processes will be in two parts.

In this section we shall prove that if a system is time reversal invariant then the scattering operator satisfies

$$O_T \mathcal{S} O_T^{-1} = \mathcal{S}^\dagger$$

From this we can derive the more useful statement: if time reversal invariance holds the transition amplitude from a state a to a state b is equal to the transition amplitude from the time reversed state of b to the time reversed state of a, i.e. the initial and final states are interchanged as well as being individually time reversed.

In §6.3 we shall discuss the time reversal properties of helicity states. The results of that discussion combined with the statement just made will give the consequences of time reversal invariance for transition amplitudes in the helicity description.

The scattering operator $\mathcal{S}$ was introduced for a system undergoing a localised interaction, as that operator which relates the states well before and well after the interaction. In order to understand the effect of the time reversal operation it is necessary to consider the time dependence of these states in more detail.

For definiteness we shall consider two particle scattering in the CM frame with neglect of spin.

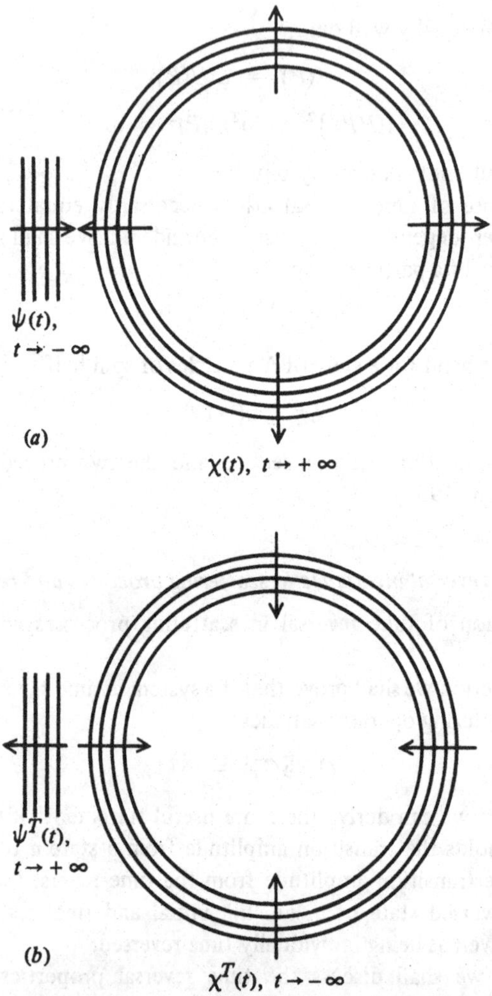

Fig. 6.2. Asymptotic forms of the states (a) $\Phi(t)$ and (b) $\Phi^T(t)$. $\Phi(t)$ corresponds to the conditions of a scattering experiment.

The development of the state of the system in time is described by a state vector $\Phi(t)$. It satisfies some equation of motion which we shall not write explicitly.

The conditions of a scattering experiment are such that $\Phi(t)$ has simple asymptotic forms as $t \to \pm\infty$. Thus as $t \to -\infty$, $\Phi(t)$ tends to a plane wave and as $t \to +\infty$, $\Phi(t)$ becomes an outgoing spherical wave,

$$\left.\begin{aligned}\Phi(t) &\to \psi(t), \quad t \to -\infty \\ &\to \chi(t), \quad t \to +\infty\end{aligned}\right\} \tag{6.36}$$

See fig. 6.2.

χ and ψ are solutions of the equations of motion without interaction. We may write them as superpositions of plane wave states ϕ_k of definite momentum and energy, which have simple time dependence. Thus

$$\psi(t) = \int d^3k\, \tilde{\psi}(k)\, \phi_k e^{-i\omega(k)t} \tag{6.37a}$$

where $\tilde{\psi}(k)$ is the momentum space wavefunction.

Similarly

$$\chi(t) = \int d^3k\, \tilde{\chi}(k)\, \phi_k e^{-i\omega(k)t} \tag{6.37b}$$

The scattering operator expresses the linear connection between the system states ψ and χ at $t \to -\infty$ and $t \to +\infty$. This may be equally well expressed in terms of the momentum space wavefunctions as follows

$$\tilde{\chi}(k) = \int d^3k'\, \mathcal{S}(k, k')\, \tilde{\psi}(k') \tag{6.38}$$

Since in practice ψ is approximately a plane wave, $\tilde{\psi}(k)$ has a δ-function peak at say $k = k_1$ so $\tilde{\chi}(k)$ which is the amplitude to find the system in the state ϕ_k as $t \to +\infty$ is just $\mathcal{S}(k, k_1)$. It will sometimes be convenient to suppress momentum integrations and write (6.38) as

$$\tilde{\chi} = \mathcal{S}\, \tilde{\psi} \tag{6.39}$$

Now let us assume that the system we are considering is invariant under time reversal. Then

$$\Phi^T(t) = O_T\, \Phi(t) = \Phi^*(-t)$$

is a possible state of the system, although difficult to prepare in practice, see fig. 6.2. From (6.36) with the substitution $t \to -t$, the asymptotic forms of $\Phi^T(t)$ are

$$\begin{aligned}\Phi^T(t) &\to \chi^*(-t) = \chi^T, \quad t \to -\infty \\ &\to \psi^*(-t) = \psi^T, \quad t \to +\infty\end{aligned}$$

The time reversal operation applied to the momentum expansion (6.37a) gives

$$\begin{aligned}\psi^*(-t) &= \int d^3k\, \tilde{\psi}^*(k)\, \phi_{-k} e^{-i\omega(k)t} \\ &= \int d^3k\, \tilde{\psi}^*(-k)\, \phi_k e^{-i\omega(k)t}\end{aligned}$$

because for a momentum eigenstate time reversal changes the sign of k. The second line then follows by a change of variable. We have derived the time reversal transformation law for a momentum space wave-function

$$\tilde{\psi}(k) \rightarrow O_T \tilde{\psi}(k) = \tilde{\psi}^*(-k)$$

So we write

$$O_T \psi(t) = \int d^3k \, \{O_T \tilde{\psi}(k)\} \phi_k e^{-i\omega(k)t}$$

and

$$O_T \chi(t) = \int d^3k \, e^{-i\omega(k)t} \phi_k O_T \tilde{\chi}(k)$$

Since $O_T \Phi(t)$ is an admissible state of the system with asymptotic forms $O_T\chi$ and $O_T\psi$ as $t \rightarrow -\infty$ and $+\infty$ respectively, the discussion which led to (6.39) must now give

$$O_T \tilde{\psi} = \mathcal{S} \, O_T \tilde{\chi} \tag{6.40}$$

From (6.39) and (6.40) the desired equation can be obtained. We multiply (6.40) by $\mathcal{S}^\dagger$ and use the unitary property

$$\mathcal{S}^\dagger \mathcal{S} = 1$$

yielding

$$\mathcal{S}^\dagger O_T \tilde{\psi} = O_T \tilde{\chi} \tag{6.41}$$

On the other hand, the application of O_T to (6.39) gives

$$\begin{aligned} O_T \tilde{\chi} &= O_T \, \mathcal{S} \tilde{\psi} \\ &= O_T \, \mathcal{S} O_T^{-1} O_T \tilde{\psi} \end{aligned} \tag{6.42}$$

On comparing (6.41) and (6.42) we deduce the condition of time reversal invariance for the $\mathcal{S}$-operator

$$O_T \mathcal{S} O_T^{-1} = \mathcal{S}^\dagger \tag{6.43}$$

For the transition operator defined by

$$\mathcal{S} = 1 + i\,\mathcal{T}$$

it follows that

$$O_T \mathcal{T} O_T^{-1} = \mathcal{T}^\dagger \tag{6.44}$$

We shall assume that (6.43) and (6.44) are the general expressions of time reversal invariance when spins are present and also for decays and reactions where the number and kinds of particles may change.

To derive the condition on $\mathcal{T}$-matrix elements we rearrange (6.44) and take the ϕ_n–ϕ_m matrix element. Then

$$(\phi_n, O_T^{-1} \mathcal{T}^\dagger O_T \phi_m) = (\phi_n, \mathcal{T} \phi_m)$$

With the aid of (6.30) we have

$$(\phi_n, O_T^{-1}\mathcal{T}^\dagger O_T \phi_m) = (O_T\phi_n, O_T O_T^{-1}\mathcal{T}^\dagger O_T \phi_m)^*$$
$$= (\phi_n^T, \mathcal{T}^\dagger \phi_m^T)^*$$
$$= (\phi_m^T, \mathcal{T}\phi_n^T)$$

Hence time reversal invariance leads to

$$(\phi_m^T, \mathcal{T}\phi_n^T) = (\phi_n, \mathcal{T}\phi_m) \tag{6.45}$$

that is, the transition amplitude from the state m to the state n is equal to the transition amplitude from the time reverse of state n to the time reverse of state m. This is sometimes called the *reciprocity theorem*. We shall discuss the consequences of this equality in §6.4, after deriving the helicity version (6.70).

Except in the case of elastic scattering where the initial and final states contain the same particles, (6.45) is a relation between the amplitudes for two different processes $m \to n$ and $n \to m$. However, there is an approximation in which (6.44) leads to a constraint on the amplitude for a single process. This is our next topic.

6.2.5 *Time reversal in first-order processes*

By 'first-order process' we mean one in which the squares of the relevant $\mathcal{T}$-matrix elements are negligible compared with the matrix elements themselves,

$$|\mathcal{T}|^2 \ll \mathcal{T} \tag{6.46}$$

The significance of this is as follows. The unitarity condition on $\mathcal{S}$

$$\mathcal{S}^\dagger\mathcal{S} = 1$$

implies that $\mathcal{T}$ defined by

$$\mathcal{S} = 1 + \mathrm{i}\mathcal{T}$$

satisfies

$$\mathcal{T} - \mathcal{T}^\dagger = \mathrm{i}\mathcal{T}\mathcal{T}^\dagger$$

Thus if the interaction is weak as we have defined it, the right-hand side can be neglected and $\mathcal{T}$ is Hermitian. Hence

$$(\phi_m^T, \mathcal{T}\phi_n^T) = (\phi_n^T, \mathcal{T}\phi_m^T)^*$$

so that (6.45) becomes

$$(\phi_n^T, \mathcal{T}\phi_m^T)^* = (\phi_n, \mathcal{T}\phi_m) \tag{6.47}$$

In quantum electrodynamics or weak interaction theory the $\mathcal{T}$-operator or its matrix element for any particular process is accurately given by the first few terms of a power series in a dimensionless coupling strength

$$\mathcal{T} = g\,\mathcal{T}_1 + g^2\mathcal{T}_2 + \ldots$$

In quantum electrodynamics the fine structure constant plays the role of g.

The first-order approximation formulated here corresponds to taking only the first term of this expansion. It is usually called the Born approximation. For weak interaction, the interaction Hamiltonian itself forms the first term

$$\mathcal{T}_1 = H_{\mathbf{wk}}$$

Since the effect of O_T on ϕ_m and ϕ_n is to transform momenta and spins according to

$$\boldsymbol{p} \to -\boldsymbol{p}, \quad \boldsymbol{S} \to -\boldsymbol{S}$$

it follows from (6.47) that

$$W(-\boldsymbol{p}, -\boldsymbol{S}) = W(\boldsymbol{p}, \boldsymbol{S}) \tag{6.48}$$

where in the notation of §5.1.5

$$W(\boldsymbol{p}, \boldsymbol{S}) = |(\phi_n, \mathcal{T}\phi_m)|^2$$

This equality only holds to the extent that the approximation (6.46) is valid. However (6.46) is very well satisfied for weak interactions, and so (6.48) may be used to test time reversal invariance in such processes. For example in the $K_{\mu 3}$ decay

$$K^{\pm} \to \pi^0 + \mu^{\pm} + \nu(\bar{\nu}) \tag{6.49}$$

(6.48) forbids the existence of a non-zero expectation of the transverse μ-spin component

$$\langle \boldsymbol{S}_\mu \cdot \boldsymbol{p}_\pi \wedge \boldsymbol{p}_\mu \rangle$$

This decay is a purely weak process, so the use of (6.48) is permitted. This has been used as a test of T-invariance in weak interactions. See §6.7.

In a decay such as

$$\Sigma^- \to n + \pi^-$$

which proceeds via the weak interaction, the decay products can interact strongly, and the approximation (6.46) which refers to the whole interaction $\mathcal{T}$-matrix, is not valid. However in these cases of weak decays with strong final state interaction, T-invariance still leads to useful restrictions via the final state theorem (see §6.4 and Feinberg (1960)).

6.3 Time reversal and the helicity description

In this section the time reversal transformation properties of helicity states will be derived and used to show the consequences of time reversal invariance for two particle scattering and reactions.

The method is broadly similar to that used for space inversion in the last chapter but it differs in detail because of the antiunitary nature of time reversal.

The results of the calculation are contained in (6.63) and (6.66), and for two particle states (6.67) and (6.68). The reader whose principal interest is in the applications may refer to these and then proceed to the following subsection where these results are applied.

6.3.1 *Time reversal transformation of one and two particle states*

The time reversal operation T in a four-dimensional notation is

$$\begin{pmatrix} x_0 \\ x_1 \\ x_2 \\ x_3 \end{pmatrix} \to \begin{pmatrix} x'_0 \\ x'_1 \\ x'_2 \\ x'_3 \end{pmatrix} = \begin{pmatrix} -1 & 0 & 0 & 0 \\ 0 & +1 & 0 & 0 \\ 0 & 0 & +1 & 0 \\ 0 & 0 & 0 & +1 \end{pmatrix} \begin{pmatrix} x_0 \\ x_1 \\ x_2 \\ x_3 \end{pmatrix}$$

The matrix on the right will be denoted by T. We have

$$T^2 = 1 \tag{6.50}$$

Since T leaves the space coordinates fixed it commutes with space displacements and space rotations. Denoting displacements by D_{a_μ} rather than T_{a_μ} in this section,

$$D_{a_\mu}: \begin{array}{l} x_0 \to x'_0 = x_0 + a_0 \\ \mathbf{x} \to \mathbf{x}' = \mathbf{x} + \mathbf{a} \end{array}$$

we have

$$TD_{(a_0, \mathbf{a})} = D_{(-a_0, \mathbf{a})}T \tag{6.51}$$

For an arbitrary rotation R

$$TR = RT \tag{6.52}$$

On the other hand T does not commute with Lorentz transformations. For a boost along the z-axis,

$$T\mathfrak{Z}_v = \mathfrak{Z}_{-v}T \tag{6.53}$$

and in general the time reversal transform of a boost with velocity $\mathbf{v}$ is a boost with velocity $-\boldsymbol{v}$,

$$TL(\boldsymbol{v})T^{-1} = L(-\boldsymbol{v})$$

These are the space–time properties of T. In a quantum-mechanical system invariant under time reversal there is a corresponding time reversal operator acting on the states of the system. If we suppose that this operator is a unitary one $U(T)$, and follow the development of §5.4.1 for the unitary parity operator $U(P)$ we should have as a consequence

$$P_0 U(T) = -U(T)P_0 \quad \text{(wrong)}$$

$$\boldsymbol{P}\,U(T) = +U(T)\boldsymbol{P} \quad \text{(wrong)}$$

corresponding to (5.40) and (5.41) where P_0 and $\boldsymbol{P}$ are the energy and momentum operators.

It follows from these equations that if $\phi_{E,\boldsymbol{p}}$ is an energy–momentum eigenstate of a spinless particle, $U(T)\phi_{E,\boldsymbol{p}}$ has energy $-E$ and momentum $\boldsymbol{p}$. $U(T)$ is thus not an acceptable time reversal operator: it turns positive energy states into states of negative energy. The resolution of this difficulty follows from the elementary discussion earlier: the quantum-mechanical operator corresponding to T must be antiunitary. We denote it by $O(T)$, rather than O_T, from this point on.

The operator equations corresponding to (6.51), (6.52) and (6.53) therefore take the form

$$U(-a_0, \boldsymbol{a})O(T) = O(T)U(a_0, \boldsymbol{a}) \tag{6.54}$$

$$U(R)O(T) = O(T)U(R) \tag{6.55}$$

$$U(\mathfrak{Z}_{-\boldsymbol{v}})O(T) = O(T)U(\mathfrak{Z}_{\boldsymbol{v}}) \tag{6.56}$$

Consider now an infinitesimal space–time displacement. Combining (4.28) and (4.29) we have

$$U(a_0, \boldsymbol{a}) = 1 + \mathrm{i}(a_0 P_0 - \boldsymbol{a}\cdot\boldsymbol{P})$$

we substitute this into (6.54) and equate the coefficients of a_0 and $\boldsymbol{a}$. On making use of the antilinear property (6.28) we obtain

$$P_0 O(T) = O(T)P_0 \tag{6.57}$$

$$\boldsymbol{P}O(T) = -O(T)\boldsymbol{P} \tag{6.58}$$

Following the argument used before, we find that $O(T)\phi_{E,\boldsymbol{p}}$ is a state of energy E and momentum $-\boldsymbol{p}$. Thus $O(T)$ is an acceptable time reversal operator.

We now consider a massive particle of spin s. Starting with the set of $(2s+1)$ rest-states $\phi_{0\lambda}$ it follows from (6.58) that the states $O(T)\phi_{0\lambda}$ are also rest-states. However they do not transform in the standard way under rotations. Rather we have

$$\begin{aligned} U(R)O(T)\phi_{0\lambda} &= O(T)U(R)\phi_{0\lambda} \\ &= O(T)\{\sum_{\lambda'}\phi_{0\lambda'}\mathfrak{D}^s_{\lambda'\lambda}(R)\} \\ &= \sum_{\lambda'}\mathfrak{D}^{s*}_{\lambda'\lambda}(R)O(T)\phi_{0\lambda'} \end{aligned}$$

with $\mathfrak{D}^*$ on the right instead of $\mathfrak{D}$.

Now from (3.70) and (3.84c) we find

$$\mathfrak{D}^{s*}_{\lambda'\lambda}(R) = (-1)^{\lambda'-\lambda}\mathfrak{D}^s_{-\lambda',-\lambda}(R)$$

which may be used to show that

$$U(R)(-1)^{-s-\lambda}O(T)\phi_{0,-\lambda} = \sum_{\lambda'}\mathfrak{D}^s_{\lambda'\lambda}(R)(-1)^{-s-\lambda'}O(T)\phi_{0,-\lambda'}$$

and hence that the set of $(2s+1)$ rest-states

$$(-1)^{-s-\lambda}O(T)\phi_{0,-\lambda}$$

also transforms in the standard way under rotations. It may be concluded that $O(T)\phi_{0,-\lambda}$ and $(-1)^{s+\lambda}\phi_{0\lambda}$ are proportional. So putting $-\lambda$ for λ, we have

$$O(T)\phi_{0\lambda} = \eta_T(-1)^{s-\lambda}\phi_{0,-\lambda} \tag{6.59}$$

Here η_T is a phase factor independent of λ.

Unlike the case of space inversion, we cannot interpret η_T as an eigenvalue since we do not have the same state vector on both sides of this equation. Even the fact that the square of T is the identity transformation (6.50) does not restrict η_T, because, by (6.27), we have

$$\begin{aligned} O(T)^2\phi_{0\lambda} &= O(T)\{\eta_T(-1)^{s-\lambda}\phi_{0,-\lambda}\} \\ &= \eta_T^*(-1)^{s-\lambda}\eta_T(-1)^{s+\lambda}\phi_{0\lambda} \end{aligned}$$

and thus

$$O(T)^2\phi_{0\lambda} = (-1)^{2s}\phi_{0\lambda} \tag{6.60}$$

because $|\eta_T| = 1$. This generalises the results for the cases $s=0$ and $s=\frac{1}{2}$ that we found earlier. η_T can in fact be changed at will by multiplying all states by a common phase factor. We shall therefore choose

$$\eta_T = 1 \tag{6.61}$$

As in the discussion of the parity operator, (6.59) may be put into a more useful form by noticing that the state $\phi_{0,-\lambda}$ can be obtained from $\phi_{0\lambda}$ by a suitable rotation, which we take to be a rotation through π about the y-axis,

$$U(Y_\pi)\phi_{0\lambda} = (-1)^{s-\lambda}\phi_{0,-\lambda}$$

Thus (6.59) with $\eta_T = 1$, becomes

$$O(T)\phi_{0\lambda} = U(Y_\pi)\phi_{0\lambda} \tag{6.62}$$

Next we consider $O(T)$ acting on a state with momentum along the z-direction. Using (6.56), we find

$$\begin{aligned} O(T)\phi_{p\,000\lambda} &= O(T)U(\mathcal{Z}_p)\phi_{000\lambda} \\ &= U(\mathcal{Z}_{-p})O(T)\phi_{000\lambda} \\ &= U(\mathcal{Z}_{-p})U(Y_\pi)\,\phi_{000\lambda} \\ &= U(Y_\pi)U(\mathcal{Z}_p)\phi_{000\lambda} \end{aligned}$$

and thus

$$O(T)\phi_{p00\lambda} = U(Y_\pi)\phi_{p00\lambda} \tag{6.63}$$

This is not unexpected: since a rotation does not change the helicity, (6.63) states that the T-transform of the momentum–helicity state is a state of opposite momentum but the same helicity (and hence opposite spin projection).

A similar calculation may be made for the state $\hat{\phi}_{p\pi0\lambda}$ defined by (4.66). The T-transform of a state of momentum along the negative z-direction with helicity λ will have momentum along $+z$ and the opposite spin (hence the same helicity). Again this state can be related to the original by a rotation Y_π. The detailed calculation only serves to check the phase factors. We have

$$\begin{aligned} O(T)\hat{\phi}_{p\pi0\lambda} &= (\mathrm{e}^{-i\pi s})^*\,U(Z_\pi)U(Y_\pi)O(T)\phi_{p00\lambda} \\ &= \mathrm{e}^{i\pi s}\,U(Z_\pi)U(Y_\pi)U(Y_\pi)\phi_{p00\lambda} \\ &= \mathrm{e}^{i\pi(s-\lambda)}(-1)^{2s}\,\phi_{p00\lambda} \end{aligned} \tag{6.64}$$

because Y_π^2 is a rotation by 2π which is represented by ± 1 for integer or half integer spin. Equation (4.66) may be inverted to read

$$\phi_{p00\lambda} = (-1)^{s-\lambda}(-1)^{2s}U(Y_\pi)\hat{\phi}_{p\pi0\lambda} \tag{6.65}$$

Combining the last two equations we have the simple result

$$O(T)\hat{\phi}_{p\pi0\lambda} = U(Y_\pi)\hat{\phi}_{p\pi0\lambda} \tag{6.66}$$

in parallel with (6.63).

To find the time reversal properties of a two particle angular momentum eigenstate we combine the results obtained so far with the integral representation (4.89).

Since the two particle CM plane wave state has the form

$$\Phi_{W00\lambda_a\lambda_b} \sim \phi_{p00\lambda_a} \hat{\phi}_{p\pi 0\lambda_b}$$

apart from factors, (6.63) and (6.66) show that

$$O(T)\,\Phi_{W00\lambda_a\lambda_b} = U(Y_\pi)\,\Phi_{W00\lambda_a\lambda_b} \tag{6.67}$$

Applying $O(T)$ to both sides of (4.89) and using (6.67) one finds

$$O(T)\,\Psi_{WJM\lambda_a\lambda_b} = (-1)^{J-M}\,\Psi_{WJ,-M\lambda_a\lambda_b} \tag{6.68}$$

($J-M=$ integer). For details the reader is referred to Jacob and Wick (1959).

6.3.2 *Consequences of time reversal invariance for reactions*

We proceed by deriving the consequences of time reversal invariance for the $\mathcal{S}$ -matrix elements in the angular momentum basis, cf. (4.93), and deduce the consequences for the plane wave amplitude $f_{\lambda_c\lambda_d,\lambda_a\lambda_b}(\theta,\phi)$ with the aid of the partial wave expansion (4.95).

We have shown above that if a system is time reversal invariant then the operators $\mathcal{S}$ and $\mathcal{T}$ satisfy

$$O(T)\,\mathcal{S}O(T)^{-1} = \mathcal{S}^\dagger \tag{6.43a}$$

$$O(T)\,\mathcal{T}O(T)^{-1} = \mathcal{T}^\dagger \tag{6.44a}$$

We rearrange the last equation as

$$\mathcal{T} = O(T)^{-1}\,\mathcal{T}^\dagger\,O(T)$$

and take the matrix element between angular momentum eigenstates,

$$\begin{aligned}
&(\Psi^{cd}_{JM\lambda_c\lambda_d},\ \mathcal{T}\Psi^{ab}_{JM\lambda_a\lambda_b})\\
&= (\Psi^{cd}_{JM\lambda_c\lambda_d}, O(T)^{-1}\,\mathcal{T}^\dagger\,O(T)\Psi^{ab}_{JM\lambda_a\lambda_b})\\
&= (O(T)\,\Psi^{cd}_{JM\lambda_c\lambda_d},\ \mathcal{T}^\dagger\,O(T)\,\Psi^{ab}_{JM\lambda_a\lambda_b})^*, \quad \text{by (6.30)}\\
&= (-1)^{2J-2M}\,(\Psi^{cd}_{J,-M\lambda_c\lambda_d},\ \mathcal{T}^\dagger\,\Psi^{ab}_{J,-M\lambda_a\lambda_b})^*, \quad \text{by (6.68)}\\
&= (\Psi^{ab}_{J,-M\lambda_a\lambda_b},\ \mathcal{T}\Psi^{cd}_{J,-M\lambda_c\lambda_d})
\end{aligned}$$

Thus since the matrix element is independent of M and $(J-M)=$ integer,

$$\mathcal{T}^J_{\lambda_c\lambda_d,\,\lambda_a\lambda_b} = \mathcal{T}^J_{\lambda_a\lambda_b,\,\lambda_c\lambda_d} \tag{6.69}$$

It is important to note that two different processes are related, unless of course we are dealing with elastic scattering $a = b, c = d$.

We substitute this result in (4.95), specialising to $\phi = 0$, and using (3.84*b*) we find

$$p_{ab} f_{\lambda_c\lambda_d,\,\lambda_a\lambda_b}(\theta) = (-1)^{\lambda_a-\lambda_b-\lambda_c+\lambda_d} p_{cd} f_{\lambda_a\lambda_b,\,\lambda_c\lambda_d}(\theta) \tag{6.70}$$

This is the consequence of time reversal invariance for the two particle reaction amplitude.

For πN elastic scattering, we obtain

$$f_{\mu\lambda}(\theta) = (-1)^{\lambda-\mu} f_{\lambda\mu}(\theta) \tag{6.71}$$

Comparing with (4.99) we find that for this case, time reversal invariance leads to no additional restrictions beyond those following from space reflection invariance.

More generally this is the case for any spin half, spin-0 reaction in which the reaction parity $\eta_a\eta_b/\eta_c\eta_d$ is $+1$, and so it holds for the scattering or reaction of any of the $\frac{1}{2}^+$ baryons with the 0^- mesons.

In the case of a *first-order reaction* process for which time reversal invariance implies

$$(O(T)\phi_n, \mathcal{T}O(T)\phi_m)^* = (\phi_n, \mathcal{T}\phi_m) \tag{6.47a}$$

it is easy to see that the helicity transition amplitude is purely real. We take the initial and final states in (6.47) as CM angular momentum eigenstates; then since the effect of $O(T)$ is to reverse the sign of M, while the $\mathcal{T}$-matrix element is independent of M (by rotational invariance), we have

$$\mathcal{T}^{J*}_{\lambda_c\lambda_d,\,\lambda_a\lambda_b} = \mathcal{T}^J_{\lambda_c\lambda_d,\,\lambda_a\lambda_b} = \text{real}$$

Then by (4.95) with $\phi = 0$, it follows that

$$f_{\lambda_c\lambda_d,\,\lambda_a\lambda_b}(\theta) = \text{real}$$

Thus the relative phases of the amplitudes for the various helicity states are fixed to be 0 or π.

This last result shows that in a reaction for which the first-order approximation is good, there can be no polarization of the final particles. This can be seen in the spin half, spin-0 case from the explicit formula (5.84) for the polarisation produced off an unpolarised target. It can also be seen more directly by noting that the polarisation corresponds to a non-zero value for $\langle \boldsymbol{S}\cdot \boldsymbol{p}_{ab} \wedge \boldsymbol{p}_{cd}\rangle$ which must vanish when (6.47) and hence (6.48) holds.

It may similarly be shown that for a first-order decay process, (6.47) applied to (4.115) and the equation following it, leads to the conclusion

$$a_{\lambda_a \lambda_b} = \text{real}$$

We may say that (6.47) fixes the relative phases of the decay amplitudes into the different helicity states to be zero, or π.

6.3.3 *Time reversal invariance and massless particles*

The final form of the transformation laws for particle states obtained in §6.3.1, and hence the results of the last section, are valid for massless particles, but the details of the derivation are different.

As in the parity discussion of §5.4.3 one must start with a standard state $\phi_{p00\lambda}$ in which the momentum is directed along the $+z$-axis.

Consider the state $O(T)\phi_{p00\lambda}$.

Since the momentum changes sign on time reversal while the helicity is invariant (cf. (6.58) and (6.33)), we find that $O(T)\phi_{p00\lambda}$ is a state with momentum along the $-z$-axis with helicity λ. So we may again put

$$O(T)\phi_{p00\lambda} = \eta_T U(Y_\pi)\phi_{p00\lambda}$$

η_T is a phase factor. One can show that η_T is independent of p by applying a Lorentz boost to both sides. As before one may arrange η_T to be $+1$ by choice of phases. The last equation is then the same as (6.63) so the rest of the discussion for massless particles follows that of §6.3.1.

The important distinction between this analysis and that in the parity case is that the requirement that the time reversal operator be applicable to massless particle states does *not* require the existence of both helicity states $\pm\lambda$ of the particle. So it is possible to construct time reversal invariant interactions involving a massless particle which exists in only one helicity state. This is the case for weak interactions involving the neutrinos.

6.4 Consequences of time reversal invariance

6.4.1 *Reciprocity theorem for cross-sections*

The relation (6.70) between the helicity transition amplitudes for the reaction

$$a + b \rightarrow c + d$$

and the reverse reaction

$$c + d \rightarrow a + b$$

can be used to derive relations between the differential cross-sections for the two processes.

The differential cross-section for the first process between definite helicity states is given by

$$\frac{d\sigma_{ab}}{d\Omega} = |f^{ab}_{\lambda_c\lambda_d,\, \lambda_a\lambda_b}(W, \theta)|^2 \tag{4.96}$$

Here the σ and f are labelled by the initial state of the process.

A similar expression holds for the reverse reaction and hence by (6.70) we have

$$p^2_{ab}\frac{d\sigma_{ab}}{d\Omega} = p^2_{cd}\frac{d\sigma_{cd}}{d\Omega} \tag{6.72}$$

where the cross-sections are evaluated at the same CM energy and angle.

If the initial beam and target are unpolarised and the detectors are insensitive to the final spins, the 'unpolarised' cross-section given by

$$\overline{\frac{d\sigma_{ab}}{d\Omega}} = \frac{1}{(2s_a+1)(2s_b+1)}\sum_{\lambda_a\lambda_b\lambda_c\lambda_d}|f^{ab}_{\lambda_c\lambda_d,\, \lambda_a\lambda_b}(W, \theta)|^2 \tag{4.97a}$$

is related to that for the reverse process by

$$(2s_a+1)(2s_b+1)p^2_{ab}\overline{\frac{d\sigma_{ab}}{d\Omega}} = (2s_c+1)(2s_d+1)p^2_{cd}\overline{\frac{d\sigma_{cd}}{d\Omega}} \tag{6.73}$$

This is called the *principle of detailed balance*. p^2_{ab}/p^2_{cd} measures the ratio of the phase space available in the two reactions.

It was observed by Marshak (1951) and Cheston (1951) that the spin of the charged pion could be determined by applying the principle of detailed balance to the reactions

$$\mathrm{p} + \mathrm{p} \rightleftarrows \pi^+ + \mathrm{d}$$

since the spins of the proton and deuteron were known. The reactions $\mathrm{pp} \rightarrow \pi\mathrm{d}$ and $\pi\mathrm{d} \rightarrow \mathrm{pp}$ were studied by Cartwright *et al.* (1953) and Durbin *et al.* (1951). The second group showed that two sets of differential cross-sections were in agreement with (6.73) only for $S_\pi = 0$.

6.4.2 *Final state theorem*

In a process such as

$$\Sigma^- \longrightarrow n + \pi^-$$

in which the primary decay interaction is weak but the final particles can interact strongly, the decay amplitude is related by time reversal invariance to the elastic scattering amplitude for the final state particles. This relation is called the final state theorem, and is due to Aidzu (1953), Watson (1954) and Fermi (1955). We shall prove it in the context of the particular process just cited. Thus this discussion extends that on hyperon decays in §5.6.4 and the notation established there will be used.

A set of particles which forms the initial or final state in a reaction is sometimes called a channel. In this case there are two channels, Σ^- and $n\pi^-$. The $n\pi^-$ channel can be divided into *subchannels* of definite angular momentum J and helicity λ. Since only an $n\pi^-$ state of $J = \frac{1}{2}$ can result from hyperon decay, we have a three channel problem:

$$\Sigma, \quad (\pi N)_{J=1/2,\,\lambda=+1/2}, \quad (\pi N)_{J=1/2,\,\lambda=-1/2}$$

The corresponding matrix elements of the $\mathcal{S}$-operator will be written

$$\mathcal{S} = \begin{bmatrix} \mathcal{S}_{11} & \mathcal{S}_{12} & \mathcal{S}_{13} \\ \mathcal{S}_{21} & \mathcal{S}_{22} & \mathcal{S}_{23} \\ \mathcal{S}_{31} & \mathcal{S}_{32} & \mathcal{S}_{33} \end{bmatrix} \tag{6.74}$$

where the row (column) refers to the final (initial) channel.

The relation between $\mathcal{S}$ and $\mathcal{T}$ now takes the form

$$\mathcal{S}_{nm} = \delta_{nm} + i\mathcal{T}_{nm}$$

Thus for $n \neq m$, $\mathcal{S}_{nm} = i\mathcal{T}_{nm}$ is exactly the transition amplitude for $m \to n$.

To start with, we consider a zeroth-order approximation in which the weak decay interaction is switched off, so that $1 \to 2$ and $1 \to 3$ are forbidden and

$$\mathcal{S}_{21} = \mathcal{S}_{31} = 0$$

Similarly

$$\mathcal{S}_{12} = \mathcal{S}_{13} = 0$$

However, $2 \leftrightarrow 3$ can still occur: it corresponds to πN scattering with helicity flip. It is more convenient if the $\mathcal{S}$-matrix is diagonal in the zeroth-order approximation. This will be so in the present case, if instead of the πN helicity states we use the parity eigenstates $\Psi_{JM}^{(\pm)}$ of

(4.106). So for channels 2 and 3 we take $\Psi^{(-)}_{1/2M}$ and $\Psi^{(+)}_{1/2M}$ which correspond to $L = 0$ and 1, and parity $\eta = -1$ and $+1$ respectively.

Then since the strong interaction conserves J and η, we have in zeroth order

$$\mathcal{S}^{(0)} = \begin{bmatrix} 1 & 0 & 0 \\ 0 & e^{2i\delta_s} & 0 \\ 0 & 0 & e^{2i\delta_p} \end{bmatrix} \tag{6.75}$$

where we have denoted $\delta_{1/2-}$ and $\delta_{1/2+}$ of (4.108) by δ_s and δ_p. We put $\mathcal{S}^{(0)}_{11} = 1$ since the Σ^- is stable in this approximation; hence if initially present with unit amplitude, it remains so finally. Note that in any problem since $\mathcal{S}$ is unitary it can always be brought to diagonal form by suitable choice of the channel labels.

When the decay interaction is switched on, our first-order approximation corresponds to being able to write

$$\mathcal{S} = \mathcal{S}^{(0)} + i\Sigma \tag{6.76}$$

where the matrix elements of Σ are small enough that their squares can be neglected. The amplitudes Σ_{21} and Σ_{31} corresponding to $1 \to 2$ and $1 \to 3$ are simply the amplitudes a_s and a_p of §5.6.4.

We now apply unitarity and time reversal invariance. The first of these requires that

$$\mathcal{S}\mathcal{S}^\dagger = \mathcal{S}^\dagger\mathcal{S} = 1$$

and thus

$$\begin{aligned} 1 &= (\mathcal{S}^{(0)} + i\Sigma)(\mathcal{S}^{(0)\dagger} - i\mathcal{S}^\dagger) \\ &\simeq \mathcal{S}^{(0)}\mathcal{S}^{(0)\dagger} + i(\Sigma\mathcal{S}^{(0)\dagger} - \mathcal{S}^{(0)}\Sigma^\dagger), \end{aligned}$$

with neglect of the $\Sigma\Sigma^\dagger$ term.

$\mathcal{S}^{(0)}$ of (6.75) is already unitary

$$\mathcal{S}^{(0)}\mathcal{S}^{(0)\dagger} = 1$$

and so we have

$$\Sigma\mathcal{S}^{(0)\dagger} = \mathcal{S}^{(0)}\Sigma^\dagger$$

Since $\mathcal{S}^{(0)}$ is diagonal this gives

$$\Sigma_{nm}\, e^{-2i\delta_m} = e^{2i\delta_n}\,\Sigma^*_{mn} \tag{6.77}$$

Time reversal invariance of the $\mathcal{S}$-operator has the same form as (6.45)

$$(\phi^T_m, \mathcal{S}\phi^T_n) = (\phi_n, \mathcal{S}\phi_m)$$

Since in the angular momentum basis $O(T)$ simply changes M to $-M$, cf. (6.68), while $\mathcal{S}$-matrix elements are independent of M, time reversal invariance is simply expressed by

$$\mathcal{S}_{mn} = \mathcal{S}_{nm}$$

and hence

$$\Sigma_{mn} = \Sigma_{nm} \tag{6.78}$$

(We also pick up $(-1)^{2J-2M}$ which is always $+1$.)

Combining (6.77) and (6.78) we find after rearrangement

$$\Sigma_{nm}\, e^{-i(\delta_n+\delta_m)} = \Sigma^*_{nm}\, e^{i(\delta_n+\delta_m)}$$

which shows that this quantity is real. We may therefore write

$$\Sigma_{nm} = R_{nm}\, e^{i(\delta_n+\delta_m)}$$

where R_{nm} is a real matrix.

This is the *final state theorem*: it relates the phases of decay amplitudes to the phases of the elastic scattering channels $2 \to 2$ and $3 \to 3$.

In this case we have

$$a_s = \Sigma_{21} = \pm|a_s|\, e^{i\delta_s} \tag{6.79a}$$

$$a_p = \Sigma_{31} = \pm|a_p|\, e^{i\delta_p} \tag{6.79b}$$

(the R_{nm} are not necessarily positive).

Since we were working at a fixed energy throughout, the phase shifts on the right must be evaluated at a πN CM total energy corresponding to the Σ^- mass.

We may note that if there were no final state interaction, T-invariance and the first-order approximation would imply that a_s and a_p are relatively real (see the end of §6.3.2).

We saw in §5.6 that the transverse polarisation $P_{y'}$ of the decay nucleon is proportional to the quantity $2\,\mathrm{Im}(a_s^*\, a_p)$ via the parameter β, defined in (5.94). Thus in the absence of final state interaction β would be zero.

Using (6.79) and (5.94) the final state theorem gives

$$\beta/\alpha = \tan(\delta_p - \delta_s)$$

This relation is valid to the extent that second-order weak interactions can be neglected, and thus forms a sensitive test of time reversal invariance.

However, for the $\Sigma^- \to \pi^- n$ decay, α is small and β/α is only poorly known. Let us consider instead the Λ^0 decay.

In the case of Σ^- decay there is a unique final charge state, but for Λ^0 decay we have

$$\Lambda^0 \rightarrow p + \pi^- \quad \text{(66 per cent)}$$

$$\rightarrow n + \pi^0 \quad \text{(33 per cent)}$$

Each of the two charge states can be in s- or p-waves giving five channels (with the Σ). Charge exchange scattering $p\pi^- \rightarrow n\pi^0$ couples the channels, so to diagonalise $\mathbb{S}$ in this case it is necessary to work with eigenstates of total isospin, I.

I can take the values $\frac{1}{2}$ and $\frac{3}{2}$, but it is found that the partial width for Λ^0 decay into the $I = \frac{3}{2}$ state is almost zero, as is shown by the branching ratios for the two modes. This is in accordance with the $|\Delta I| = \frac{1}{2}$ rule for non-leptonic weak decays. Thus the relevant πN phase shifts are those corresponding $I = \frac{1}{2}$, and of course $J = \frac{1}{2}$; i.e. P_{11} and S_{11} phase shifts in the standard notation $L_{2I,\,2J}$. From the πN phase shift analysis of Roper and Wright (1965) we find

$$\delta(P_{11}) - \delta(S_{11}) = +7.8°$$

at the CM energy corresponding to the Λ^0 mass. Using the average values of the Λ^0 decay parameters quoted by the Particle Data Group (1973) we obtain

$$\arctan(\beta/\alpha) = 7.5°$$

with an estimated error of 3°. Thus we obtain satisfactory agreement with the final state theorem.

Another important application of the final state theorem which we shall not discuss is to relate the phase shifts in pion photo-production

$$\gamma + N \rightarrow \pi + N$$

to the πN scattering phase shifts.

6.4.3 *Static electric dipole moments*

Let us first show that time reversal invariance leads to the vanishing of the electric dipole moment of a stationary state. The system may be an atom, nucleus or in our case an elementary particle.

Let ϕ_m denote the state of the particle at rest with spin projection on the z-axis equal to m (i.e. ϕ_{0m} in the notation of §6.3). If the particle has spin zero, then of course the stationary state at rest is rotationally invariant and all the static electric and magnetic moments of the particle must be zero. We consider the expectation value of the dipole moment operator $\boldsymbol{d}$. From the definition

$$\boldsymbol{d} = \sum_i e\boldsymbol{r}^i$$

and the fact that the coordinate operator is unchanged under time reversal, it follows that $\boldsymbol{d}$ satisfies

$$O(T)\boldsymbol{d}\,O(T)^{-1} = +\boldsymbol{d}$$

Consider the expectation value of this equation with respect to the state $O(T)\phi_m$. Using (6.29) and the Hermitian property of $\boldsymbol{d}$, we have

$$\begin{aligned}(O(T)\phi_m, \boldsymbol{d}\,O(T)\phi_m) &= (O(T)\phi_m, O(T)\boldsymbol{d}\,O(T)^{-1}\,O(T)\,\phi_m)\\ &= (O(T)\phi_m, O(T)\,\boldsymbol{d}\,\phi_m)\\ &= (\phi_m, \boldsymbol{d}\,\phi_m)^*\\ &= (\phi_m, \boldsymbol{d}\,\phi_m)\end{aligned}$$

Now, by (6.62), $O(T)$ acting on a rest-state has the same effect as a rotation by π, so

$$\begin{aligned}(O(T)\phi_m, \boldsymbol{d}\,O(T)\phi_m) &= (U(Y_\pi)\phi_m, \boldsymbol{d}\,U(Y_\pi)\phi_m)\\ &= (\phi_m, U(Y_\pi)^{-1}\,\boldsymbol{d}\,U(Y_\pi)\phi_m)\\ &= -(\phi_m, \boldsymbol{d}\,\phi_m)\end{aligned}$$

because a rotation by π changes the sign of the vector operator $\boldsymbol{d}$. Thus

$$(\phi_m, \boldsymbol{d}\phi_m) = -(\phi_m, \boldsymbol{d}\phi_m) = 0$$

Experiments performed over many years have placed extremely low limits on the electric dipole moment (EDM) of the neutron. Smith, Purcell and Ramsey (1957) describe an experiment performed in 1950 with a resonance-beam apparatus. Neutrons from a reactor are polarised by reflection from magnetised iron. They pass through a region of uniform parallel electric and magnetic fields. Transitions between the two spin states can be induced by applying a radio-frequency magnetic field at the Larmor frequency corresponding to the magnetic moment interaction energy. If an electric moment interaction is present as well, the frequency at which the maximum number of transitions is induced will be shifted. In this way, the EDM was found to be

$$d = -(0.1 \pm 2.4) \times 10^{-20}\ e\,\text{cm}$$

where e is the electronic charge.

Subsequent experiments by the Harvard group and others are summarised in table 6.1. The review paper of Golub and Pendlebury (1972) contains a survey of the experimental techniques and further references.

We may ask what is the significance of these small limits on the EDM of the neutron? Because the existence of a non-zero EDM require

Table 6.1. *Measurements of electric dipole moment of the neutron*

Reference	Dipole moment (e cm)
Smith, Purcell and Ramsey (1957)	$(0.1 \pm 2.4) \times 10^{-20}$
Miller *et al.* (1967)	$(-2 \pm 3) \times 10^{-22}$
Shull and Nathans (1967)	$(2.4 \pm 3.9) \times 10^{-22}$
Baird *et al.* (1969)	5×10^{-23}
Dress, Miller and Ramsey (1973)	$(3.2 \pm 7.5) \times 10^{-23}$

that both P and T be violated the theoretical interpretation is not clear. We have noted that a parity-violating strangeness-conserving nucleon–nucleon force with strength given by the Fermi coupling constant G_F is expected to exist. If we suppose a T-violating electromagnetic interaction to exist with matrix elements of order f, then an EDM of order

$$d \sim f G_F \frac{M_p}{\hbar^2}$$

might be expected (Feinberg, 1965). This follows at least on dimensional grounds, since

$$G_F = 10^{-49} \text{ erg cm}^3 = 10^{-5} M_p c^2 \left(\frac{\hbar}{M_p c}\right)^3$$

so that

$$d \sim 10^{-5} \frac{\hbar}{M_p c} f \sim 10^{-19} F \; e \text{ cm}$$

where M_p is the proton mass and $F = f/e$ is the ratio of T-violating to T-conserving electromagnetic matrix elements. The present experimental values show that

$$F < 10^{-3}$$

A similar method has been used to measure the EDM of neutral atoms with spin half in order to put a limit on the EDM of the electron. Stein *et al.* (1969) used caesium and found

$$d(\text{Cs}) < 3 \times 10^{-21} \; e \text{ cm}$$

If this result is interpreted in terms of an EDM of the electron, then according to the calculations of Sandars (1968) it follows that

$$d(e^-) < 2.5 \times 10^{-23} \; e \text{ cm}$$

Player and Sandars (1970) using xenon atoms found

$$d(e^-) = (0.7 \pm 2.2) \times 10^{-24} \; e \text{ cm}$$

Again the interpretation is not clear cut because both *P*- and *T*-violation are required to give a non-zero value for the EDM. At least the results confirm the *P*- and *T*-conserving character of the electrodynamic interaction of the electron which is already suggested by the accurate agreement of quantum-electrodynamic calculations with experiment such as the Lamb shift.

6.5 Tests of time reversal invariance in strong interactions

As with parity, we distinguish the experiments according to the part of the particle interaction Hamiltonian whose invariance is being tested. In this section strong interactions are considered.

6.5.1 *Detailed balance tests*

According to the principle of detailed balance (6.73), the differential cross-sections for a reaction and its inverse are in the ratio of phase space factors if *T*-invariance holds.

To present the results quantitatively we must suppose that there is a small amount of *T*-violating amplitude. This may be represented by writing the $\mathcal{T}$-matrix element as

$$\mathcal{T} = \mathcal{T}_e + \mathcal{T}_o$$

where $\mathcal{T}_e$ satisfies (6.44*a*) but $\mathcal{T}_o$ satisfies

$$O(T)\,\mathcal{T}_o\,O(T)^{-1} = -\mathcal{T}_o^\dagger$$

Then it is easy to see that deviations from (6.73) will come from the term $2\,\mathrm{Re}(\mathcal{T}_e\,\mathcal{T}_o^*)$ in $|\mathcal{T}|^2$.

Thus if *L* and *R* denote the left and right sides of (6.73) then

$$\frac{L-R}{L+R} \approx \frac{2\,\mathrm{Re}(\mathcal{T}_e\,\mathcal{T}_o^*)}{|\mathcal{T}_e|^2 + |\mathcal{T}_o|^2} \approx \left|\frac{\mathcal{T}_o}{\mathcal{T}_e}\right|$$

In this way the amount of *T*-violating amplitude present can be estimated.

Because of the difficulties in accurately measuring absolute differential cross-sections, it is convenient to make a relative comparison, in which the ratio of the differential cross-sections for $ab \to cd$ at two different energies or angles is compared with the ratio of the two corresponding cross-sections for $cd \to ab$. In order to be detectable the time reversal non-invariant effect must be a function of energy or angle.

Table 6.2. *Tests of detailed balance*

Test	Reaction $ab \to cd$	Energy (MeV) of $ab \to cd$	$\lvert\mathfrak{T}_o/\mathfrak{T}_e\rvert$ (per cent)	Reference
Absolute	$p + t \to d + d$	6.3	2	Rosen and Brolley (1959)
Relative in angle	$\alpha + {}^{12}C \to d + {}^{14}N$	40	3	Bodansky *et al.* (1959)
Relative in angle	$d + {}^{24}Mg \to p + {}^{25}Mg$	10	0.4	Bodansky *et al.* (1966)
Relative in energy	$\alpha + {}^{24}Mg \to p + {}^{27}A$	10–15	0.2–0.4	Von Witsch *et al.* (1967)
Relative in angle	$d + {}^{24}Mg \to p + {}^{25}Mg$	10	0.3	Weitkamp *et al.* (1968)
Absolute	$d + {}^{16}O \to \alpha + {}^{14}N$	4	0.3	Thornton *et al.* (1968)

Thornton *et al.* (1968) have made an absolute comparison of differential cross-sections for the processes

$$d + {}^{16}O \rightleftharpoons {}^{14}N + \alpha$$

and found detailed balance to be satisfied to within ±0.5 per cent. Some other tests are summarised in table 6.2. The paper referred to gives criteria for choosing a reaction that may be sensitive to T-violation.

Henley and Jacobsohn (1959) have made a critical examination of detailed balance as a test of T-invariance. They point out that in certain cases detailed balance holds regardless of T-invariance. For example, if $\mathfrak{T}$ is Hermitian as in the first-order approximation of §6.2.5, then

$$(\phi_n, \mathfrak{T}\phi_m) = (\phi_m, \mathfrak{T}\phi_n)^*$$

In the $J\lambda$ basis this gives

$$\mathfrak{T}^{J(cd \leftarrow ab)}_{\lambda_c\lambda_d,\, \lambda_a\lambda_b} = \mathfrak{T}^{J(ab \leftarrow cd)*}_{\lambda_a\lambda_b,\, \lambda_c\lambda_d}$$

from which (6.70) follows without involving T-invariance. Thus although a failure of detailed balance implies breakdown of T-invariance, the converse is not necessarily true.

6.5.2 *Polarisation-asymmetry equality*

This is a generalisation of a result proved in §5.5.5 for spin half–spin-0 scattering as a consequence of P-invariance. In the case of spin half particles scattering off a target of spin $s \neq 0$ the polarisation-asymmetry equality does not follow from P-invariance. It does follow from T-invariance as we shall now show.

We first consider the left–right asymmetry in the scattering of a fully transversely polarised proton off an unpolarised target. The amplitude is

$$f_{\lambda_c\lambda_d,\,\lambda_a\lambda_b}(\theta,\phi) = f_{\lambda_c\lambda_d,\,\lambda_a\lambda_b}(\theta)\,e^{i(\lambda_a-\lambda_b)\phi}$$

where λ_a (λ_c) and λ_b (λ_d) denote the initial (final) helicities of the proton p and target particle x of spin s. The ϕ-dependence has been made explicit on the right.

The state

$$2^{-1/2}(\phi_{p,\,\lambda_a=+1/2} + i\,\phi_{p,\,\lambda_a=-1/2})$$

represents a proton fully polarised in the y-direction. Hence the scattering probability to the left ($\phi = 0$) off an unpolarised target is

$$\mathcal{I}_L(\theta) = \frac{1}{2(2s+1)}\sum_{\lambda_b\lambda_c\lambda_d} |f_{\lambda_c\lambda_d,+1/2,\,\lambda_b}(\theta) + i f_{\lambda_c\lambda_d,-1/2,\lambda_b}(\theta)|^2 \tag{6.80}$$

This can be written more concisely as

$$\mathcal{I}_L(\theta) = \mathrm{Tr}\{f(\theta)\rho^i f(\theta)^\dagger\}$$

where $f(\theta)$ is a matrix with rows and columns labelled by the pairs (λ_c, λ_d) and (λ_a, λ_b) of helicity indices respectively. ρ^i denotes the initial state density matrix,

$$\rho^i_{\lambda_a\lambda_b,\,\lambda'_a\lambda'_b} = \rho_{\lambda_a\lambda'_a}\,\rho_{\lambda_b\lambda'_b} \tag{6.81}$$

with

$$\rho_{\lambda_b\lambda'_b} = \frac{1}{(2s_b+1)}\,\delta_{\lambda_b\lambda'_b} \tag{6.82}$$

$$\rho_{\lambda_a\lambda'_a} = \tfrac{1}{2}(1+\sigma_y)_{\lambda_a\lambda'_a}$$

corresponding to an unpolarised x and a transversely polarised proton.

Right scattering corresponds to $\phi = \pi$, and differs from (6.80) in that $f_{\lambda_c\lambda_d,\,\lambda_a\lambda_b}(\theta)$ is replaced by

$$f_{\lambda_c\lambda_d,\,\lambda_a\lambda_b}(\theta)\,e^{i(\lambda_a-\lambda_b)\pi}$$

This can be incorporated in the matrix notation as follows

$$f(\theta,\phi=\pi) = f(\theta)R$$

where R is a diagonal matrix

$$(R)_{\lambda_a\lambda_b,\,\lambda'_a\lambda'_b} = (-1)^{\lambda_a-\lambda_b}\delta_{\lambda_a\lambda'_a}\delta_{\lambda_b\lambda'_b} \tag{6.83}$$

So

$$\mathcal{I}_{\mathrm{R}}(\theta) = \mathrm{Tr}\{fR\rho^{\mathrm{i}}(fR)^\dagger\}$$
$$= \mathrm{Tr}\{fR\rho^{\mathrm{i}}R^\dagger f^\dagger\}$$

The left–right asymmetry is defined to be

$$\epsilon(\theta) = \frac{\mathcal{I}_{\mathrm{L}}(\theta)-\mathcal{I}_{\mathrm{R}}(\theta)}{\mathcal{I}_{\mathrm{L}}(\theta)+\mathcal{I}_{\mathrm{R}}(\theta)}$$

which can be put in the form

$$\epsilon(\theta) = \frac{\mathrm{Tr}\{fNf^\dagger\}}{\mathrm{Tr}\{fDf^\dagger\}}$$

where

$$\left.\begin{matrix} D \\ N \end{matrix}\right\} = \rho^{\mathrm{i}} \pm R\rho^{\mathrm{i}}R^\dagger$$

From (6.81), (6.82) and (6.83) we easily find

$$\rho^{\mathrm{i}} \pm R\rho^{\mathrm{i}}R^\dagger = \begin{cases} 1 \\ \sigma_y^{\mathrm{a}} \end{cases}$$

where σ_y^{a} denotes the Pauli matrix for the proton, i.e.

$$(\sigma_y^{\mathrm{a}})_{\lambda_a\lambda_b,\,\lambda'_a\lambda'_b} = (\sigma_y)_{\lambda_a\lambda'_a}\delta_{\lambda_b\lambda'_b}$$

thus we have

$$\epsilon(\theta) = \frac{\Sigma f_{\lambda_c\lambda_d,\,\lambda_a\lambda_b}(\sigma_y)_{\lambda_a\lambda'_a}f^*_{\lambda_c\lambda_d,\,\lambda'_a\lambda_b}}{\Sigma|f_{\lambda_c\lambda_d,\,\lambda_a\lambda_b}|^2} \tag{6.84}$$

where the sums go over all helicity indices.

Next the transverse polarisation of the proton produced in scattering off an unpolarised x target is given by

$$\mathcal{I}(\theta)P_{y'} = \frac{1}{2(2s+1)}\sum_{\lambda_c\lambda'_c\lambda_a\lambda_b\lambda_d}(\sigma_y)_{\lambda'_c\lambda_c}f^*_{\lambda'_c\lambda_d,\,\lambda_a\lambda_b}f_{\lambda_c\lambda_d,\,\lambda_a\lambda_b} \tag{6.85}$$

where $\mathcal{I}(\theta)$ is the differential cross-section

$$\mathcal{I}(\theta) = \frac{1}{2(2s+1)}\sum_{\lambda_c\lambda_d,\,\lambda_a\lambda_b}|f_{\lambda_c\lambda_d,\,\lambda_a\lambda_b}(\theta)|^2 \tag{6.86}$$

Now T-invariance for elastic scattering (6.70) gives

$$f_{\lambda_c\lambda_d,\,\lambda_a\lambda_b}(\theta) = (-1)^{\lambda_a-\lambda_b-\lambda_c-\lambda_d} f_{\lambda_a\lambda_b,\,\lambda_c\lambda_d}(\theta)$$

Thus noting that the phase factor is ± 1 we can transform (6.85)

$$\begin{aligned} \mathcal{I}(\theta)P_{y'} &= \frac{1}{2(2s+1)}\sum(\sigma_y)_{\lambda'_c\lambda_c} f^*_{\lambda_a\lambda_b,\lambda'_c\lambda_d} f_{\lambda_a\lambda_b,\lambda_c\lambda_d} \\ &\quad \times(-1)^{\lambda_a-\lambda_b-\lambda'_c-\lambda_d}(-1)^{-\lambda_a+\lambda_b+\lambda_c-\lambda_d} \\ &= \frac{1}{2(2s+1)}\sum(\sigma_y)_{\lambda_c\lambda'_c} f^*_{\lambda_a\lambda_b,\,\lambda'_c\lambda_d} f_{\lambda_a\lambda_b,\,\lambda_c\lambda_d} \end{aligned} \tag{6.87}$$

where we have used the identity

$$(-1)^{\lambda_c-\lambda'_c}(\sigma_y)_{\lambda'_c\lambda_c} = (\sigma_y)_{\lambda_c\lambda'_c}$$

Comparing (6.87) and (6.84) we have the *polarisation-asymmetry equality*

$$P_{y'}(\theta) = \epsilon(\theta) \tag{6.88}$$

Wolfenstein and Ashkin (1952) and Dalitz (1952) originally derived this result for the spin half–spin half case by examining the most general transition operator invariant under rotations, space inversion and time reversal. Bell and Mandl (1958*a*, *b*) pointed out that it holds independently of space inversion invariance. If instead of using the helicity convention, we quantise spins along the normal to the scattering plane, a slightly more elementary proof can be given as in the work of Bell and Mandl.

It follows from the remarks at the beginning of this section that in order to use (6.88) as a test of T-invariance we must examine p–p scattering or scattering of protons off nuclei with non-zero spin. Several groups have tested the polarisation asymmetry equality in p–p scattering. The measurements of Hillman, Johansson and Tibell (1958), Abashian and Hafner (1958) and Hwang *et al.* (1960) made in the range 150 to 210 MeV and those of Zulkarneyev *et al.* (1967) at 600 MeV found the equality to be satisfied to within experimental errors. The amount of T-violating amplitude was estimated to be at most a few per cent.

6.6 Tests of time reversal invariance in electromagnetic interactions

We may consider separately the electromagnetic interactions of hadrons and leptons. This is symbolised by writing

$$H_{\mathrm{em}} = H_{\mathrm{em}}^{\mathrm{had}} + H_{\mathrm{em}}^{\mathrm{lep}}$$

The electromagnetic interaction of charged leptons (electron and muon) is accurately described by quantum electrodynamics. The Hamiltonian of this theory is formally invariant under time reversal and this provides indirect evidence of the time reversal invariance of $H_{\mathrm{em}}^{\mathrm{lep}}$.

The evidence for time reversal invariance in the electromagnetic interaction of hadrons is rather limited. The smallness of $e^2/\hbar c$ means that an electromagnetic process is accurately represented as a first-order process as we have defined it in §6.2.5. Simple detailed balance tests of reactions of the form

$$\gamma + a \rightleftarrows b + c$$

are therefore not sensitive to T-violations. It is necessary to make detailed spin–momentum analyses to test reciprocity.

Similarly, electron–proton scattering is well described by the one-photon exchange process shown in fig. 6.3. The blob representing the interaction of the virtual photon with the proton is given by the matrix element of the electromagnetic current operator j_μ between two proton states. Considerations of Lorentz invariance and parity conservation show that this may be written as

$$\langle p' | j_\mu | p \rangle = \bar{u}(p')\{\gamma_\mu F_1 + \sigma_{\mu\nu} q_\nu F_2 + q_\mu F_3\} u(p)$$

where $u(p)$ and $\bar{u}(p)$ are Dirac spinor wavefunctions. The Fs are form factors which depend only on the invariant four-momentum squared, q^2, where q_μ is the momentum transfer to the nucleon. The Hermitian property of the current operator requires the form factors to be real. On the other hand it can be shown that time reversal invariance requires F_1 and F_2 to be real and F_3 to be purely imaginary. Thus $F_3 = 0$ if T-invariance holds. However, it also follows from the conservation law of the charge-current density

$$\sum_\mu \frac{\partial j_\mu}{\partial x_\mu} = 0$$

that $F_3 = 0$, and thus no test of T-invariance is possible in e-scattering (Bernstein, Feinberg and Lee, 1965).

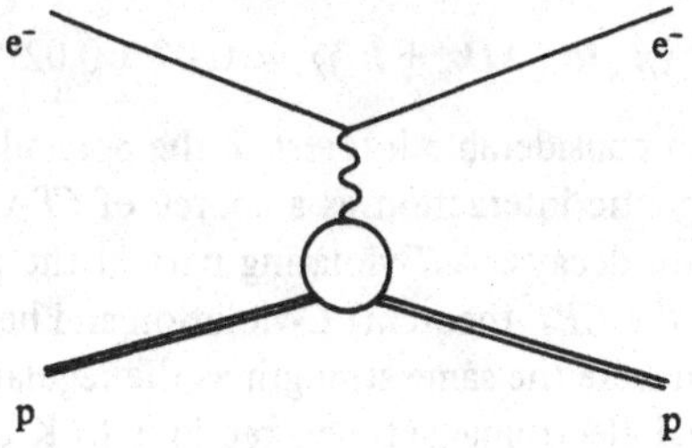

Fig. 6.3. Electron–proton scattering in the one photon approximation.

This theoretical restriction is avoided if we consider inelastic processes (Christ and Lee, 1966*a*, *b*).

Chen *et al.* (1968) studied the process

$$e^- + N \to N^* + e^-$$

and observed inelastically scattered electrons from a polarised proton target with energy loss corresponding to excitation of the N* (1236) and higher resonances. The F_3 term is not now excluded by current conservation and gives rise to a contribution to the predicted intensity which changes with the sign of the target polarisation. To within the experimental error (5–10 per cent) no change in the intensity was found, in accordance with *T*-invariance.

Tests of *T*-invariance with higher accuracy at lower energies are possible in studies of nuclear γ-transitions. These are reviewed by Henley (1969). However the limit on the *T*-odd part of the interaction is at best $< 10^{-3}$.

A different type of test is afforded by the Σ^0 electromagnetic decay

$$\Sigma^0 \to \Lambda^0 e^+ e^-$$

and theoretically analysed in detail by Bernstein, Feinberg and Lee (1965).

If $\hat{\boldsymbol{p}}$ and $\hat{\boldsymbol{k}}_\pm$ denote unit vectors along the momenta of the Λ and $e^\pm$, a non-zero expectation

$$\langle \boldsymbol{\sigma}_\Lambda \cdot \hat{\boldsymbol{p}} \wedge (\hat{\boldsymbol{k}}_+ + \hat{\boldsymbol{k}}_-) \rangle \neq 0$$

representing the Λ^0-polarisation normal to the decay plane would be a clear indication of *T*-violation cf. (6.48), while consistent with *P*-conservation.

Glasser *et al.* (1966) analysed 907 events of this decay, using the proton momentum as indicator of the Λ-spin, and found

$$\langle p_N \cdot \hat{p}_\Lambda \wedge (\hat{k}_+ + \hat{k}_-)\rangle = 0.02 \pm 0.02$$

There has been considerable interest in the possibility of *T*-violation in the electromagnetic interaction as a source of *CP*-violation observed in neutral K-meson decays. A *T*-violating part of the parity conserving H_{em} implies (by the *CPT* theorem) *C*-violation and hence *CP*-violation. If this interaction were the same strength as the regular electromagnetic interaction, virtual electromagnetic corrections to K decays could give a *CP*-violation of the observed magnitude. However the associated *C*-violation has not been observed. We shall return to this when discussing *CP*-violation.

6.7 Tests of time reversal invariance in weak interactions

We may write

$$H_{wk} = H_{wk}^{ll} + H_{wk}^{lh} + H_{wk}^{hh}$$

where the three terms represent respectively the parts of the weak interaction responsible for (1) purely leptonic processes e.g. μ decay; (2) leptonic decays of hadrons, e.g. β decay or $\pi \to \mu\nu$; (3) non-leptonic decays of hadrons e.g. $\Lambda \to N\pi$, $K \to 2\pi$, 3π. Tests of *T*-invariance may be distinguished according to the part of H_{wk} involved.

Although it may eventually be possible to test *T*-invariance in neutrino scattering the tests discussed here all involved decay processes. Thus the considerations of §6.2.5 apply.

Muon decay. This is a purely leptonic weak process to the extent that radiative corrections can be neglected. Observations on muon decay are consistent with the current–current form of interaction which is manifestly *T*-invariant.

Leptonic decays of hadrons. A test of *T*-invariance is provided by the electron–neutrino correlation in the decay of polarised neutrons. A non-zero value for

$$\langle \boldsymbol{\sigma}_n \cdot \boldsymbol{p}_e \wedge \boldsymbol{p}_\nu \rangle$$

indicates *T*-violation.

Burgy *et al.* (1958) measured the quantity

$$\langle \boldsymbol{p}_p \cdot \boldsymbol{p}_e \wedge \boldsymbol{\sigma}_n \rangle$$

which by momentum conservation is equivalent to the previous expression. The result may be presented as a phase angle between the vector and the axial vector decay coupling constants, G_V and G_A, since according to T-invariance their ratio is real implying $\phi = 0°$ or $180°$. The experiment quoted gave the result

$$\phi < 8°$$

Calaprice *et al.* (1967) have measured the phase angle between G_V and G_A in ^{19}Ne decay and found

$$\phi = 180.2° \pm 1.6°$$

In $K_{\mu 3}$ decays measurements have been made of $\langle \boldsymbol{\sigma}_\mu \cdot \boldsymbol{p}_\mu \wedge \boldsymbol{p}_e \rangle$ which should vanish by T-invariance. Camerini *et al.* (1965) measured this correlation directly in the decay

$$K^+ \rightarrow \pi^0 + \mu^+ + \nu$$

of stopped K^+ in a heavy liquid bubble chamber. They obtained 690 events in which both gammas from the π^0 convert in the chamber and also the e^+ from the μ^+ decay is seen. It was found that

$$\langle \boldsymbol{\sigma}_\mu \cdot \boldsymbol{n} \rangle = 0.04 \pm 0.35$$

where

$$\boldsymbol{n} = \frac{\boldsymbol{p}_\mu \wedge \boldsymbol{p}_e}{|\boldsymbol{p}_\mu \wedge \boldsymbol{p}_e|}$$

Other less direct tests on this decay have been made by comparing predictions of the usual T-conserving theory with experiment. The results are summarised by Rowe and Squires (1969).

Non-leptonic decays of hadrons. We have seen above that the final state theorem applied to non-leptonic hyperon decays shows that T-invariance is satisfied within the rather large experimental errors.

Many of the T-tests described here were prompted by the discovery in 1964 of CP-violation. If the CPT theorem holds then such a violation must be accompanied by T-violation.

Subsequent studies of the neutral kaon system have shown that independently of one another, time reversal invariance is violated and CPT-invariance is satisfied. This will be discussed further in §8.5.

CHAPTER 7

CHARGE INDEPENDENCE, ISOSPIN AND STRANGENESS

7.1 Evidence for charge independence of strong interactions

Two types of nucleus, in which each contains the same number of protons as the other does neutrons, are sometimes called mirror nuclei. Well-known pairs of mirror nuclei, differing in the nature of a single nucleon, are $^{7}_{4}Be$, $^{7}_{3}Li$ and $^{11}_{6}C$, $^{11}_{5}B$. In both these pairs we find that the symmetry of structure is reflected in a correspondence of general properties. For example, the binding energies of ^{11}C and ^{11}B in their ground states differ by an amount which corresponds to the Coulomb energy of the extra proton in ^{11}C, within the limits of accuracy set by our uncertainty about the detailed distributions of the nuclear charges.

This correspondence extends to the excited states, which occur at energies showing marked qualitative, and sometimes quantitative, similarities. For example, the first three excited states of ^{11}B and ^{11}C occur at energies given by Ajzenberg-Selove and Lauritsen (1968) as

$$^{11}C: \quad 2.00, \quad 4.31, \quad 4.79 \quad \text{MeV}$$

$$^{11}B: \quad 2.12, \quad 4.44, \quad 5.02 \quad \text{MeV}$$

While similarities of this order prove nothing, even in contrast to the wide spread observed in the corresponding properties of unrelated nuclei, taken together they provide an impressive weight of evidence for the charge symmetry of nuclear forces, at least for the strong interactions which are responsible for nuclear binding energies. By charge symmetry we mean equivalence of the proton–proton and neutron–neutron forces. For evidence about charge independence, a wider concept including also the neutron–proton force, we have to search farther.

The first such evidence is provided by sets of mirror nuclei in which two nucleons are different, for example $^{14}_{6}C$ and $^{14}_{8}O$. Here again, the binding energies of the ground states are consistent with charge symmetry of the strong interaction, $^{14}_{6}C$ being considered as a core of $^{12}_{6}C$ with two extra neutrons, and $^{14}_{8}O$ as the same core with two extra protons. But when the discussion is extended to include $^{14}_{7}N$, with a neutron–proton pair attached to the same core, the correspondence becomes less obvious. The ground state has a higher binding energy, i.e. a lower potential energy; the difference is in fact approximately equal to the excitation energy (2.31 MeV) of the first excited state of $^{14}_{7}N$. Thus this excited state may be considered as the centre member of a triplet of nuclear states with equivalent binding energies (after correction for different Coulomb energies), the outer members being the ground states of $^{14}_{6}C$ and $^{14}_{8}O$. Such an equivalence, if significant, would imply similarity of the neutron–proton interactions in these three nuclei with the proton–proton and neutron–neutron interactions. The lower potential energy of $^{14}_{7}N$ in its ground state would have to be attributed to a difference in the states accessible to nucleons in this type of nucleus.

As an indication that such a difference may be genuine, we note the following peculiarity of the 2.31 MeV level in $^{14}_{7}N$, a peculiarity which distinguishes it from the ground state and most of the other excited states of $^{14}_{7}N$: it is formed extremely weakly, if at all, in the reaction

$$^{16}_{8}O + ^{2}_{1}H \rightarrow ^{14}_{7}N + ^{4}_{2}He \qquad (7.1)$$

although it is formed copiously along with the other states of $^{14}_{7}N$ in the processes

$$^{14}_{7}N + p \rightarrow ^{14}_{7}N^{*} + p \quad \text{(inelastic scattering)} \qquad (7.2)$$

$$^{13}_{6}C + ^{2}_{1}H \rightarrow ^{14}_{7}N + n \quad \text{((d, n) reaction in } ^{13}_{6}C\text{)} \qquad (7.3)$$

$$^{14}_{8}O \rightarrow ^{14}_{7}N + e^{+} \quad \text{(positron decay)} \qquad (7.4)$$

Before discussing the interpretation of these effects in terms of conservation of isospin, we shall give a brief account of the background from which this idea grows.

7.2 The concept of isospin

The now-familiar idea that the neutron and the proton should be treated as two charge-states of a single particle, the nucleon, was in fact first put forward by Heisenberg in 1932. This suggestion followed closely the realisation of the part played by neutrons along with protons in the structure of nuclei. The formalism of the idea, in terms analogous to those used for ordinary quantum-mechanical spin, was further developed by Wigner (1937), who suggested the name isotopic spin for the variable whose eigenvalues were to give the charge quantum numbers.

The importance of 'isotopic spin' increased as its relevance in the properties of light nuclei became understood (Inglis, 1953). After some years the original name was abandoned in favour of the more accurate 'isobaric spin', a term which is now more often contracted to isospin or i-spin. Nowadays, however, it is in the theory of elementary particles that the concept of isospin finds its most advanced applications, and it is for use in this context that we now develop it.

The proton p and neutron n are considered as two states of a single particle, the nucleon N. There is a mathematical analogy here with the spin degree of freedom of a particle with spin half. However, there is no physical connection between ordinary spin and isospin. It only happens they they use the same operator mathematics. A complete specification of a state of a nucleon must give its space and spin coordinates and also indicate whether it is a proton or a neutron.

We write for a proton in the space–spin state $\psi(\boldsymbol{r},\boldsymbol{\sigma})$

$$\psi(\boldsymbol{r},\boldsymbol{\sigma})|\mathrm{p}\rangle$$

and for a neutron

$$\psi(\boldsymbol{r},\boldsymbol{\sigma})|\mathrm{n}\rangle$$

$|\mathrm{p}\rangle$ and $|\mathrm{n}\rangle$ are the two basic charge states of a nucleon. They are the analogues of the basic spin states of a spin half particle. Henceforth we shall suppress the space–spin state where it is not essential to the discussion.

The general nucleon state may be written

$$|\mathrm{N}\rangle = a\,|\mathrm{p}\rangle + b\,|\mathrm{n}\rangle$$

where a and b are complex coefficients. A nucleon in such a state has a probability of $|a|^2$ of being a proton and $|b|^2$ of being a neutron. We may adopt a matrix notation in which

$$|\mathrm{p}\rangle \to \begin{pmatrix}1\\0\end{pmatrix} \qquad |\mathrm{n}\rangle \to \begin{pmatrix}0\\1\end{pmatrix}$$

so that

$$|\mathrm{N}\rangle \to a\begin{pmatrix}1\\0\end{pmatrix} + b\begin{pmatrix}0\\1\end{pmatrix} = \begin{pmatrix}a\\b\end{pmatrix}$$

We introduce isospin operators acting on these states by analogy with the angular momentum operators for a spin half particle

$$I_1 = \frac{1}{2}\begin{pmatrix}0 & 1\\1 & 0\end{pmatrix}, \quad I_2 = \frac{1}{2}\begin{pmatrix}0 & -\mathrm{i}\\\mathrm{i} & 0\end{pmatrix}, \quad I_3 = \frac{1}{2}\begin{pmatrix}1 & 0\\0 & -1\end{pmatrix} \tag{7.5}$$

The matrices on the right without the $\frac{1}{2}$ are simply the Pauli spin matrices which in this context are denoted by τ_1, τ_2 and τ_3.

I_1, I_2, and I_3 may be regarded as components of a vector $\boldsymbol{I}$ in an *isospin space*. Both p and n are eigenstates of the square of the total isospin defined by

$$I^2 = I_1^2 + I_2^2 + I_3^2$$

The corresponding eigenvalue is $\frac{1}{2}(\frac{1}{2}+1) = \frac{3}{4}$. The eigenvalue of I_3 is $+\frac{1}{2}$ for p and $-\frac{1}{2}$ for n.

We shall also use $\boldsymbol{\tau}^2$ and τ_3 which have the eigenvalues 3 and ± 1 respectively.

Finally we define the shift operators τ_+ and τ_- as

$$\tau_+ = 2^{-1/2}(\tau_1 + \mathrm{i}\tau_2), \quad \tau_- = 2^{-1/2}(\tau_1 - \mathrm{i}\tau_2) \tag{7.6}$$

with the properties

$$\tau_+|\mathrm{p}\rangle = 0, \quad \tau_+|\mathrm{n}\rangle = 2^{1/2}|\mathrm{p}\rangle$$

$$\tau_-|\mathrm{p}\rangle = 2^{1/2}|\mathrm{n}\rangle, \quad \tau_-|\mathrm{n}\rangle = 0$$

If we add to the set τ_1, τ_2, τ_3 the unit operator

$$1 = \begin{pmatrix} 1 & 0 \\ 0 & 1 \end{pmatrix}$$

then any operator acting on the general one nucleon wavefunction $|\mathrm{N}\rangle$ may be expressed as a linear combination of these four operators.

Thus the electric charge operator Q whose eigenstates are the states of definite charge (+1 for p and 0 for n) is seen to be given by

$$Q = \tfrac{1}{2}\cdot 1 + I_3 = \tfrac{1}{2}(1+\tau_3) = \begin{pmatrix} 1 & 0 \\ 0 & 0 \end{pmatrix} \tag{7.7}$$

All the preceding is quite formal, but its physical significance becomes visible when we consider two or more particles and in particular the interactions between them.

Before doing this we note that we may regard $|\mathrm{p}\rangle$, $|\mathrm{n}\rangle$ and $|\mathrm{N}\rangle$ as functions of a charge coordinate, q. This must be a two valued (dichotomic) variable since the only possible outcomes of the measurement of the internal state of the nucleon are p and n, so q takes the value +1 and −1 (or 1 and 2) and serves to label the first and second rows in the column vector notation.

Thus in a one nucleon wavefunction we may display the space and spin coordinates (collectively denoted by ξ) and insert a charge coordinate q

$$\Psi(\xi, q) = \psi(\xi)|\mathrm{N}(q)\rangle$$

The natural generalisation to two nucleons is to write a wavefunction

$$\Psi(\xi_1, q_1; \xi_2, q_2) \tag{7.8}$$

involving the coordinates of both nucleons 1 and 2. Correspondingly we introduce isospin operators $\boldsymbol{I}_{(1)}$ and $\boldsymbol{I}_{(2)}$ for the two particles.

Let us consider some special cases of (7.8). For a state of two protons

$$\Psi = \psi_1(\xi_1, \xi_2')|\mathrm{p}_1\rangle|\mathrm{p}_2\rangle \tag{7.9}$$

and for two neutrons

$$\Psi = \psi_2(\xi_1, \xi_2')|\mathrm{n}_1\rangle|\mathrm{n}_2\rangle \tag{7.10}$$

For a proton–neutron system we might write

$$\Psi = \psi_3(\xi_1, \xi_2)|\mathrm{p}_1\rangle|\mathrm{n}_2\rangle \tag{7.11}$$

or alternatively

$$\psi_4(\xi_1, \xi_2)|\mathrm{n}_1\rangle|\mathrm{p}_2\rangle \tag{7.12}$$

This brings us to an important point of principle. Now that we have chosen to regard the proton and neutron as different states of the same particle, labelled by a charge coordinate q, we should extend the Pauli principle to cover the charge coordinate as follows:

Generalised Pauli principle: The two nucleon wavefunction Ψ must be antisymmetric under exchange of space, spin and charge coordinates, i.e.

$$\Psi(\xi_1, q_1; \xi_2, q_2) = -\Psi(\xi_2, q_2; \xi_1, q_1)$$

What does this imply for the various special cases considered above? For the two proton and two neutron systems the charge part of the wavefunction is symmetrical under $q_1 \leftrightarrow q_2$ exchange; hence the space–spin wavefunction must be antisymmetrical, which is the usual requirement of the Pauli principle for two identical particles. Neither of the proton–neutron wavefunctions (7.11) and (7.12) has the required antisymmetry. However, the correctly symmetrised isospin wavefunction can be set up by analogy with the treatment of two spin half wavefunctions in, say, the quantum mechanics of the helium atom.

We form

$$|\chi_s\rangle = 2^{-1/2}\{|\mathrm{p}_1\mathrm{n}_2\rangle + |\mathrm{n}_1\mathrm{p}_2\rangle\} \tag{7.13}$$

$$|\chi_a\rangle = 2^{-1/2}\{|\mathrm{p}_1\mathrm{n}_2\rangle - |\mathrm{n}_1\mathrm{p}_2\rangle\} \tag{7.14}$$

which are respectively symmetrical and antisymmetrical under the exchange $q_1 \leftrightarrow q_2$. On multiplication by space–spin wavefunctions which are respectively antisymmetrical and symmetrical, we obtain admissible wavefunctions Ψ. Thus there is essentially no restriction on the space–spin wavefunction for the proton–neutron system. However, we can no longer say that the first particle is a proton and the second a neutron. This is not surprising when we remember that exchange potentials due to charged meson exchange may make it impossible to keep track of which particle is a proton and which is a neutron during an interaction.

We define the total isospin $\boldsymbol{I}$ of the two nucleon system as

$$\boldsymbol{I} = \tfrac{1}{2}\boldsymbol{\tau}_{(1)} + \tfrac{1}{2}\boldsymbol{\tau}_{(2)}$$

Eigenstates of I^2 and I_3 can be constructed by the rules developed for the vector addition of angular momentum. Thus the two nucleon states constructed above correspond to eigenvalues of total isospin and third component as follows

$$\begin{aligned} I = 1,\ I_3 = {} & +1\colon |\mathrm{p}_1\mathrm{p}_2\rangle \\ & 0\colon \tfrac{1}{2}(|\mathrm{p}_1\mathrm{n}_2\rangle + |\mathrm{n}_1\mathrm{p}_2\rangle) \\ & -1\colon |\mathrm{n}_1\mathrm{n}_2\rangle \\ I = 0,\ I_3 = {} & 0\colon \tfrac{1}{2}(|\mathrm{p}_1\mathrm{n}_2\rangle - |\mathrm{n}_1\mathrm{p}_2\rangle) \end{aligned}$$

For a system containing a number B of nucleons, the eigenvalue of I_3 is related to the total charge Q by

$$Q = I_3 + \tfrac{1}{2}B \tag{7.15}$$

while I may have any value from $\frac{1}{2}B$, by unit steps down to $\frac{1}{2}$ or 0 according to whether B is odd or even.

Extending this idea from positive values of B to the case $B = -1$, we find that the antinucleons can be covered by the same formalism, with the eigenvalues listed in table 7.1.

With these assignments, (7.15) covers all systems of nucleons and antinucleons, with pions also if the pion is assigned a baryon number $B = 0$.

The three observed types of pion, π^+, π^0 and π^- are treated as three charged-states of a single particle, the pion, which is assigned a total isospin $I = 1$; the three possible eigenvalues of I_3 are $+1$, 0 and -1, representing charges Q of $+1$, 0 and -1. Thus we may describe the properties of systems containing pions and nucleons in terms of a total isospin I which is the vector sum of individual isospins of $\frac{1}{2}$ for

Table 7.1. *Nucleons and antinucleons*

Particle	B	I	I_3	Q
p	1	$\frac{1}{2}$	$\frac{1}{2}$	1
n	1	$\frac{1}{2}$	$-\frac{1}{2}$	0
$\bar{\text{n}}$	−1	$\frac{1}{2}$	$\frac{1}{2}$	0
$-\bar{\text{p}}$[a]	−1	$\frac{1}{2}$	$-\frac{1}{2}$	−1

[a] The − sign in the label $-\bar{\text{p}}$ indicates that the antiproton has to be assigned a phase opposite to that given by operating on $\bar{\text{n}}$ by the shift operator I_- (see appendix C).

each nucleon or antinucleon, and 1 for each pion, with eigenvalue of I_3 related to the total charge by (7.15).

The rules for vector addition of isospins are identical with those given in chapter 3 for vector addition of angular momenta, and the same tables of Clebsch–Gordan coefficients apply.

7.3 Conservation of isospin

As a physical hypothesis connecting the above formalism with experimentally testable predictions we propose that total isospin I is conserved by the strong interaction.

$$[H_{st}, I] = 0 \tag{7.16}$$

By analogy with conservation of angular momentum we can deduce the following consequences:

(*a*) Energy eigenstates of a system may be labelled by a definite value of the total isospin and its third component. States with the same I and different I_3 are degenerate in energy.

(*b*) The total isospin and the third component remain constant during a transition.

As evidence for the hypothesis (7.16) there is an instructive calculation given for example by Fermi (1955). We recall the experimental law of charge independence. This states that the p–p and n–n interaction potentials are equal to one another and to the n–p potential in antisymmetric space–spin states. This last proviso is necessary because the p–n system can exist in symmetric space–spin states forbidden to the p–p and n–n systems by the Pauli principle. One can show that the general form of the nucleon–nucleon interaction which embodies charge independence is

$$H_{\text{int}} = U + V\boldsymbol{\tau}_{(1)} \cdot \boldsymbol{\tau}_{(2)} \tag{7.17}$$

where $\boldsymbol{\tau}_{(1)}$ and $\boldsymbol{\tau}_{(2)}$ are (twice) the isospin operators of the constituent nucleons and U and V depend only on the space and spin coordinates.

If it is to satisfy the condition (7.16), H_{int} cannot depend on any of the components of I separately. However we have

$$\begin{aligned} I^2 &= \tfrac{1}{4}(\boldsymbol{\tau}_{(1)} + \boldsymbol{\tau}_{(2)})^2 \\ &= \tfrac{1}{4}(\boldsymbol{\tau}_{(1)}^2 + \boldsymbol{\tau}_{(2)}^2 + 2\,\boldsymbol{\tau}_{(1)}\cdot\boldsymbol{\tau}_{(2)}) \\ &= \tfrac{1}{2}(3 + \boldsymbol{\tau}_{(1)}\cdot\boldsymbol{\tau}_{(2)}) \end{aligned} \tag{7.18}$$

since

$$\tau_{(2)}^2 = \tau_{(2)}^2 = 3$$

Thus a term in H of the type $\boldsymbol{\tau}_{(1)}\cdot\boldsymbol{\tau}_{(2)}$ must commute with $\boldsymbol{I}^2$, and therefore also with the components of $\boldsymbol{I}$.

The two degrees of freedom associated with the two potentials in (7.17) are related to the two values $I = 0, 1$ of the total isospin. Indeed from (7.18) we see that in the basis in which $\boldsymbol{I}^2$ is diagonal so is $\boldsymbol{\tau}_{(1)}\cdot\boldsymbol{\tau}_{(2)}$. In the states of $I = 0$ or 1 the eigenvalues are

$$\begin{aligned} \boldsymbol{\tau}_{(1)}\cdot\boldsymbol{\tau}_{(2)} &= 2I^2 - 3 = 2I(I+1) - 3 \\ &= -3 \quad (I = 0) \\ &\text{or}\quad 1 \quad (I = 1) \end{aligned}$$

Thus the nucleon–nucleon interaction represented by (7.17) has the effective values

$$\begin{aligned} H_{\text{int}} &= U - 3V \quad \text{in state with } I = 0 \\ &\text{or } U + V \quad \text{in state with } I = 1 \end{aligned}$$

Total isospin should be conserved if no interactions occur except the charge-independent strong interaction. However, the total nucleon–nucleon interaction includes a Coulomb term in the case of two protons. This extra term is outside the scope of the isospin formalism, and may lead to partial non-conservation of isospin. If the total isospin is conserved in a nuclear reaction, the total final isospin, obtained from the vector sum of the isospins of the final particles will be the same as the total initial isospin, obtained similarly as the vector sum of the isospins of the initial particles. An example fitting this is the reaction (7.1) in which we remarked that the 2.31 MeV level of ^{14}N was formed weakly if at all.

The initial particles, ^{16}O (ground state) and ^{2}H are expected to have zero isospin, as is the α-particle. Thus the only states of ^{14}N which can be produced with an α-particle in an isospin-conserving reaction will be those with $I = 0$.

If the triplet of nuclear states discussed in §7.1 is a genuine isospin triplet, the 2.31 MeV level of ^{14}N must have $I = 1$, along with the ground states of ^{14}C and ^{14}O. The ground state of ^{14}N, being more tightly bound than those of ^{14}O and ^{14}C, is likely to have $I = 0$, a value not accessible to nuclei with unbalanced pairs of nucleons.

Thus the ground state of ^{14}N, and any excited states with $I = 0$, would be expected to be formed freely in reaction (7.1), while the 2.31 MeV level would be formed not at all, or weakly if isospin is not perfectly conserved.

7.4 Application to strange particles

The adjective 'strange' was applied to the kaons and hyperons in recognition of the fact that they lay outside the 'non-strange' family of nucleons and pions which were sufficient to explain the basic properties of nuclei. The strange particles were observed to be produced in pairs in interactions between pions and nucleons at energies above about 1 GeV, according to a law of associated production. In modern terms, this law allows reactions such as

$$\pi^+ + p \to \Sigma^+ + K^+ \tag{7.19}$$

$$\pi^- + p \to \Lambda^0 + K^0 \tag{7.20}$$

$$\pi^- + p \to K^- + K^+ + n \tag{7.21}$$

in which a Σ- or a Λ-hyperon is created in association with a positive or a neutral kaon, or oppositely charged kaons are produced as a pair. But, on the other hand, negatively charged kaons are not produced in association with Σ- or Λ-hyperons.

To describe this law in terms of conservation of a quantum number, the K^0- and K^+-mesons were given a strangeness $S = 1$, while the Σ- and Λ-hyperons were allocated a value $S = -1$. The K^-, being the antiparticle of the K^+, also had $S = -1$. The complete picture of the kaons (see chapter 9) requires a fourth kaon, the $\bar{K}^0$ which is the antiparticle of the K^0 and has strangeness -1. Pions and protons, not being strange in the above sense, are given strangeness zero.

Strangeness, defined as above, is conserved in the production reactions (7.19)–(7.21), and conservation of strangeness may be quoted as something equivalent to the old law of associated production in forbidding, for example, the production of $\Sigma^+ + K^-$ from the interaction of π^- with p. We may thus say that strangeness is conserved in the strong interactions which lead to the production reactions, but in saying this we note that strangeness is not conserved in the weak processes by which hyperons decay to non-strange particles:

$$\Lambda^0 \rightarrow p + \pi^- \tag{7.22}$$

$$\Sigma^+ \rightarrow p + \pi^0 \tag{7.23}$$

$$\searrow n + \pi^+ \tag{7.24}$$

At this point we may mention the cascade hyperons Ξ^- and Ξ^0, which have to be given strangeness $S = -2$; they are formed with conservation of strangeness in reactions like

$$K^- + p \rightarrow \Xi^- + K^+ \tag{7.25}$$

which start with total strangeness -1, and they decay to non-strange particles in two steps each with unit change of strangeness, e.g.:

$$\left.\begin{array}{l} \Xi^- \rightarrow \Lambda^0 + \pi^- \\ \qquad\quad \searrow p + \pi^- \end{array}\right\} \tag{7.26}$$

When we come to describe these particles and their different charge states in terms of isospin, we find that the cascade hyperons and the two pairs of kaons form isospin doublets, while the singly strange hyperons form an isospin triplet (Σ^+, Σ^0 and Σ^-) and a singlet (Λ^0).

The list of particles, in terms of strangeness and isospin quantum numbers, now takes the form shown in table 7.2. For systems involving strange particles, the baryon number B of (7.15), which has so far been merely the number of nucleons, has to be extended to have values zero for all mesons and 1 for nucleons and hyperons. With these values, B is rigidly conserved in all known processes, as is total electric charge.

Study of table 7.1 reveals a relation between S, B, Q and I_3:

$$S + B = 2(Q - I_3) \tag{7.27}$$

This means that (7.15) may be rewritten to cover strange particles as:

$$Q = I_3 + \tfrac{1}{2}Y \tag{7.28}$$

where $Y = S + B$ is the quantity also known as the hypercharge. Equation (7.28) is known as the Gell-Mann–Nishijima formula.

The hypercharge is thus the mean charge of the particles constituting an isospin multiplet, and the strangeness is the amount by which the the hypercharge differs from the baryon number.

Table 7.2. *Quantum numbers of baryons and mesons*

Particle	Q	S	B	I	I_3	Y
p n	1 0	0	1	$\frac{1}{2}$	$\frac{1}{2}$ $-\frac{1}{2}$	1
Σ^+ Σ^0 Σ^-	1 0 −1	−1	1	1	1 0 −1	0
Λ^0	0	−1	1	0	0	0
Ξ^0 Ξ^-	0 −1	−2	1	$\frac{1}{2}$	$\frac{1}{2}$ $-\frac{1}{2}$	−1
π^+ π^0 π^-	1 0 −1	0	0	1	1 0 −1	0
K^+ K^0	1 0	1	0	$\frac{1}{2}$	$\frac{1}{2}$ $-\frac{1}{2}$	1
$\bar{K}^0$ K^-	0 −1	−1	0	$\frac{1}{2}$	$\frac{1}{2}$ $-\frac{1}{2}$	−1

7.5 Pion–nucleon scattering

The vector addition of isospins, and the conservation of total isospin, provide a basis for discussing the properties of the pion–nucleon system, and the processes of pion–proton scattering. The three scattering processes which are easily observed and need to be described, are elastic scattering:

$$\pi^+ + p \rightarrow \pi^+ + p \tag{7.29}$$

$$\pi^- + p \rightarrow \pi^- + p \tag{7.30}$$

and charge exchange scattering:

$$\pi^- + p \rightarrow \pi^0 + n \tag{7.31}$$

Without distinguishing between these three processes, the differential cross-sections were calculated in §4.8 by the method of helicity amplitudes. Our present task is to show how the differential cross-sections for the three processes are related to each other by the requirement for conservation of isospin.

The differential cross-section for any of the scattering processes may be written as

$$\frac{d\sigma}{d\Omega} = |f(\theta)|^2 \tag{7.32}$$

Here we have omitted the indices which were used to specify the helicity state in (4.98), and instead we specify the isospin state by writing

$$f(\theta) = (\psi(I_3^{N'}, I_3^{\pi'}), \mathcal{T}\psi(I_3^{N}, I_3^{\pi})) \tag{7.33}$$

The initial and final isospin states may be expressed in terms of states of definite total isospin $\phi(I, I_3)$ by means of Clebsch–Gordan coefficients.

The hypothesis that isospin is conserved by the interaction, which is expressed by

$$[I, \mathcal{T}] = 0$$

leads to the conclusions that (*a*) both I and I_3 are conserved, and (*b*) the transition matrix element is independent of I_3. The proof of this is exactly the same as in the case of angular momentum which was treated in §3.3.2 and in analogy to (3.61) we obtain

$$(\phi(I', I_3'), \mathcal{T}\phi(I, I_3)) = \delta_{I'I}\delta_{I_3'I_3}\mathcal{T}^I$$

The total I for the nucleon–pion system has two possible values, $\frac{3}{2}$ and $\frac{1}{2}$. To these values correspond two independent transition matrix elements, namely

$$M_1 = (\phi(\tfrac{1}{2}, I_3), \mathcal{T}\phi(\tfrac{1}{2}, I_3)) \tag{7.34}$$

$$M_3 = (\phi(\tfrac{3}{2}, I_3), \mathcal{T}\phi(\tfrac{3}{2}, I_3)) \tag{7.35}$$

The $p\pi^+$ elastic scattering process (7.29) can go only through the $I = \frac{3}{2}$ state, so we may put

$$\left(\frac{d\sigma}{d\Omega}\right)_{p\pi^+} = K|M_3|^2 \tag{7.36}$$

In the processes (7.30) and (7.31) involving a $p\pi^-$ initial state, both total isospin states can contribute, and we must consider the initial and final states as superpositions of $I = \frac{3}{2}$ and $I = \frac{1}{2}$ states in proportions given by the Clebsch–Gordan coefficients of table 3.2.

For $p\pi^-$ elastic scattering

$$\psi_{\text{initial}} = \psi_{\text{final}} = (\tfrac{1}{3})^{1/2}\phi(\tfrac{3}{2}, -\tfrac{1}{2}) + (\tfrac{2}{3})^{1/2}\phi(\tfrac{1}{2}, -\tfrac{1}{2}) \tag{7.37}$$

The differential cross-section is therefore

$$\left(\frac{d\sigma}{d\Omega}\right)_{p\pi^-\text{el}} = |(\psi_{\text{final}}, \mathcal{T}\psi_{\text{initial}})|^2 = K|\tfrac{1}{3}M_3 + \tfrac{2}{3}M_1|^2 \tag{7.38}$$

But for charge exchange scattering (7.30) the final state is a different superposition

$$\psi_{\text{final}} = (\tfrac{2}{3})^{1/2}\,\phi(\tfrac{3}{2},-\tfrac{1}{2}) - (\tfrac{1}{3})^{1/2}\,\phi(\tfrac{1}{2},-\tfrac{1}{2}) \tag{7.39}$$

With an initial state given by (7.38), this gives a differential cross-section

$$\left(\frac{d\sigma}{d\Omega}\right)_{\text{ch.ex.}} = K\,|\tfrac{1}{3}\cdot 2^{1/2}M_3 - \tfrac{1}{3}\cdot 2^{1/2}M_1|^2 \tag{7.40}$$

It follows from the above that the differential cross-sections for these processes are in the ratio

$$\left(\frac{d\sigma}{d\Omega}\right)_{p\pi^+} : \left(\frac{d\sigma}{d\Omega}\right)_{p\pi^- \text{el}} : \left(\frac{d\sigma}{d\Omega}\right)_{\text{ch.ex.}}$$

$$= |M_3|^2 : |\tfrac{1}{3}M_3 + \tfrac{2}{3}M_1|^2 : |\tfrac{1}{3}\cdot 2^{1/2}M_3 - \tfrac{1}{3}\cdot 2^{1/2}M_1|^2$$

$$= 9:1:2 \quad \text{if } M_3 \gg M_1$$

or

$$9:3:0 \quad \text{if } M_3 = M_1$$

or

$$9:1:8 \quad \text{if } M_3 = -M_1$$

or

$$0:2:1 \quad \text{if } M_3 \ll M_1$$

At a total energy around 1236 MeV, the observed ratio is close to 9 : 1 : 2, indicating that transitions are predominantly through the $I=\frac{3}{2}$ channel; this fact provides clear evidence that the resonance at this energy has isospin $\frac{3}{2}$. At higher energies contributions from both channels are important.

7.6 The isospin invariance group, $SU(2)$

In chapter 3 we explored the relation between angular momentum and rotational invariance. It is useful similarly to base the isospin concept on an invariance principle, or symmetry group.

Since isospin has been introduced by analogy with angular momentum, we might choose to regard the isospin operators I as associated with the group of rotations about the three axes in a three-dimensional isospin space different from but having the same structure as real space. However we choose instead to conceive of the symmetry transformation as acting in a two-dimensional complex space. In this section we shall give a precise definition of these transformations and their connection with the isospin formalism as we defined it in preceding sections. The reason for this apparently more abstract definition of the isospin symmetry group is one of mathematical convenience, but it will also facilitate the generalisation to higher symmetries such as $SU(3)$.

If we can imagine the electromagnetic interaction of particles to be 'switched off' then there is nothing to distinguish a proton from a neutron and the isospin symmetry becomes exact. We may express this by saying that instead of taking the one proton and one neutron states as the basic states of the nucleon we could equally well choose new linear combinations of these states, as basic states, for example

$$\left.\begin{aligned} |\mathrm{p}'\rangle &= a|\mathrm{p}\rangle + b|\mathrm{n}\rangle \\ |\mathrm{n}'\rangle &= -b^*|\mathrm{p}\rangle + a^*|\mathrm{n}\rangle \end{aligned}\right\} \tag{7.41}$$

where a and b are two arbitrary complex numbers. The coefficients in the second equation have been chosen so that the states $|\mathrm{n}'\rangle$ and $|\mathrm{p}'\rangle$ are orthogonal.

Thus

$$\langle \mathrm{p}'| = a^*\langle \mathrm{p}| + b^*\langle \mathrm{n}|$$

and hence

$$\begin{aligned} \langle \mathrm{p}'|\mathrm{n}'\rangle &= -a^*b^*\langle \mathrm{p}|\mathrm{p}\rangle + (a^*)^2\langle \mathrm{p}|\mathrm{n}\rangle - (b^*)^2\langle \mathrm{n}|\mathrm{p}\rangle + b^*a^*\langle \mathrm{n}|\mathrm{n}\rangle \\ &= 0 \end{aligned}$$

as a consequence of the normalisation and orthogonality of the original basis

$$\langle \mathrm{p}|\mathrm{p}\rangle = \langle \mathrm{n}|\mathrm{n}\rangle = 1$$

$$\langle \mathrm{p}|\mathrm{n}\rangle = \langle \mathrm{n}|\mathrm{p}\rangle = 0$$

Similarly by evaluating $\langle \mathrm{p}'|\mathrm{p}'\rangle$ we find that if we require a and b to satisfy

$$|a|^2 + |b|^2 = 1 \tag{7.42}$$

then the transformed states will be normalised

$$\langle \mathrm{p}'|\mathrm{p}'\rangle = \langle \mathrm{n}'|\mathrm{n}'\rangle = 1$$

Any pair of states defined by equations of the form (7.41) with coefficients satisfying (7.42) is a suitable pair of basis states in the limit of exact isospin symmetry. This suggests that the set of all such transformations is the symmetry group underlying isospin symmetry. If this is the case then, according to the general discussion of chapter 2, we should by a consideration of the infinitesimal symmetry transformations be able to derive the conserved Hermitian generators which in this case are the isospin operators I_1, I_2, and I_3.

Before doing this it is necessary to define the unitary operator U

which acts on $|p\rangle$ and $|n\rangle$ states to produce the transformation (7.41). U is labelled by the parameters a and b of the transformation, so that we have

$$U(a,b)|p\rangle = |p'\rangle = a|p\rangle + b|n\rangle$$

$$U(a,b)|n\rangle = |n'\rangle = -b^*|p\rangle + a^*|n\rangle$$

The identity isospin transformation which changes nothing is obtained by setting $a = 1, b = 0$, so that

$$U(1,0) = 1 = \text{unit operator}$$

An infinitesimal transformation is represented by

$$a = 1 - \tfrac{1}{2}i\epsilon_3, \quad b = \tfrac{1}{2}(\epsilon_2 - i\epsilon_1)$$

where ϵ_1, ϵ_2 and ϵ_3 are infinitesimal real parameters.

We first set $\epsilon_1 = \epsilon_2 = 0$, and consider the resulting infinitesimal transformation

$$U(\epsilon_3)|p\rangle = |p\rangle - \tfrac{1}{2}i\epsilon_3|p\rangle$$

$$U(\epsilon_3)|n\rangle = |n\rangle + \tfrac{1}{2}i\epsilon_3|n\rangle$$

The Hermitian generator G of a transformation is defined by

$$U(\epsilon_3) = 1 - i\epsilon_3 G$$

On substituting this into the preceding equation we find that G must act as follows

$$G|p\rangle = \tfrac{1}{2}|p\rangle$$

$$G|n\rangle = -\tfrac{1}{2}|n\rangle$$

i.e. $G = I_3$.

If instead we put $\epsilon_1 = \epsilon_3 = 0$

$$U(\epsilon_2) = 1 - i\epsilon_2 G$$

to generate the transformation

$$U(\epsilon_2)|p\rangle = |p'\rangle = |p\rangle + \tfrac{1}{2}\epsilon_2|n\rangle$$

$$U(\epsilon_2)|n\rangle = |n'\rangle = |n\rangle - \tfrac{1}{2}\epsilon_2|p\rangle$$

then we find

$$G|p\rangle = \tfrac{1}{2}i|n\rangle$$

$$G|n\rangle = -\tfrac{1}{2}i|p\rangle$$

i.e. $$G = I_2$$

Finally the generator of the transformation with ϵ_1 non-zero is found to be I_1.

We summarise what has been done in the language of group theory. For a deductive group theoretical presentation the reader is referred elsewhere. We have identified the isospin symmetry group with the set of all complex matrices

$$A = \begin{pmatrix} a & -b^* \\ b & a^* \end{pmatrix} \tag{7.43}$$

for which

$$|a|^2 + |b|^2 = 1$$

The successive application of two transformations of the type (7.41) corresponds to multiplication of the corresponding matrices (7.43). The matrices A are unitary,

$$A^\dagger A = 1$$

and have determinant equal to 1,

$$\det(A) = 1$$

The multiplicative group of such 2 x 2 matrices is denoted as $SU(2)$, in which U indicates unitary (|determinant| = 1), and S (special) indicates a limitation to determinant = +1.

CHAPTER 8

CHARGE CONJUGATION

8.1 Symmetry under charge conjugation

8.1.1 *The operator C*

Symmetry under space inversion (P) and under time reversal (T) having been considered in earlier chapters, we now turn to symmetry under charge conjugation (C). The operation C was originally conceived as one which would change the sign of electric charges and interchange positrons with electrons. It was later generalised to replace all particles by their antiparticles, changing the signs of all internal quantum numbers such as baryon number B, lepton number L, strangeness S and hypercharge Y as well as charge Q and isospin component I_3. We may describe the effect of charge conjugation on a state with momentum $\boldsymbol{p}$ and spin λ by the relation

$$U_C \phi_{Qp\lambda} = \phi_{-Qp\lambda} \tag{8.1}$$

where U_C is the unitary operator corresponding to C, and Q is used to denote all the additive internal quantum numbers.

Under charge conjugation, electrically charged systems transform into oppositely charged systems, and thus are not eigenstates of the operator U_C. In fact an eigenstate of U_C must have zero values of all the internal quantum numbers. For other types of state, C can be useful in relating the properties of a system to those of its charge conjugate system.

8.1.2 *Eigenstates of C*

Neutral mesons with zero strangeness can be eigenstates of U_C with eigenvalues $\eta_C = +1$ or -1. η_C is sometimes called 'charge conjugation parity', and ± 1 are the only permitted values, since operation with U_C twice in succession has to restore the initial state, i.e.

$$U_C^2 = 1$$

For a system with several components, each of which is an eigenstate of C, the overall symmetry under charge conjugation will be the product of the values of η_C for the component parts of the system, just as overall parity is given by multiplication of individual parities. We expect that strong and electromagnetic interactions will be invariant under charge conjugation, i.e. that U_C and the Hamiltonian H satisfy the relation

$$U_C H U_C^{-1} = H$$

This invariance requires that η_C is conserved in transitions between states which are eigenstates of C, and that a transition between states X and Y which are not eigenstates of C must have an amplitude equal to that of the charge conjugate transition between $\overline{X}$ and $\overline{Y}$.

The photon must have $\eta_C = -1$, since all components of the electromagnetic field change sign under charge conjugation, and the product field × current must be invariant under C. A system of n photons thus has eigenvalue $\eta_C = (-1)^n$. In particular this leads us to assign a value $\eta_C = +1$ to the neutral pion, which decays to two photons by strong or electromagnetic interaction. The importance of C-invariance is indicated by the fact that the decay $\pi^0 \rightarrow 3\gamma$ is not observed; on purely electromagnetic arguments this decay would have been expected with relative probability of 1/137.

8.1.3 *Positronium*

Just as an electron can move in an orbit around a proton in a hydrogen atom, an electron and a positron can move around their common centre of mass, in the system known as positronium. This system, having zero net charge and zero lepton number as well as zero baryon number and strangeness, can be an eigenstate of the operator U_C.

Let us consider a positronium atom, consisting of an electron and a positron with total spin S and orbital angular momentum l. Simultaneous interchange of the space, spin and charge coordinates is equivalent to interchange of the two particles, under which a system of two fermions must be antisymmetric. Interchange of the separate coordinates gives symmetry as follows:

$$\text{Space: } (-1)^l$$

$$\text{Spin: } (-1)^s$$

$$\text{Charge: } \eta_C$$

The product of these must be -1, so we conclude that

$$\eta_C = (-1)^{l+S}$$

The ultimate fate of a positronium atom is annihilation into two or more photons (one photon cannot conserve energy and momentum); a final state of n photons will have charge conjugation symmetry $\eta_C = (-1)^n$, so if η_C is to be conserved in this annihilation, two photon annihilation requires $l + S$ to be even. On the other hand, states with $l + S$ odd can annihilate only to states containing an odd number of photons.

The shortest-lived state of positronium is in fact $^1S_0$ which has $l + S = 0$ and decays by two photon annihilation. The next is $^3S_0$, with $l + S = 1$, which is observed to give three photons.

8.2 Tests of charge conjugation invariance

As with P- and T-invariances, we distinguish the experiments according to the kind of interaction whose invariance under charge conjugation is being tested.

8.2.1 *C in strong interactions*

A direct test of C-invariance would be to compare the differential cross-section, energy distribution, polarisation, etc., for a process

$$a + b \to c + d + e + \ldots$$

with the corresponding charge conjugate process

$$\bar{a} + \bar{b} \to \bar{c} + \bar{d} + \bar{e} + \ldots$$

However, there are very few systems for which both the initial states can be easily prepared. The only one which has been used is $\bar{p}p$ annihilation either at rest or in flight.

Let us consider for example annihilation in flight into $\pi^+\pi^-$ and $n - 2$ particles of definite kinds. C-invariance relates the CM amplitude for $\bar{p}p$ annihilation into a π^+ and $n - 1$ other particles to that for $p\bar{p}$ annihilation into π^- and $n - 1$ other particles (the charge conjugates of the previous $n - 1$). This is illustrated in fig. 8.1. The initial states differ by a spatial rotation through π, and rotational invariance is assumed, so that after squaring and summing over all variables save one, we have

$$\left(\frac{d\sigma}{d\cos\theta_+}\right)_{\theta_+=\theta} = \left(\frac{d\sigma}{d\cos\theta_-}\right)_{\theta_-=\pi-\theta}$$

The angles are measured with respect to the p-direction in both cases. Similarly for the differential energy spectra of π^+ and π^-,

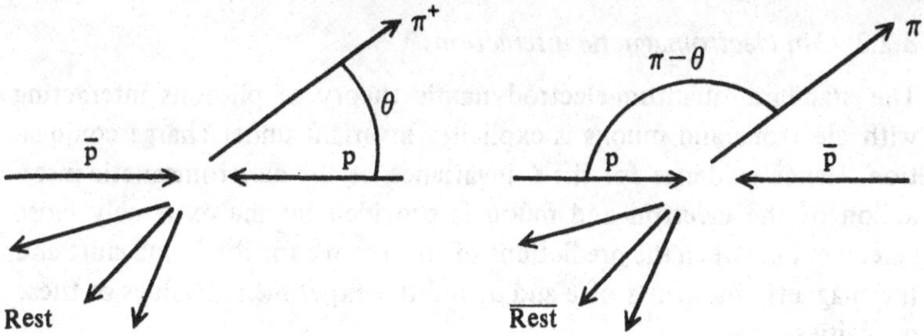

Fig. 8.1. Two p̄p annihilations in flight related by charge conjugation.

$$\frac{\mathrm{d}\sigma}{\mathrm{d}E_+} = \frac{\mathrm{d}\sigma}{\mathrm{d}E_-}$$

Relations such as these were given by Pais (1959). Similarly one can show that the angular distribution of a π^0 should be symmetrical about $\theta = \pi/2$.

Xuong, Lynch and Hinrichs (1961) and Maglić, Kalbfleisch and Stevenson (1961) studied annihilations in flight, and found no statistically significant evidence for deviations from these equations. Dobrzynski *et al.* (1966) in a study of several different annihilation channels estimated the relative *C*-violating amplitude to be of order 1 per cent.

In the case of annihilations at rest we may assume that the p̄ is captured from an atomic orbit, so that the initial state has definite charge conjugation parity η_C. Then *C*-invariance states that the amplitude for annihilation into, say, a π^+ and $n-1$ other mesons is equal to η_C times the amplitude for decay into π^- and $n-1$ mesons (charge conjugates of the preceding ones) and leads to the same predictions as before.

In the experiment of Baltay *et al.* (1965) only energy distributions were tested (single particle distributions and two particle invariant mass distributions). The effect of integrating over angles is to eliminate any interference terms which may result if there is coherence between initial states of opposite charge conjugation parity. Here it is assumed that capture takes place from s-states and so states with $\eta_C = +1$ and -1 have different total angular momentum ($^1S_0$ and $^3S_1$ respectively).

These workers examined separately purely pionic decays 34 811 events (dominant channel $\pi^+\pi^+\pi^-\pi^-\pi^0$) and kaonic decays 4663 events ($K^0K^+\pi^-\pi^0$ and $K^0K^-\pi^+\pi^0$). The ratio of *C*-violating amplitude to the *C*-conserving amplitude was estimated to be at most 1 per cent for pionic modes and 2 per cent for kaonic modes.

8.2.2 *C in electromagnetic interactions*

The standard quantum-electrodynamic theory of photons interacting with electrons and muons is explicitly invariant under charge conjugation. Hence evidence for the *C*-invariance of the electromagnetic interaction of the electron and muon is provided by the extremely close agreement between the predictions of this theory for the Lamb shift and the magnetic moments of e and μ, and the experimental values of these quantities.

Hadrons. A hadronic process which goes through electromagnetic interaction rather than by strong interaction is provided by the decay of the η^0 meson. This has $J^P = 0^-$, which forbids decay to 2π, and even *G*-parity; its mass is too low for decay to 4π, so the only available decay processes (see §9.5) are *G*-violating decay by electromagnetic interaction to $\pi^+\pi^-\gamma$ and to $\pi^+\pi^-\pi^0$. *C*-violation in one of these decay processes would lead to a difference between the energy distributions of the π^+ and the π^-.

Many experiments have been carried out to search for differences in these distributions. Of these, an early experiment at Columbia (Baltay *et al.*, 1966) seemed to indicate a slight excess of the mean π^+ energy over the mean π^- energy as measured in the rest-frame of the η^0. However, a CERN experiment (Cnops *et al.*, 1966) and further work at Columbia (Layter *et al.*, 1973) have found no further evidence for *C*-violation, the latest value of the charge asymmetry for $\eta \to 3\pi$ being quoted as $\alpha = -0.005 \pm 0.0022$, and for $\eta \to \pi\pi\gamma$ as $\alpha = 0.005 \pm 0.006$.

8.2.3 *C in weak interactions*

According to an analysis of Lee, Oehme and Yang (1957) the experiments of Wu *et al.* (1957), Garwin, Lederman and Weinrich (1957) and Friedman and Telegdi (1957) which demonstrated parity violation in β-decay and in muon decay, also show that *C* is violated in these decays. More direct evidence is provided by the leptonic pion decays

$$\pi^- \to \mu^- + \bar{\nu}_\mu \tag{8.2a}$$

$$\pi^+ \to \mu^+ + \nu_\mu \tag{8.2b}$$

In the π^--decay, the μ^- is found to be polarised in the direction of its momentum as described in §5.6.3. In the π^+-decay the μ^+ has its spin antiparallel to its momentum (the spin is measured in the pion rest-frame in either case). The comparison of these charge conjugate processes clearly shows the violation of charge conjugation invariance.

It is interesting to note, however, that these observations are compatible with invariance under the combined operation *CP* which in the case of the pion decays, (8.2), exchanges particle with antiparticle and changes the sign of the helicities of the leptons. In particular, for the picture of the neutrino distinguished from its antiparticle by its leptonic charge as described in §5.6.3, the operation *CP* takes a left-handed neutrino into a right-handed antineutrino and vice versa. The separate operations of *C* and *P* are undefined for physical neutrinos since they would require the existence of the right-handed neutrino and left-handed antineutrino.

Similarly non-leptonic decays such as those of the neutral kaons are invariant under *CP* to a high degree of accuracy; indeed *CP*-invariance was thought to be exact up to the time of the experiment of Christensen, Cronin, Fitch and Turley in 1964.

8.3 Invariance under *CP*

8.3.1 *The operator CP*

Having noted the occurrence of parity violation in decays involving the weak interaction, we may ask whether some more general symmetry is valid in these processes. Before the experiment of Christensen *et al.* (1964), it was thought that invariance under the combined operation *CP* held universally. Under *CP*-invariance, any parity violation observed with one system of particle is observed with reversed sense with the antiparticles of this system; inversion of space coordinates by the operation *P*, together with particle–antiparticle inversion by the charge conjugation operation *C*, give a new operation *CP*, with an associated unitary operation U_{CP} whose eigenvalue is conserved even when those of its component operators are not. The most familiar application of this idea was to the neutral kaons, to which it is now known to apply as an approximation valid to ± 2 per cent, as follows.

8.3.2 *Decay of neutral kaons*

The two neutral kaons, K^0 with strangeness + 1, and $\bar{K}^0$ with strangeness − 1, are antiparticles of each other, with wavefunctions which should be related by the operator U_C or U_{CP} but cannot themselves be eigenstates of U_C or U_{CP}. The systems of pions to which neutral kaons decay may be eigenstates of U_{CP}: if the decay processes involve weak interactions which may violate *P* while conserving (exactly or approximately) *CP*, we must look for possible initial states which are eigenstates of U_{CP}. These normalised initial states are

$$|K_1^0\rangle = 2^{-1/2}(|K^0\rangle - |\overline{K}^0\rangle) \qquad (8.3)$$

$$|K_2^0\rangle = 2^{-1/2}(|K^0\rangle + |\overline{K}^0\rangle) \qquad (8.4)$$

If we insert the conditions

$$U_{CP}|K^0\rangle = -|\overline{K}^0\rangle \qquad (8.5)$$

and

$$U_{CP}|\overline{K}^0\rangle = -|K^0\rangle \qquad (8.6)$$

we may check that K_1^0 and K_2^0 indeed represent eigenstates of *CP* with eigenvalues + 1 and − 1 respectively, so that

$$U_{CP}|K_1^0\rangle = +|K_1^0\rangle$$

and

$$U_{CP}|K_2^0\rangle = -|K_2^0\rangle$$

The minus sign in (8.5) and (8.6) arises in our convention from the negative parity of the kaons. In some treatments the arbitrary relative phase of the hypercharge eigenstates K^0 and $\overline{K}^0$ is chosen so as to absorb this sign with consequent changes in the definitions (8.3) and (8.4).

If we now consider a final state consisting of two pions ($\pi^+\pi^-$ or $2\pi^0$), we see that this must have eigenvalue of $CP = +1$. It can therefore be formed from K_1^0, but not from K_2^0. The rest-mass of the K^0 and $\overline{K}^0$ is enough to allow decay to either two or three pions. The available phase space is greater for decay to two pions, so we may expect the transition probability to be higher when the initial state is K_1^0, from which decay to two pions is allowed; when the initial state is K_2^0, decay to three pions is the only available channel, and the overall decay rate is correspondingly lower. The observed mean lives are in fact 0.86×10^{-10} s (K_1^0) and 5.3×10^{-8} s (K_2^0).

8.3.3 *The development of a neutral kaon beam*

Let us consider a beam of neutral kaons produced by the strangeness conserving strong interaction

$$\pi^- + p \rightarrow K^0 + \Lambda^0$$

At first the beam consists entirely of K^0, with $S = 1$; this is not an eigenstate of *CP*, but may be described as a mixture with equal amplitudes of the two *CP*-eigenstates, K_1^0 and K_2^0, since (8.3) and (8.4) may be combined to give

$$|K^0\rangle = 2^{-1/2}(|K_1^0\rangle + |K_2^0\rangle) \tag{8.7}$$

$$|\overline{K}^0\rangle = 2^{-1/2}(|K_2^0\rangle - |K_1^0\rangle) \tag{8.8}$$

As time passes, however, the K_1^0-component decays with a mean life about 10^{-10} s, leaving the relatively stable K_2^0. After a few mean lives of K_1^0, the residual beam is almost pure K_2^0, which is a fifty–fifty mixture of K^0 and $\overline{K}^0$. The strangeness violating decay thus leads to spontaneous conversion of either K^0 or $\overline{K}^0$ to a mixture in equal parts having the relative phase and mean life characteristic of K_2^0.

Conversely, we may pass a beam of K_2^0 through material in which strong interactions can occur selectively according to the strangeness of the incident particle; for example

$$\overline{K}^0 + n \rightarrow \pi^- + \Sigma^-$$

has a much higher cross-section than any reaction between K^0 and n. The result is to deplete the $S = -1$ component in the beam, leaving a mixture which must be described as K_2^0 with some K_1^0. Thus the effect of the material is to regenerate some of the short lived *CP*-eigenstate K_1^0. Downstream from the regenerator material, the K_1^0 will decay with mean life 10^{-10} s, leaving the K_2^0 to continue with reduced intensity.

The time evolution of the different amplitudes in a neutral kaon beam may be described by using different time dependences $e^{-i\lambda_1 t}$ and $e^{-i\lambda_2 t}$ for K_1^0, and K_2^0, where

$$\lambda_1 = m_1 - \tfrac{1}{2}i\gamma_1 \tag{8.9}$$

and

$$\lambda_2 = m_2 - \tfrac{1}{2}i\gamma_2 \tag{8.10}$$

γ_1 and γ_2 are the decay rates for K_1^0 and K_2^0, γ_1 being the larger. m_1 and m_2 are the ordinary masses of K_1^0 and K_2^0, with a difference $\Delta m = m_2 - m_1$ which will control any interference effects between K_1^0 and K_2^0.

After a proper time t, a beam which was initially K^0 will be represented by a wavefunction given by inserting time dependences in (8.7), to get

$$\psi(t) = 2^{-1/2}(|K_1^0\rangle e^{-i\lambda_1 t} + |K_2^0\rangle e^{-i\lambda_2 t}) \tag{8.11}$$

Breaking this up in terms of K^0 and $\overline{K}^0$, by means of (8.3) and (8.4) we get

$$\psi(t) = \tfrac{1}{2}\{|K^0\rangle(e^{-i\lambda_1 t} + e^{-i\lambda_2 t}) + |\overline{K}^0\rangle(e^{-i\lambda_2 t} - e^{-i\lambda_1 t})\}$$

The intensity of the $\overline{K}^0$-component at time t is therefore

$$\tfrac{1}{4}|e^{-i\lambda_2 t} - e^{-i\lambda_1 t}|^2 = \tfrac{1}{4}\{e^{-\gamma_1 t} + e^{-\gamma_2 t} - 2e^{-1/2(\gamma_1+\gamma_2)t}\cos(\Delta m t)\} \tag{8.12}$$

A corresponding calculation gives the intensity of K^0 as

$$\tfrac{1}{4}\{e^{-\gamma_1 t} + e^{-\gamma_2 t} + 2e^{-1/2(\gamma_1+\gamma_2)t}\cos(\Delta m t)\} \tag{8.13}$$

The main features of these time dependences are determined by the relative magnitudes of the mass difference Δm and the larger decay constant γ_1. Bubble chamber measurements have shown that Δm and γ_1 are approximately equal, observed values for $\Delta m/\gamma_1$ ranging from 0.6 to 1.9. For $\Delta m \sim \gamma_1$, the intensities of K^0 and $\bar{K}^0$ are shown as functions of time in fig. 8.2. For large Δm, the initial oscillations are very pronounced, while smaller Δm gives a damped approach to the fifty–fifty mixture, without oscillation.

As was mentioned earlier, the above theory assumes *CP*-invariance. The departures from it are discussed in §8.5.

8.4 *CPT*-invariance

The symmetry operation *CPT* consisting of the successive operations of time reversal, space inversion and charge conjugation has a significance over and above that of the three component symmetries. This is because of the *CPT theorem*, discovered by Pauli (1955) and Luders and Zumino (Luders, 1954). This theorem states that in a quantum field theory, any Hamiltonian which is invariant under the proper Lorentz transformations is necessarily also invariant under the combined operation *CPT*, whether or not it is invariant under *C*, *P* or *T* separately. In modern constructive quantum field theory as expounded for example in the book of Streater and Wightman (1964), the proof of the *CPT* theorem has been so refined that the only ingredients of the proof are relativistic invariance and the local field concept.

A proof of the theorem is beyond the scope of this book. We shall be content with an examination of some of its consequences, and of the experimental evidence for its validity.

In view of the general nature of the assumptions on which the *CPT* theorem is based, the predictions to which it leads are also general, i.e. not unexpected. Indeed it is often assumed that *CPT*-invariance is in the nature of an absolute symmetry principle, on the level of conservation of charge or baryon number. However, it can and should be tested. Since all dynamical models for particle processes are formulated with the aid of quantum field theory or some equivalent thereof,

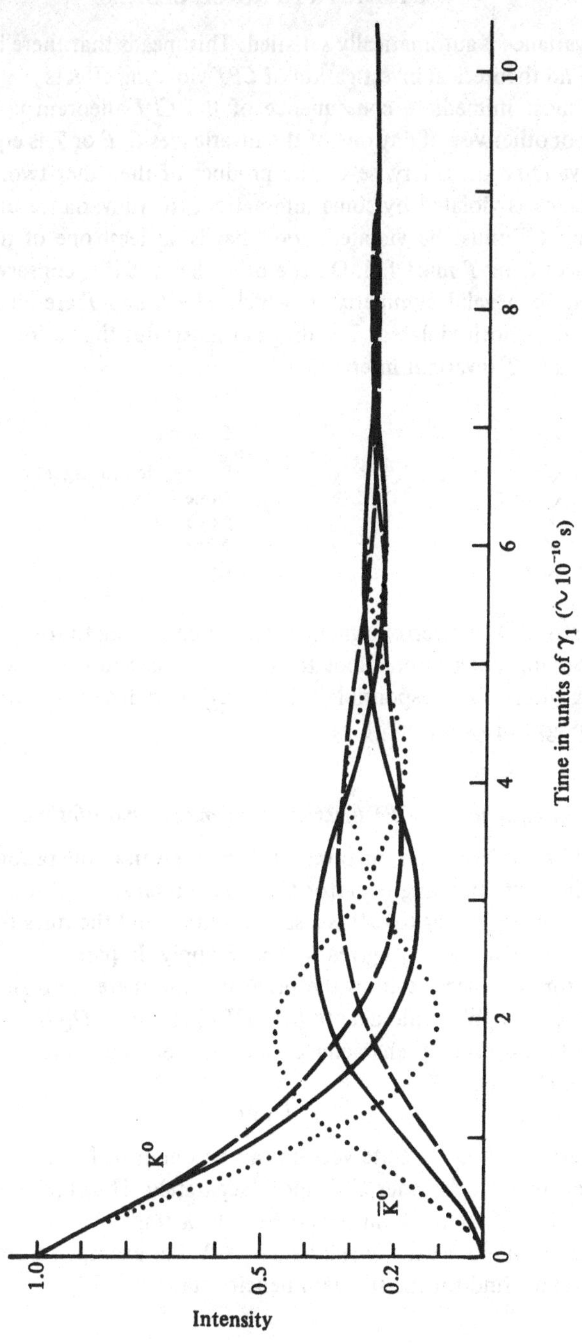

Fig. 8.2. Intensity of K^0 and $\bar{K}^0$, as functions of time, in a beam initially pure K^0. Full line: $\Delta m = \gamma_1$; broken line: $\Delta m = 0.75\,\gamma_1$; dotted line: $\Delta m = 1.5\,\gamma_1$.

CPT-invariance is automatically satisfied. This means that there has been little or no theoretical investigation of *CPT*-violating effects.

The most immediate consequence of the *CPT* theorem is that the validity or otherwise of any one of the invariances *C*, *P* or *T* is equivalent to the validity or otherwise of the product of the other two. Thus if *P*-invariance is violated by some interaction, then invariance under the operation *CT* must be violated, too: that is, at least one of the other invariances *C* or *T* must fail. On the other hand if *P* is conserved then *CT* must be a valid symmetry in which case *C* and *T* are either both conserved or both violated. The diagram illustrates the various possibilities for a *CPT*-invariant interaction.

P	*C*	*T*	Example
√	√	√	Strong, electromagnetic
√	×	×	None
×	×	√	β-decay
×	√	×	None
×	×	×	$K_L^0 \to 2\pi$

The strong and electromagnetic interactions belong to row 1 and the *CP*-conserving weak interactions to row 3. On present evidence the *CP*-violating interaction responsible for the $K_L^0 \to 2\pi$ decay conserves *CPT* and belongs to row 5 (see §8.5).

8.4.1 *Consequences of CPT-invariance for masses and lifetimes*

The *CPT*-transformation involves the time reversal operation *T* for which the corresponding operator O_T is antiunitary. It follows that the operator corresponding to *CPT* is also antiunitary and the rules for manipulation of such operators given in §6.2.3 apply. In particular although *CPT* performed twice restores the original state, there is no such thing as a *CPT*-parity. We could denote the *CPT*-operator by O_{CPT}. However, to indicate the primary character of the joint operation we shall use a special symbol, Θ.

$$\Theta = O_{CPT}$$

The effect of Θ on state vectors can be obtained from their transformation properties under *C*, *P* and *T* separately. Thus under Θ, a one particle state of momentum $\boldsymbol{p}$ and helicity λ transforms into an antiparticle state of momentum $\boldsymbol{p}$ and helicity $-\lambda$. For example, from (5.46) and (6.63) we find for the standard helicity state

$$\Theta\phi^a_{p00\lambda} = U_C U_P O_T \phi^a_{p00\lambda} = \eta_P(-1)^{s-\lambda}\phi^{\bar{a}}_{p00,-\lambda} \tag{8.14}$$

where a denotes the particle label, $\bar{a}$ the corresponding antiparticle with the momentum $\boldsymbol{p}$ is directed along the z-axis.

Invariance of the Hamiltonian H under CPT is expressed by

$$\Theta^{-1}H\Theta = H$$

If we take the matrix element of this equation between any two states ϕ_b and ϕ_a we have

$$(\phi_b, H\phi_a) = (\phi_b, \Theta^{-1}H\Theta\phi_a)$$
$$= (\Theta\phi_b, \Theta\Theta^{-1}H\Theta\phi_a)^*$$

by (6.30), and hence

$$(\phi_b, H\phi_a) = (\phi_{\bar{b}}, H\phi_{\bar{a}})^* \tag{8.15}$$

where $\phi_{\bar{a}}$ denotes the CPT-transform of ϕ_a. If we take for ϕ_a and ϕ_b the state of a single particle at rest, then the diagonal matrix element $(\phi_a, H\phi_a)$ is just the mass of the particle and (8.15) gives

$$m_a = m_{\bar{a}}$$

Thus invariance under CPT implies that the mass of particle and antiparticle are equal, irrespective of whether H is invariant under C alone. This result may be understood directly as follows.

Θ applied to a single particle a with spin projection λ *at rest* yields the antiparticle $\bar{a}$ with spin projection $-\lambda$, and by a rotation we can make the latter state into an antiparticle $\bar{a}$ with spin $+\lambda$. Thus Θ combined with a rotation has the same effect as C *on a single particle state* without assuming C-invariance. An unstable particle may be regarded as an eigenstate of H with a complex eigenvalue

$$m - \tfrac{1}{2}i\Gamma$$

where m is the mass and Γ is the total decay width, or the reciprocal of the lifetime. CPT-invariance in this case requires that both the mass and lifetime of the particle are equal to those of the antiparticle:

$$m_a = m_{\bar{a}}, \quad \Gamma_a = \Gamma_{\bar{a}}$$

(Luders and Zumino, 1957).

Some results on the measurements of particle and antiparticle lifetimes are given in table 8.1. All are consistent with CPT-invariance to within the sensitivity of order 10^{-3}.

The best evidence for CPT-invariance comes from the neutral kaon system. For the states $|K^0\rangle$ and $|\bar{K}^0\rangle$ related by charge conjugation, CPT-invariance requires that

Table 8.1. *Comparison of particle–antiparticle lifetimes as a test of CPT-invariance*

Particle	$\dfrac{\tau^+ - \tau^-}{\tau^-}$	Reference
$\mu^+\mu^-$	0.000 ± 0.001	Meyer *et al.* (1963)
K^+K^-	-0.0009 ± 0.0008	Lobkowicz *et al.* (1966)
$\pi^+\pi^-$	0.004 ± 0.0018	Lobkowicz *et al.* (1966)
$\pi^+\pi^-$	$0.000\,64 \pm 0.000\,69$	Ayres *et al.* (1968)

$$m(K^0) = m(\bar{K}^0)$$

where

$$m(K^0) = \langle K^0 | H | K^0 \rangle$$

$$m(\bar{K}^0) = \langle \bar{K}^0 | H | \bar{K}^0 \rangle$$

and

$$H = H_{st} + H_{em} + H_{wk}$$

$|K^0\rangle$ and $|\bar{K}^0\rangle$ are eigenstates of H_{st}, but not of the total Hamiltonian. The physical states $|K^0_S\rangle$ and $|K^0_L\rangle$ which are superpositions of $|K^0\rangle$ and $|\bar{K}^0\rangle$ have definite masses (and lifetimes). One may show that the difference $m(K^0) - m(\bar{K}^0)$ cannot be greater than $m(K_S) - m(K_L)$, the mass difference between the short lived and long lived neutral kaons. The latter is known to very great accuracy, measurements giving

$$\left| \frac{m(K_S) - m(K_L)}{m(K)} \right| \sim 10^{-14}$$

This gives an upper limit to the *CPT*-violation in the strong interaction. The simplest electromagnetic contributions to the mass involve emission and re-absorption of a virtual photon, and are therefore expected to be of order α relative to the strong interaction. A slightly stronger *CPT*-violating electromagnetic term could be present, of strength 10^{-12} relative to the *CPT* conserving H_{st}. By the same kind of argument we may conclude that the upper limit to the *CPT*-violation in the weak interaction is of order 10^{-7}. Only the $S = 0$ part of the weak interaction is being tested here. Although *CPT*-invariance implies that the total decay widths of particle and antiparticle are equal, it does not require equality of the particle width into charge conjugate channels, for example

$$n \to pe^-\bar{\nu}$$

and

$$\bar{n} \to \bar{p}e^+\nu$$

If, however, the decay is describable by the first-order approximation in which the transition operator $\mathcal{T}$ is Hermitian, the most important case being weak decays when $H_{wk} = \mathfrak{J}_{wk}$, then the considerations of §6.2.5 generalised to *CPT* enable one to show that

$$(\phi_b, \mathfrak{J}\phi_a) = (\Theta\phi_b, \mathfrak{J}\Theta\phi_a)^*$$

Since the effect of Θ is to turn particles into antiparticles, and reverse spins and helicities but not momenta, it follows that the squared amplitudes or probabilities satisfy

$$W_{a\to b}(p, S) = W_{\bar{a}\to\bar{b}}(p, -S)$$

On summing over spins we find that the decay distributions of the processes $a \to b$ and $\bar{a} \to \bar{b}$ are predicted to be the same. Consequently the partial widths for the processes are also equal (Lee, Oehme and Yang, 1957). Examples of pairs for which these results are applicable are

$$K^+ \to \mu^+ + \nu, \quad K^- \to \mu^- + \bar{\nu}$$

and

$$K^+ \to \pi^0 + \mu^+ + \nu, \quad K^- \to \pi^0 + \mu^- + \bar{\nu}$$

In the case of processes such as

$$\Lambda^0 \to p + \pi^- \quad \bar{\Lambda}^0 \to \bar{p} + \pi^+$$

or

$$K^\pm \to \pi^+ + \pi^- + l^\pm + \nu(\bar{\nu})$$

the strong final state interactions mean that the condition of a first-order process is not fulfilled. However the kind of analysis made in the final state theorem (§6.4.2) can be used to show that invariance under *CPT* leads to relations between the phases of two charge conjugate decay amplitudes and the phase shift for the final state rescattering. The application to the case of $K^\pm_{l4}$ decays and their relation to $\pi\pi$ rescattering in the final state is discussed in chapter 8.IV of Lee and Wu (1966).

8.4.2 *CPT-invariance and magnetic dipole moments*

By considering the effect of a *CPT*-transformation on the motion of a particle in a given electromagnetic field, it can be shown that the magnetic moments of particle and antiparticle are equal in magnitude. Similarly the corresponding electromagnetic form factors have the same functional form for both particle and antiparticle.

We shall consider only the magnetic moment equality for a spin-$\frac{1}{2}$ particle using the simple argument given by Grawert, Luders and Rollnik (1959).

We produce a magnetic field by means of a current loop and place the particle at the centre with its spin up, fig. 8.3. The *CPT*-transformation

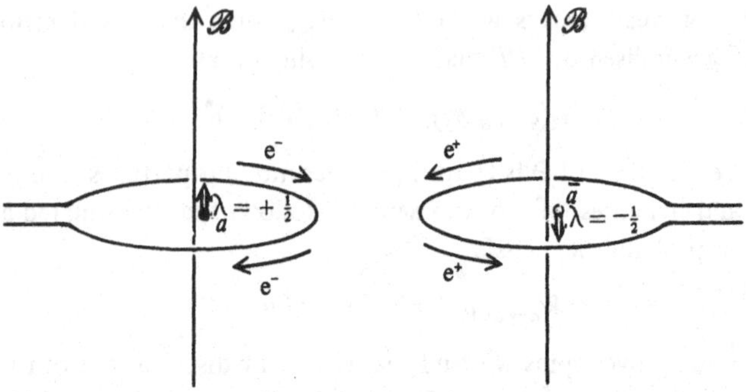

Fig. 8.3. Effect of the *CPT*-transformation on a particle in the field of a current-carrying coil.

is now applied to the whole system, coil and particle, taking the centre of the coil as the origin of space inversion. Then the electron current becomes a positron current in the opposite direction, so that the electric current and hence the magnetic field are unchanged. The particle a at rest with spin up becomes the antiparticle $\bar{a}$ at rest with spin down. *CPT*-invariance requires that the energy of both situations is the same. Hence the magnetic dipole moment must be the same in both cases. It follows that the magnetic dipole moment relative to the spin direction is equal and opposite for particle and antiparticle.

The magnetic moments of electrons and muons have been measured to considerable accuracy. The results are conveniently quoted in terms of the gyromagnetic ratio g.

From the results of Rich and Crane (1966) one finds that

$$g(e^+) - g(e^-) = (3 \pm 4)\alpha^2/\pi^2$$

where α is the fine structure constant. Experiments on muons in the CERN muon storage ring (Bailey *et al.*, 1968) show that

$$g(\mu^+) - g(\mu^-) = (1 \pm 1.5)\alpha^2/\pi^2$$

Thus the electromagnetic interactions of leptons appear to be *CPT*-invariant.

8.5 Violation of *CP*-invariance

The *CP*-invariant model of the behaviour of neutral kaons, as presented in §8.2, was accepted as exact until 1964. In that year, the experiment of Christensen *et al.* showed that a small proportion of the long lived neutral kaons decayed to two pions, violating the simple *CP*-invariant model. The amount of this anomalous decay (to $\pi^+\pi^-$) was 0.2 per cent of all decays to charged particles. Subsequent experiments have shown that there is a similar *CP*-violating decay to $2\pi^0$, and that the leptonic decays

$$K_L^0 \to \pi^- l^+ \nu \tag{8.16}$$

$$K_L^0 \to \pi^+ l^- \bar{\nu} \tag{8.17}$$

give even more striking evidence for *CP*-violation. The neutral particle K_L^0 decays preferentially by process (8.16) into positive leptons.

CP-invariance for this system implies that (*a*) the K_L^0 of definite lifetime is the same as the *CP*-eigenstate

$$|K_2^0\rangle = 2^{-1/2}(|K^0\rangle + |\bar{K}^0\rangle)$$

and (*b*) relations hold between the K^0- and $\bar{K}^0$-decay amplitudes. Denoting the final state by its charged lepton for brevity, *CP*-invariance requires

$$\langle l^+, p|\mathcal{T}|K^0\rangle = -\langle l^-, -p|\mathcal{T}|\bar{K}^0\rangle$$

$$\langle l^-, p|\mathcal{T}|K^0\rangle = -\langle l^+, -p|\mathcal{T}|\bar{K}^0\rangle$$

From (*a*) and (*b*) we conclude that the rates for (8.16) and (8.17) should be equal.

Experimental studies of these relatively rare modes show clear evidence for a small but non-zero charge asymmetry measured by

$$\delta = \frac{\Gamma(l^+) - \Gamma(l^-)}{\Gamma(l^+) + \Gamma(l^-)}$$

The current average value given by the Particle Data Group (1974) is

$$\delta = (3.26 \pm 0.1) \times 10^{-3}$$

Attempts to describe and explain these observations have used many different techniques and have had differing degrees of success (see e.g. Kabir, 1968), but no single explanation stands out uniquely. Recently the description in terms of the superweak interaction (Wolfenstein 1964, 1966; Lee and Wolfenstein, 1965) has been increasingly favoured.

It is usual to describe the situation in terms of short lived and long lived state K_S^0 and K_L^0, which are roughly but not exactly identical with the *CP*-eigenstates K_1^0 and K_2^0. The observations to be explained are then the last two entries in table 8.2, expressed as fractions of the *CP*-conserving decay constants for either K_L^0 or K_S^0.

Table 8.2. *Decay of neutral kaons, listed in order of frequency*

Initial state	Decay channel	Decay constant (s^{-1})	Comment
K_S^0	$\pi^+\pi^-$	0.799×10^{10}	*CP*-conserving hadronic decays, rapid
	$\pi^0\pi^0$	0.361×10^{10}	
K_L^0	$\pi e \nu$	7.54×10^6	Leptonic decays
	$\pi \mu \nu$	5.18×10^6	
	$\pi^0\pi^0\pi^0$	4.13×10^6	*CP*-conserving hadronic decays
	$\pi^+\pi^-\pi^0$	2.43×10^6	
	$\pi^+\pi^-$	3.03×10^4	*CP*-violating hadronic decays
	$\pi^0\pi^0$	1.81×10^4	

8.5.1 *The decay formalism*

To accommodate the *CP*-violating terms, the time dependence of a neutral kaon beam may be described as follows: instead of the two simple time dependences $e^{-i\lambda t}$ of §8.3.3, we assume a time dependence given by

$$i\frac{d}{dt}\begin{pmatrix}a_1\\a_2\end{pmatrix} = \begin{pmatrix}\Lambda_{11} & \Lambda_{12}\\ \Lambda_{21} & \Lambda_{22}\end{pmatrix}\begin{pmatrix}a_1\\a_2\end{pmatrix} \tag{8.18}$$

where a_1 and a_2 are the coefficients defining a general state,

$$|\psi(t)\rangle = a_1(t)\,|K^0\rangle + a_2(t)\,|\bar{K}^0\rangle \tag{8.19}$$

In general, the time dependence (8.18) gives a changing value of the ratio a_1/a_2 which defines the relative proportions of K^0 and $\bar{K}^0$ in the beam. Algebraic manipulation shows, however, that there are two values of a_1/a_2 for which the passage of time involves no change in the beam composition, but only exponential decay of its intensity. These are the two states of the form (8.19) for which $\begin{pmatrix}a_1\\a_2\end{pmatrix}$ is an eigenvector of the complex mass matrix $\mathbf{\Lambda}$, and they correspond to the short and long lived kaons K_S and K_L.

We denote the eigenvalues by

$$\lambda_L = m_L - \tfrac{1}{2}i\gamma_L, \quad \lambda_S = m_S - \tfrac{1}{2}i\gamma_S \tag{8.20}$$

and write the eigenstates as follows

$$\left.\begin{aligned} |K_L\rangle &= [2(1+|\epsilon_L|^2)]^{-1/2}[(1+\epsilon_L)|K^0\rangle + (1-\epsilon_L)|\bar{K}^0\rangle] \\ |K_S\rangle &= [2(1+|\epsilon_S|^2)]^{-1/2}[(1+\epsilon_S)|K^0\rangle - (1-\epsilon_S)|\bar{K}^0\rangle] \end{aligned}\right\} \tag{8.21}$$

The complex parameters ϵ_L and ϵ_S have been defined so that ϵ_L measures the amplitude of K_1^0 relative to K_2^0 in K_L, while ϵ_S is the amplitude of K_2^0 relative to K_1^0 in K_S. To the extent that *CP* is approximately conserved, ϵ_L and ϵ_S are expected to be small.

It is useful to define

$$\left.\begin{aligned} \epsilon &= \tfrac{1}{2}(\epsilon_S + \epsilon_L) \\ \text{and} \quad \delta &= \tfrac{1}{2}(\epsilon_S - \epsilon_L) \end{aligned}\right\} \tag{8.22}$$

In an interval of time t, K_L and K_S undergo exponential decay according to

$$|K_L\rangle \to e^{(-im_L t - 1/2\gamma_L t)}|K_L\rangle \tag{8.23a}$$

$$|K_S\rangle \to e^{(-im_S t - 1/2\gamma_S t)}|K_S\rangle \tag{8.23b}$$

The relation between the parameters ϵ_L and ϵ_S and the elements of $\boldsymbol{\Lambda}$ may be obtained most easily by writing the general state of the neutral kaon both as a superposition of K^0 and $\bar{K}^0$ and of K_L and K_S, and considering its change in a small time δt as expressed by (8.18) and (8.23) (Bell and Steinberger, 1966; Steinberger, 1970). After straightforward algebra one obtains

$$\left.\begin{aligned} \Lambda_{11} &= \tfrac{1}{2}(\lambda_L + \lambda_S) + \frac{\epsilon_L - \epsilon_S}{2(1 - \epsilon_L\epsilon_S)}(\lambda_L - \lambda_S) \\ \Lambda_{22} &= \tfrac{1}{2}(\lambda_L + \lambda_S) - \frac{\epsilon_L - \epsilon_S}{2(1 - \epsilon_L\epsilon_S)}(\lambda_L - \lambda_S) \\ \Lambda_{12} &= \frac{(1 + \epsilon_L)(1 + \epsilon_S)}{2(1 - \epsilon_L\epsilon_S)}(\lambda_L - \lambda_S) \\ \Lambda_{21} &= \frac{(1 - \epsilon_L)(1 - \epsilon_S)}{2(1 - \epsilon_L\epsilon_S)}(\lambda_L - \lambda_S) \end{aligned}\right\} \tag{8.24}$$

We wrote (8.18) down as a phenomenological equation, but it can be derived as a partial solution to the time dependent Schrödinger equations for the system consisting of $|K^0\rangle|\bar{K}^0\rangle$ and all those states $|\alpha\rangle$ into which K^0 and $\bar{K}^0$ can make transitions (real or virtual) under the influence of the total Hamiltonian

$$H = H_{st} + H_{em} + H_{wk}$$

This is the Weisskopf–Wigner approximation which has been described many times (see e.g. Kabir, 1968, appendix A). From it one obtains expressions for the elements of $\boldsymbol{\Lambda}$ in terms of the decay amplitudes as follows. We write the 2×2 matrix in terms of a Hermitian mass matrix M and a Hermitian decay matrix $\boldsymbol{\Gamma}$,

$$\Lambda = M - \tfrac{1}{2}\mathrm{i}\Gamma$$

Then the matrix elements of Γ are determined by the real (or on-energy-shell) transition matrix elements of H_{wk} by the familiar formulae of the Golden rule,

$$\left.\begin{aligned}\Gamma_{11} &= 2\pi \int \mathrm{d}\alpha\rho(E_\alpha)\,|\langle\alpha|H_{\mathrm{wk}}|\mathrm{K}^0\rangle|^2\delta(E_\alpha - m_0)\\ \Gamma_{22} &= 2\pi \int \mathrm{d}\alpha\rho(E_\alpha)\,|\langle\alpha|H_{\mathrm{wk}}|\overline{\mathrm{K}}^0\rangle|^2\delta(E_\alpha - m_0)\end{aligned}\right\} \tag{8.25}$$

and by

$$\Gamma_{21} = \Gamma_{12}^* = 2\pi \int \mathrm{d}\alpha\rho(E_\alpha)\langle\overline{\mathrm{K}}^0|H_{\mathrm{wk}}|\alpha\rangle\langle\alpha|H_{\mathrm{wk}}|\mathrm{K}^0\rangle\delta(E_\alpha - m_0)$$

and the matrix elements of M depend on the virtual (or off-energy-shell) transitions by formulae reminiscent of second-order perturbation theory,

$$\left.\begin{aligned}M_{11} &= m_0 + \langle\mathrm{K}^0|H_{\mathrm{wk}}|\mathrm{K}^0\rangle + P\int \mathrm{d}\alpha\rho(E_\alpha)\frac{|\langle\alpha|H_{\mathrm{wk}}|\mathrm{K}^0\rangle|^2}{m_0 - E_\alpha}\\ M_{22} &= m_0 + \langle\overline{\mathrm{K}}^0|H_{\mathrm{wk}}|\overline{\mathrm{K}}^0\rangle + P\int \mathrm{d}\alpha\rho(E_\alpha)\frac{|\langle\alpha|H_{\mathrm{wk}}|\overline{\mathrm{K}}^0\rangle|^2}{m_0 - E_\alpha}\end{aligned}\right\} \tag{8.26}$$

$$M_{12} = M_{21}^* = \langle\mathrm{K}^0|H_{\mathrm{wk}}|\overline{\mathrm{K}}^0\rangle + P\int \mathrm{d}\alpha\rho(E_\alpha)\frac{\langle\mathrm{K}^0|H_{\mathrm{wk}}|\alpha\rangle\langle\alpha|H_{\mathrm{wk}}|\overline{\mathrm{K}}^0\rangle}{m_0 - E_\alpha} \tag{8.27}$$

Here m_0 denotes the mass of the K^0 and $\overline{\mathrm{K}}^0$, assumed to be the same in the absence of H_{wk}. $\int \mathrm{d}\alpha$ denotes the integral over all eigenstates of $H_{\mathrm{st}} + H_{\mathrm{em}}$ and $\rho(E_\alpha)$ denotes the density of states. P denotes the principal value, i.e. states with energy $E_\alpha = m_0$ are excluded.

These expressions are a convenient starting point for the discussion of the consequences of the *CPT*-, *T*- and *CP*-symmetries.

8.5.2 *CPT- and T-invariance*

The assumption of validity of the symmetries *CPT*, *T* or *CP* leads to relations between the elements of the mass matrix Λ. We shall now derive these.

CPT-invariance. According to (8.14) the *CPT* operator Θ acts on a state of a neutral kaon of spin zero and negative parity at rest as follows

$$\Theta|K^0\rangle = -|\bar{K}^0\rangle$$

$$\Theta|\bar{K}^0\rangle = -|K^0\rangle$$

CPT-invariance of H_{wk}, (8.15), then gives

$$\langle\alpha|H_{wk}|K^0\rangle = -\langle\bar{K}^0|H_{wk}|\alpha^{\Theta}\rangle$$

$$\langle\alpha|H_{wk}|\bar{K}^0\rangle = -\langle K^0|H_{wk}|\alpha^{\Theta}\rangle$$

where α^{Θ} denotes the *CPT*-transformed state of α. We then find from (8.25) and (8.26) that

$$M_{11} = M_{22}, \quad \Gamma_{11} = \Gamma_{22} \quad (CPT) \tag{8.28}$$

and from (8.24) that

$$\epsilon_L = \epsilon_S \quad \text{or} \quad \delta = 0 \quad (CPT) \tag{8.29}$$

Thus if *CPT*-invariance holds, the admixture of K_1^0 in K_L is equal to the admixture of K_2^0 in K_S. It should be noted that $|\alpha\rangle$ represents final states such as $\pi\pi$, $\pi\pi\pi$, $\pi\mu\nu$, and so on, and when final state interactions are present $|\alpha^{\Theta}\rangle$ is different from $|\alpha\rangle$. However, if the set of $|\alpha\rangle$ is complete, then so is the set of $|\alpha^{\Theta}\rangle$ and this suffices for the derivation of (8.28). A similar remark applies in the case of *T*-invariance.

T-invariance. T-invariance of the Hamiltonian leads to

$$\langle\alpha|H_{wk}|K^0\rangle = \langle K^0|H_{wk}|\alpha^T\rangle$$

$$\langle\alpha|H_{wk}|\bar{K}^0\rangle = \langle\bar{K}^0|H_{wk}|\alpha^T\rangle$$

From (8.25) and (8.26) we find that

$$M_{21} = M_{12}, \quad \Gamma_{21} = \Gamma_{12} \quad (T) \tag{8.30}$$

which when combined with the Hermitian property, gives

$$M_{21} = M_{12} = \text{real}, \quad \Gamma_{21} = \Gamma_{12} = \text{real} \quad (T) \tag{8.31}$$

It follows that

$$\Lambda_{21} = \Lambda_{12}$$

Equation (8.24) in turn gives

$$\epsilon = 0 \quad (T) \tag{8.32}$$

Finally consider the case in which *CPT*, *T* and *CP* are conserved. This combines the consequences of (*a*) and (*b*) so we have both the conditions $\delta = 0$ and $\epsilon = 0$ or $\epsilon_L = 0$ and $\epsilon_S = 0$, and then by (8.24)

$$\Lambda_{11} = \Lambda_{22} = \tfrac{1}{2}(\lambda_L + \lambda_S)$$

$$\Lambda_{12} = \Lambda_{21} = \tfrac{1}{2}(\lambda_L - \lambda_S)$$

so the eigenstates are

$$|K_L\rangle \to 2^{-1/2}[|K^0\rangle + |\bar{K}^0\rangle] = |K_2^0\rangle$$

$$|K_S\rangle \to 2^{-1/2}[|K^0\rangle - |\bar{K}^0\rangle] = |K_1^0\rangle$$

and as expected the treatment of §8.3.3 is recovered.

8.5.3 *Tests of CPT and T in* K^0 *decay*

We do not have space here to make even a partial survey of the wealth of information on the neutral kaons. However, sufficient data are being accumulated for independent determinations of the parameters ϵ and δ (see for example, Schubert *et al.*, 1970). Starting from the Bell–Steinberger (1966) unitarity relation

$$[i(m_L - m_S) + \tfrac{1}{2}(\gamma_L + \gamma_S)]\,\langle K_S^0 | K_L^0\rangle$$
$$= 2\pi \int d\alpha \rho(E_\alpha) \langle \alpha | \mathcal{T} | K_S\rangle^* \langle \alpha | \mathcal{T} | K_L\rangle$$

inserting experimental values for $m_L - m_S$ and $\gamma_S + \gamma_L$, and by considering the experimental data for the various final states on the right, the overlap $\langle K_S | K_L\rangle$ can be evaluated. From (8.21) this is given by

$$\langle K_S | K_L\rangle = 2(\text{Re}\,(\epsilon) - i\,\text{Im}\,(\delta))$$

with the neglect of $|\epsilon_L|^2$ and $|\epsilon_S|^2$ which will be justified by the results. Thus Re (ϵ) and Im (δ) can be extracted. From the data on $K_L \to 2\pi$ decays, following the analysis of Wu and Yang (1964), Im (ϵ) and Re $(\bar{\delta})$ can be obtained.

$\bar{\delta}$ differs from δ only by a quantity relating to the *CPT*-violating part of the $K \to 2\pi$ decay, so $\bar{\delta}$ like δ should vanish if *CPT* is conserved. It was found by Schubert *et al.* (1970) that

$$\text{Re}\,(\epsilon) = (1.68 \pm 0.30) \times 10^{-3}$$

$$\text{Im}\,(\epsilon) = (1.45 \pm 0.30) \times 10^{-3}$$

$$\text{Re}\,(\bar{\delta}) = (0.07 \pm 0.43) \times 10^{-3}$$

$$\text{Im}\,(\bar{\delta}) = (-0.30 \pm 0.45) \times 10^{-3}$$

Thus the conclusion is that in kaon decays, *T* is violated along with *CP*, but that *CPT* remains valid.

CHAPTER 9

HADRONIC DECAYS OF MESONS

9.1 *G*-parity

In discussing the charge conjugation operator C, we mentioned that eigenstates of C had to have zero total charge, as well as zero strangeness and baryon number. Thus we can list values of the eigenvalue η_c among the observable properties of the non-strange neutral mesons.

The quantum number η_C can in fact be generalised to apply to isospin states provided their other quantum numbers such as baryon number B and strangeness S are zero. Examples of such states are non-strange mesons and baryon–antibaryon pairs. The resulting quantity is called *G-parity*, introduced by Michel (1953) and by Lee and Yang (1956*b*). Isotopic parity, as it was called by Michel, would be a more descriptive name.

To see how the generalisation can be made, we recall that the effect of charge conjugation on the pion states is

$$\left.\begin{aligned} U_C|\pi^+\rangle &= |\pi^-\rangle \\ U_C|\pi^0\rangle &= |\pi^0\rangle \\ U_C|\pi^-\rangle &= |\pi^+\rangle \end{aligned}\right\} \qquad (9.1)$$

Thus U_C changes the I_3 value of a state. We require a compensating isospin rotation which has the same effect. A rotation by π about the second axis has the desired effect. It is written (see appendix C.2 for a detailed discussion) as

$$P_i = e^{-i\pi I_2} \qquad (9.2)$$

The effect of a finite isospin transformation on the states of any isospin multiplet can be deduced from the analogy with angular momentum, viz. (3.69), provided that the same phase conventions are used in the two cases.

Then, from (3.68)

$$e^{-i\pi I_2}|I,I_3\rangle = (-1)^{I-I_3}|I,-I_3\rangle \qquad (9.3)$$

where the value of $d^I_{I_3' I_3}(\pi)$ has been substituted from (3.86*b*).

As explained in appendix C, the correct choice for the pion triplet is

$$|I = 1, I_3 = +1\rangle = -|\pi^+\rangle, \quad |1,0\rangle = |\pi^0\rangle, \quad |1,-1\rangle = |\pi^-\rangle$$

So we obtain

$$\left.\begin{aligned} e^{-i\pi I_2}|\pi^+\rangle &= -|\pi^-\rangle \\ e^{-i\pi I_2}|\pi^0\rangle &= -|\pi^0\rangle \\ e^{-i\pi I_2}|\pi^-\rangle &= -|\pi^+\rangle \end{aligned}\right\} \tag{9.4}$$

We now define the operator of G-parity as

$$U_G = e^{-i\pi I_2}U_C = U_C e^{-i\pi I_2} \tag{9.5}$$

which from (9.1) and (9.4) acts on one pion states as follows

$$\left.\begin{aligned} U_G|\pi^+\rangle &= -|\pi^+\rangle \\ U_G|\pi^0\rangle &= -|\pi^0\rangle \\ U_G|\pi^-\rangle &= -|\pi^-\rangle \end{aligned}\right\} \tag{9.6}$$

Thus, all three charge states of the pion are eigenstates of U_G; the corresponding eigenvalue is called the *G-parity*, $\eta_G = -1$.

The fact that G-parity is the same for different members of an isomultiplet suggests that U_G commutes with the isospin operators:

$$[U_G, \boldsymbol{I}] = 0$$

That this is indeed the case is proved in appendix C.

To sum up: for a system which satisfies isospin and charge conjugation invariance, the charge conjugation operator U_C which does not commute with all isospin transformations can be replaced by U_G which does.

Turning to the nucleon doublet, we recall that U_C transforms nucleons into antinucleons, while, from (9.3) and (3.86*b*), we have

$$e^{-i\pi I_2}|p\rangle = |n\rangle$$

$$e^{-i\pi I_2}|n\rangle = -|p\rangle$$

Hence

$$U_G|p\rangle = |\bar{n}\rangle \tag{9.7a}$$

$$U_G|n\rangle = -|\bar{p}\rangle \tag{9.7b}$$

Although nucleons are not eigenstates of U_G a nucleon–antinucleon pair can be, because it has zero baryon number and strangeness. This is considered in the next section, and as a preliminary step we consider the effect of U_G on antinucleon states.

The standard isospin states for antinucleons are (appendix C)

$$|\bar{N}, \tfrac{1}{2} + \tfrac{1}{2}\rangle = -|\bar{n}\rangle, \quad |\bar{N}, \tfrac{1}{2} - \tfrac{1}{2}\rangle = |\bar{p}\rangle$$

and so from (9.3) we have

$$e^{-i\pi I_2}|\bar{n}\rangle = -|\bar{p}\rangle$$

$$e^{-i\pi I_2}|\bar{p}\rangle = |\bar{n}\rangle$$

and therefore

$$U_G|\bar{n}\rangle = -|p\rangle \tag{9.8a}$$

$$U_G|\bar{p}\rangle = |n\rangle \tag{9.8b}$$

A similar analysis holds for the kaon system

$$\left.\begin{aligned} U_G|K^+\rangle &= |\bar{K}^0\rangle \\ U_G|K^0\rangle &= -|K^-\rangle \\ U_G|\bar{K}^0\rangle &= -|K^+\rangle \\ U_G|K^-\rangle &= |K^0\rangle \end{aligned}\right\} \tag{9.9}$$

We note that it follows from (9.8) and (9.7) that

$$U_G^2|x\rangle = -|x\rangle$$

for $x = $ p, n, $\bar{p}$ or $\bar{n}$ and similarly for $K^\pm$, K^0 or $\bar{K}^0$. On the other hand, for the pions

$$U_G^2|\pi^c\rangle = +|\pi^c\rangle \quad c = +1, 0 \text{ or } -1$$

These are special cases of

$$U_G^2 = (-1)^{2I} \tag{9.10}$$

which follows from (9.5). Equation (9.10) is sometimes written

$$U_G^2 = (-1)^Y$$

because of the Gell-Mann–Nishijima formula and the fact that all known particles have integral charges.

The chief importance of G-parity lies in the fact that a system of n pions has

$$\eta_G = (-1)^n$$

Mesons of high mass may be given G-parity $+1$ or -1, according to whether they decay through strong interaction to an even or an odd number of pions. G-parity is assumed to be conserved by the strong interactions, but not by the weak or electromagnetic interactions, for which isospin is not conserved and the rotation operator $e^{-i\pi I_2}$ is without clear significance.

9.2 Generalised Pauli principle

As quoted in §7.2, this required that the total wavefunction of a pair of fermions must be antisymmetric under simultaneous exchange of space, spin and charge coordinates. The validity of this followed from the fact that simultaneous interchange of these three types of coordinates was equivalent to interchange of the identities of the two particles.

We may now extend this principle to cover the case of a fermion–antifermion pair, for example $p+\bar{n}$, $p+\bar{p}$, $n+\bar{n}$ or $n+\bar{p}$. Antisymmetry under exchange of the two particles is now equivalent to antisymmetry under simultaneous interchange of space coordinates, spin coordinates, I_3 and baryon number B. The symmetry under interchange of baryon number is given by the eigenvalue η_G of the G-parity operator G, since the latter gives change of B without change of I. Interchange of isospin components I_3 gives symmetry $(-)^{I+1}$ while interchange of spin coordinates gives symmetry $(-)^{S+1}$, I and S being the total isospin and spin respectively. Under interchange of space coordinates, which is equivalent to inversion about the origin, orbital angular momentum l gives symmetry $(-)^l$, while the opposite intrinsic parities of the nucleon and antinucleon gives a further factor $(-)$. The overall symmetry is therefore

$$\eta_G(-)^{l+S+I+1}$$

and this is the quantity which is required to be $(-)$ by the generalised Pauli principle.

We conclude that for a nucleon–antinucleon pair, η_G must be given by

$$\eta_G = (-)^{l+S+I} \tag{9.11}$$

Mesons which satisfy (9.11) may be considered as having the possibility of virtual transitions into nucleon–antinucleon pairs, and also to quark–antiquark pairs (see chapter 11) since the statistics of a $q\bar{q}$ pair are similar to those of a $N\bar{N}$ pair.

9.3 Normal and abnormal *C*

In the case of neutral mesons which satisfy the condition (9.11), we may consider also the symmetry under charge conjugation, given by

$$\eta_G = \eta_C(-)^I$$

With (9.11), this gives

$$\eta_C = (-)^{l+S} \tag{9.12}$$

whence the symmetry under the operation *CP* is

$$\eta_{CP} = (-)^{S+1} \tag{9.13}$$

Neutral mesons with quantum numbers satisfying (9.12) and (9.13) are sometimes described as having 'normal *C*', meaning behaviour compatible with that of a q$\bar{\text{q}}$ pair or an N$\bar{\text{N}}$ pair.

If we consider also the total angular momentum *J*, given by the vector sum

$$J = l + S$$

we see that 'normal *C*' implies the list of possible combinations of quantum numbers given in (table 9.1).

Table 9.1. *Mesons with 'Normal C', i.e. which can be made from* q$\bar{\text{q}}$

S	l	J	$\eta_C = (-)^{l+S}$	$\eta_P = -(-)^l$	$\eta_{CP} = -(-)^S$	I^G	J^P
0	0	0	+	−	−	0^+ or 1^-	0^- (A)
1	0	1	−	−	+	0^- or 1^+	1^- (N)
0	1	1	−	+	−	0^- or 1^+	1^+ (A)
1	1	0, 1, 2	+	+	+	0^+ or 1^-	0^+ (N), 1^+ (A), 2^+ (N)

The right-hand columns in the table give the quantum numbers in the conventional short-hand notation, I^G showing I and η_G, and J^P showing J and η_P. Listed with J^P is a letter N or A indicating normal or abnormal parity. This is an idea which must not be confused with the normal or abnormal *C* discussed above. It simply reflects the fact that when angular momentum is all orbital as in ordinary spherical harmonics, the values $J^P = 0^+, 1^-, 2^+$ etc. are allowed.

In fact all known mesons and meson resonances are found to have normal *C*, as shown in table 9.2.

9.4 Final states

In the decay of a non-strange meson to a final state consisting of n pions, the strongest selection rule, after conservation of charge, energy and total angular momentum, will normally be conservation of parity. For a state of two pions, there is a single orbital angular momentum l, and the overall parity is $(-)^l$. For three pions, the total angular momentum j may be the vector sum of the orbital angular momentum

Table 9.2. *Mesons with $S = 0$. Principal properties and decay modes*

Name	Mass (MeV)	Full width (MeV)	$I^G J^P C$ (neutral)	Decay mode	Branching ratio (per cent)
π^+	139.6	0		$\mu\nu$	100
π^0	135.0	8 eV	$1^- 0^- +$	$\gamma\gamma$	98.84
				$\gamma e^+ e^-$	1.16
η	548.8	2.6 keV	$0^+ 0^- +$	All neutral	71
				$\pi^-\pi^+\pi^0 + \pi^+\pi^-\gamma$	29
ρ	765	135	$1^+ 1^- -$	$\pi\pi$	100
ω	783.9	10	$0^- 1^- -$	$\pi^+\pi^-\pi^0$	89.7
				$\pi^+\pi^-$	1.2
				$\pi\gamma$	9.0
ϕ	1019	4	$0^- 1^- -$	K^+K^-	49.1
				$K_L L_S$	30.7
				$\pi^+\pi^-\pi^0$	17.5
				$\eta\gamma$	2.6

l within one pair, and the orbital angular momentum of the pair with respect to the third pion. Thus if $J = 0$, $l = L$, and the overall parity of the 3π-system (including the odd intrinsic parity of the individual pion) is odd. But $J = 1$ can be the resultant of $(l, L) = (0, 1), (1, 0)$ or $(1, 1)$, so a $J = 1$ state of three pions can have either odd or even parity. Similarly, either parity is possible for $J = 1, n > 3$, and for $n = 3, J > 1$.

After parity, a further quantum number to be conserved in a decay by strong interaction is the G-parity for which the final state has the value

$$G = (-)^n$$

The hadronic decays of non-strange mesons may be expected to be governed by the availability of the final states allowing conservation of the above mentioned quantum numbers.

Mesons with non-zero strangeness, however, decay by weak interactions, with neither G-parity nor parity conserved.

9.5 Actual mesons

Some of the best-known mesons are listed in table 9.2, with their main properties. To explain the most important of the listed decay modes, we comment as follows:

η, with zero J and odd parity, requires a final state of at least three pions to conserve both these quantities. It has insufficient energy to give 4π, and therefore decays to 3π with non-conservation of G; the transition rate is thus relatively low, and the η is observed as a narrow resonance.

ρ, on the other hand, has $J^P = 1^-$ and $G = +$ and therefore decays readily to 2π with conservation of J, P and G. It is consequently observed as a resonance of large width (135 MeV).

ω, with $J^P = 1^-$, but $G = -$, has decay to 3π as its principal mode, with a G-violating decay to 2π as a weaker mode with branching ratio 1.2 per cent.

ϕ, having the same quantum numbers as the ω, has enough energy to produce a pair of kaons. Its predominant decay modes are thus to K^+K^- and to K_LK_S, with $\pi^+\pi^-\pi^0$ contributing 17.5 per cent.

The distributions of momentum and direction of the individual pions in final states of three or more pions are governed by considerations of phase space. They are commonly displayed as Dalitz plots (see, e.g. Gasiorowicz, 1966), from which important quantum numbers have on occasion been deduced.

CHAPTER 10

$SU(3)$

The successful extension of the isospin concept to the hyperons and K-mesons in the 1950s led to searches for an underlying symmetry which would relate the increased number of strongly interacting particles. Such a symmetry going beyond isospin is often referred to as a *higher symmetry*.

Several possible theoretical schemes were investigated at the beginning of the 1960s, but in this book we shall describe only the successful one, the $SU(3)$ scheme of Gell-Mann (1961, 1962) and Ne'eman (1962). We start the discussion with some general remarks on higher symmetries.

10.1 The concept of higher symmetry

The following questions arise: what is the empirical evidence, if any, for a higher symmetry of the strongly interacting particles, and how do we formulate such a symmetry mathematically? In answering them, it is convenient to look again at isospin which is the prototype of all internal symmetries.

It was postulated that there exist three isospin operators I_+, I_- and I_3 which obey the commutation rules

$$[I_3, I_\pm] = \pm I_\pm, \quad [I_+, I_-] = 2I_3 \tag{10.1}$$

From these commutation relations alone, it followed by mathematical argument (which was spelt out in the mathematically identical case of angular momentum) that particle states fall into multiplets. A multiplet is labelled by the total isospin I or more precisely by the eigenvalue $I(I+1)$ of I^2, and the individual states are distinguished by the eigenvalue of I_3. It is further assumed that $I_\pm$ and I_3 commute with the strong interaction Hamiltonian and with the operators of other space–time attributes such as spin and parity. This implies that all the particle states comprising an isospin multiplet should have the same mass, spin and parity.

With these ideas of isospin in mind, we can say that direct empirical evidence for a higher symmetry would be the observation of *supermultiplets* of hadron states of nearly the same mass and the same spin and parity, but possibly with different isospin and strangeness. This higher symmetry may be formulated mathematically by finding additional operators which commute with H_{st}, and postulating an 'interesting' set of commutation relations between them.

As in the case of isospin, we can start at a deeper level by postulating a group of transformations in an internal space. The generators of this group furnish the operators, and their commutation relations follow from the group structure. This will be the starting point of our analysis in §10.2. We return now to the question of empirical evidence for supermultiplets.

The status of the hadron spectrum at the beginning of the 1960s was roughly as follows. The eight baryons stable against strong decay were known: p, n, Σ^+, Σ^0, Σ^-, Λ^0, Ξ^0, Ξ^-. N, Λ and Σ were known to have spin $\frac{1}{2}$, while that of Ξ was unknown. The strongly stable mesons $\pi^\pm$, π^0, K^+, K^0, $\bar{K}^0$, K^- were known to have spin 0. The πN relative parity was known to be odd, and there was no strong evidence against odd ΛNK and ΣNK parity, so eight $J^P = \frac{1}{2}^+$ baryons and seven 0^- mesons could be identified.

If we assign the eight baryons to a supermultiplet of some higher symmetry then this symmetry must be badly violated because the deviations of the masses of these states from their mean is of the order of 10 per cent. Thus

$$\frac{\Lambda - \mathrm{p}}{\Lambda + \mathrm{p}} \simeq 6 \text{ per cent}$$

to be compared with an isospin multiplet

$$\frac{\mathrm{n} - \mathrm{p}}{\mathrm{n} + \mathrm{p}} \simeq 0.06 \text{ per cent}$$

It is not clear *a priori* that in such circumstances a higher symmetry is going to have any value. It is therefore a remarkable and not as yet wholly explained fact that although it is badly broken, enough regularities of the *SU*(3) symmetry 'show through' to make it useful. It is even possible to quantify the symmetry-breaking mechanism, as in the Gell-Mann–Okubo mass formula.

If the η^0-meson had been discovered earlier the octets of $\frac{1}{2}^+$ baryons and 0^- mesons might have led theorists more directly to the *SU*(3) scheme. Its absence along with the inequalities in masses, made the route more circuitous.

The 'global symmetry' scheme of Gell-Mann and Schwinger, and the four-dimensional isospin space of d'Espagnat and Prentki have only an historical interest now, but we must mention briefly the Sakata model (Sakata, 1956) because it was in the mathematical analysis of this model (Ikeda, Ogawa and Ohnuki, 1959, 1961) that the $SU(3)$ group was introduced into particle physics.

The Sakata model extends the old idea of Fermi and Yang (1949) that pions are bound states of a nucleon and an antinucleon. If we want to account for kaons similarly, then we need a heavy constituent which carries a unit of strangeness, and we choose the Λ^0. Sakata therefore proposed that p^+, n^0 and Λ^0 are the fundamental particles, from which the others are made up. Thus

$$\pi^+ \approx (p\bar{n}), \quad K^+ \approx (p\bar{\Lambda}^0), \quad K^- \approx (\Lambda^0\bar{p})$$

In this scheme the Σ- and Ξ-baryons are also composite,

$$\Sigma^+ \approx (\Lambda^0 p\bar{n}), \qquad \Xi^- \approx (\Lambda^0\Lambda^0\bar{p})$$

and thus are not equivalents of the basic triplet (p, n, Λ).

In the absence of realistic dynamical calculations predicting the binding energies of these systems, attention was mainly focussed on the isospin and strangeness content of the bound systems. This can be treated in a systematic way by considering the idealised case in which the members of the basic triplet have equal masses. We then make the following postulate of symmetry: the interactions between the p, n and Λ are invariant under linear transformations of these three objects among themselves

$$\left.\begin{aligned} p \to p' &= a_{11}p + a_{12}n + a_{13}\Lambda \\ n \to n' &= a_{21}p + a_{22}n + a_{23}\Lambda \\ \Lambda \to \Lambda' &= a_{31}p + a_{32}n + a_{33}\Lambda \end{aligned}\right\} \tag{10.2}$$

More precisely, if a Lagrangian field theory is set up, and p, n and Λ represent the fields corresponding to these particles, then (10.2) represents transformations of the fields among themselves. It is necessary to restrict the matrix of coefficients

$$A = \| a_{ij} \|$$

to be unitary in order to preserve the Hermitian property of the Lagrangian,

$$A^\dagger A = 1$$

It can be shown that the imposition of the unimodular condition

$$\det(A) = +1$$

leads to no essential restriction.

The set of all 3 × 3 matrices A satisfying these two conditions forms a group called the $SU(3)$ group.

This postulate of symmetry leads to definite predictions which we shall quote without proof. The composite states should occur in supermultiplets. Only certain dimensionalities for these supermultiplets are permitted, 1, 3, 6, 8, 10, 15 etc., and each supermultiplet has a definite isospin–strangeness content. We use the symbol $\{S, I\}$ to denote multiplet of isospin I and strangeness S.

Then bound states of one triplet $t = (\text{p}, \text{n}, \Lambda)$ and one antitriplet member $\bar{t} = (\bar{\text{p}}, \bar{\text{n}}, \bar{\Lambda})$ must form an octet or a singlet.

The singlet is $\{0, 0\}$ and the content of the octet is

$$8: \{1, \tfrac{1}{2}\}, \{0, 1\}, \{0, 0\}, \{-1, \tfrac{1}{2}\}$$

The K^-, π^-, η- and $\bar{\text{K}}$-mesons fit naturally here.

The Σ and Ξ can be realised as $(tt\bar{t})$ bound states. $(tt\bar{t})$ bound states must form supermultiplets of 15, 6 or 3 members. The 15 content is

$$15: \{1, 1\}, \{0, \tfrac{3}{2}\}, \{0, \tfrac{1}{2}\}, \{-1, 1\}, \{-1, 0\}, \{-2, \tfrac{1}{2}\}$$

It can accommodate Σ and Ξ, but then there should be ten other baryonic states of spin parity $\frac{1}{2}^+$ and similar mass. $\{0, \frac{1}{2}\}$ and $\{0, \frac{3}{2}\}$ might be nucleon resonances and $\{-1, 0\}$ a Λ-resonance.

These assignments are unsatisfactory for it is clear we have *eight* $\frac{1}{2}^+$ baryons, which would fit naturally into an octet like the mesons. To do this the p, n and Λ must be taken from their preferred positions as fundamental particles. However, we want to retain the $SU(3)$ invariance group, since it was the mechanism giving the octet as one of the supermultiplets.

This leads us to the octet model or 'eightfold way' proposed independently by Gell-Mann and Ne'eman. This can be developed in close analogy to the Sakata model by postulating a basic triplet of particles or fields on which the $SU(3)$ transformations act. These are the 'quarks' or 'aces'. The observed physical particles are formed out of quarks and/or antiquarks. However, we prefer to take first of all a more conservative if abstract point of view, and return in the next chapter to the quark model, which with stronger assumptions leads to additional predictions.

The number of review articles on $SU(3)$ symmetry and its applications is enormous. The books of Carruthers (1966) and Gourdin (1967) are wholly devoted to $SU(3)$. Gell-Mann and Ne'eman (1964) is a useful compendium of reprints.

The basic theory described here is a working out in a special case of the theory of Lie algebras and their representations, due to Cartan and Weyl, and described by Racah (1965) and Behrends *et al.* (1962).

Other reviews describing alternative approaches, calculational techniques or applications not discussed here are Behrends (1968), Berman (1965), de Swart (1963), Matthews (1967) and Smorodinskii (1965).

10.2 Conservative $SU(3)$

A conservative approach to $SU(3)$ which bypasses the question of what it is that the 3×3 transformations A act on, may be made by drawing on the general principles of symmetry in quantum theory (see §2.2).

As was seen in §2.2.3, to each transformation T of the symmetry group, there must correspond a unitary operator $U(T)$ acting in the space of particle states. For an infinitesimal transformation T, $U(T)$ takes the form

$$U = 1 + \mathrm{i}\epsilon G$$

where G is a Hermitian generator. Invariance of the system with respect to the symmetry group is expressed by

$$U(T)HU(T)^{-1} = H$$

for every T, and by

$$[G,H] = 0$$

where H is the Hamiltonian.

With this in mind we make the postulate that there exists a set of unitary operators $U(A)$ acting on particle states, which are in one-to-one correspondence with the matrices A of the group $SU(3)$, and satisfying the multiplication rule

$$U(A_2)U(A_1) = U(A_2A_1)$$

Invariance of the strong interaction under this group is expressed by

$$U(A)\,H_{\mathrm{st}}\,U(A)^{-1} = H_{\mathrm{st}} \tag{10.3}$$

The physical identification is made by considering infinitesimal transformations.

A 3×3 unitary matrix differing infinitesimally from the unit matrix can be written

$$A = 1 + \mathrm{i}\epsilon M \tag{10.4}$$

where M is 3×3 Hermitian matrix

$$M = M^{\dagger}$$

and terms of order ϵ^2 are neglected.

The condition

$$\det(A) = +1 \tag{10.5}$$

is satisfied if

$$\mathrm{Tr}(M) = 0 \tag{10.6}$$

because (10.4) can be written

$$A = e^{i\epsilon M}$$

to first order in ϵ. Now the general traceless 3×3 Hermitian matrix can be parametrised as follows

$$M = \tfrac{1}{2}\begin{pmatrix} a_3 + 3^{-1/2}a_8 & a_1 - ia_2 & a_4 - ia_5 \\ a_1 + ia_2 & -a_3 + 3^{-1/2}a_8 & a_6 - ia_7 \\ a_4 + ia_5 & a_6 + ia_7 & -2\cdot 3^{-1/2}a_8 \end{pmatrix}$$

where $a_1, a_2, \ldots, a_8$ are real quantities. By writing

$$M = \tfrac{1}{2}(a_1\boldsymbol{\lambda}_1 + a_2\boldsymbol{\lambda}_2 + \ldots + a_8\boldsymbol{\lambda}_8)$$

we define the Hermitian matrices $\boldsymbol{\lambda}_i$ introduced by Gell-Mann. They are

$$\boldsymbol{\lambda}_1 = \begin{pmatrix} 0 & 1 & 0 \\ 1 & 0 & 0 \\ 0 & 0 & 0 \end{pmatrix} \qquad \boldsymbol{\lambda}_2 = \begin{pmatrix} 0 & -i & 0 \\ i & 0 & 0 \\ 0 & 0 & 0 \end{pmatrix}$$

$$\boldsymbol{\lambda}_3 = \begin{pmatrix} 1 & 0 & 0 \\ 0 & -1 & 0 \\ 0 & 0 & 0 \end{pmatrix} \qquad \boldsymbol{\lambda}_4 = \begin{pmatrix} 0 & 0 & 1 \\ 0 & 0 & 0 \\ 1 & 0 & 0 \end{pmatrix}$$

$$\boldsymbol{\lambda}_5 = \begin{pmatrix} 0 & 0 & -i \\ 0 & 0 & 0 \\ i & 0 & 0 \end{pmatrix} \qquad \boldsymbol{\lambda}_6 = \begin{pmatrix} 0 & 0 & 0 \\ 0 & 0 & 1 \\ 0 & 1 & 0 \end{pmatrix}$$

$$\boldsymbol{\lambda}_7 = \begin{pmatrix} 0 & 0 & 0 \\ 0 & 0 & -i \\ 0 & i & 0 \end{pmatrix} \qquad \boldsymbol{\lambda}_8 = 3^{-1/2}\begin{pmatrix} 1 & 0 & 0 \\ 0 & 1 & 0 \\ 0 & 0 & -2 \end{pmatrix} \tag{10.7}$$

The $\boldsymbol{\lambda}$-matrices satisfy commutation relations which characterise the group *SU*(3). The general form is

$$[\boldsymbol{\lambda}_i, \boldsymbol{\lambda}_j] = 2\mathrm{i} \sum_{k=1}^{8} f_{ijk} \boldsymbol{\lambda}_k \tag{10.8}$$

where f_{ijk} are the *structure constants*. By direct calculation one finds the values for f_{ijk} listed in table 10.1

From the postulate made above it follows that to each of the infinitesimal transformations

$$A = 1 + \mathrm{i}\epsilon \sum_i \tfrac{1}{2} a_i \boldsymbol{\lambda}_i$$

there corresponds the operator equation

$$U(A) = 1 + \mathrm{i}\epsilon \sum_i a_i F_i$$

where U and F_i are operators acting on particle states.

It also follows that the F_i obey commutation rules analogous to (10.8)

$$[F_i, F_j] = \mathrm{i} \sum_{k=1}^{8} f_{ijk} F_k \tag{10.9}$$

Invariance under $SU(3)$ is expressed with the aid of the generators

$$[H_{\mathrm{st}}, F_i] = 0, \quad i = 1, 2, \ldots, 8 \tag{10.10}$$

Finally we make the physical identifications of the F_i. One can see that $\boldsymbol{\lambda}_1$, $\boldsymbol{\lambda}_2$ and $\boldsymbol{\lambda}_3$ are the isospin matrices augmented by a row and column of zeros. It follows that they and therefore F_1, F_2 and F_3 obey the commutation relations of isospin. So we put

$$F_1 = I_1, \quad F_2 = I_2, \quad F_3 = I_3 \tag{10.11}$$

The only matrix which commutes with $\boldsymbol{\lambda}_1$, $\boldsymbol{\lambda}_2$ and $\boldsymbol{\lambda}_3$ is $\boldsymbol{\lambda}_8$; hence F_8 is the only operator which commutes with the operators (10.11), and we may assume that it is proportional to Y, the other additive quantum number for hadrons.

We shall put

$$F_8 = M = \tfrac{1}{2} \cdot 3^{1/2} Y \tag{10.12}$$

because it will be convenient to work with M rather than Y.

It should be noted here that the baryon number B is left outside the $SU(3)$ scheme. As a consequence, the baryon number of a supermultiplet can be assigned independently. Thus, unlike the Sakata model, octets of both baryons and mesons can be accommodated in the present scheme.

Since I_3 and M (i.e. Y) commute, we can choose our states to be simultaneous eigenstates of these operators. None of the other operators

Table 10.1. *Values of the structure constants f_{ijk}. f_{ijk} vanishes for all triplets of values of ijk not listed and is totally antisymmetrical in the three indices*

ijk	f_{ijk}	*ijk*	f_{ijk}
123	1	345	$\frac{1}{2}$
147	$\frac{1}{2}$	367	$-\frac{1}{2}$
156	$-\frac{1}{2}$	458	$\frac{1}{2}\cdot 3^{1/2}$
246	$\frac{1}{2}$	678	$\frac{1}{2}\cdot 3^{1/2}$
257	$\frac{1}{2}$		

commute with both I_3 and M, so no additional additive quantum numbers are predicted by the postulated symmetry. This is in accord with empirical observation. The operators $F_4, \ldots, F_7$ are new constants of motion for H_{st}. They have, in particular, matrix elements between states of different Y, and we shall see that this leads to relations between the properties of states of different Y. Thus we can expect our symmetry to bring states with different Y (and I) into supermultiplets with similar properties.

Our first task is to determine the structure of these supermultiplets, and to do this we rewrite the commutation relations (10.9) in a more useful form.

We define *shift operators*

$$\left.\begin{aligned} I_\pm &= F_1 \pm \mathrm{i}F_2 \\ U_\pm &= F_6 \pm \mathrm{i}F_7 \\ V_\pm &= F_4 \mp \mathrm{i}F_5 \end{aligned}\right\} \tag{10.13}$$

We gather I_3 and M into a two component vector $G = (I_3, M)$. The motivation for this is that an eigenstate $\phi(I_3', M')$ of I_3 and M can be represented by a point with position vector $g = (I_3', M')$ in a plane.

We define a numerical vector $i = (1, 0)$ so that the commutation relations

$$[I_3, I_\pm] = \pm I_\pm, \quad [M, I_\pm] = 0$$

can be combined into

$$[G, I_\pm] = \pm i I_\pm \tag{10.14a}$$

and the other isospin relation in (10.1) becomes

$$[I_+, I_-] = 2i\cdot G \tag{10.14b}$$

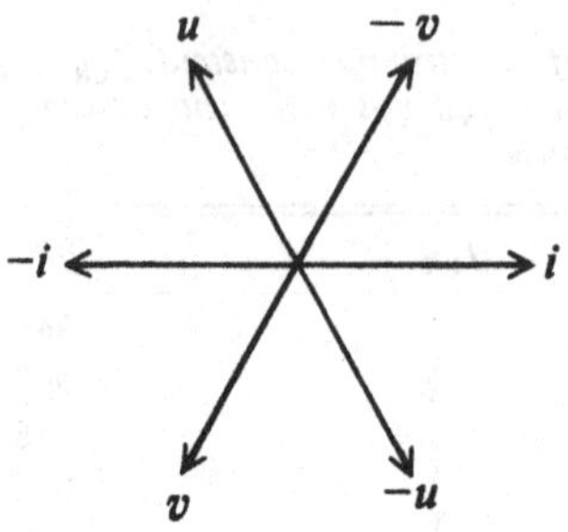

Fig. 10.1. Vectors associated with the shift operators in $SU(3)$.

Similarly we define

$$\boldsymbol{u} = (-\tfrac{1}{2}, \tfrac{1}{2}\cdot 3^{1/2}) \quad \boldsymbol{v} = (-\tfrac{1}{2}, -\tfrac{1}{2}\cdot 3^{1/2})$$

which with $\boldsymbol{i}$ have a hexagonal symmetry shown in fig. 10.1. Then from (10.8) we find

$$[G, U_\pm] = \pm \boldsymbol{u}\, U_\pm \tag{10.14c}$$

$$[U_+, U_-] = 2\boldsymbol{u}\cdot G \tag{10.14d}$$

$$[G, V_\pm] = \pm \boldsymbol{v}\, V_\pm \tag{10.14e}$$

$$[V_+, V_-] = 2\boldsymbol{v}\cdot G \tag{10.14f}$$

Further (taking the upper or lower sign throughout each equation)

$$[I_\pm, U_\pm] = \pm V_\mp \tag{10.14g}$$

$$[U_\pm, V_\pm] = \pm I_\mp \tag{10.14h}$$

$$[V_\pm, I_\pm] = \pm U_\mp \tag{10.14i}$$

and the commutator of every other pair of operators is equal to zero.

Equations (10.14*a*–*i*) which are a rewriting of (10.8), constitute the commutation relations of $SU(3)$.

In order not to lose one's way among this large set of relations, it is useful to note the natural correspondence $I_+ \to \boldsymbol{i}$, $U_+ \to \boldsymbol{u}$, $V_- \to -\boldsymbol{v}$ and so on. Then corresponding to the valid relation (fig. 10.1)

$$\boldsymbol{i} + \boldsymbol{u} = -\boldsymbol{v}$$

we have

$$[I_+, U_+] = V_-$$

but $\boldsymbol{i} - \boldsymbol{u}$ is not equal to any one of the six vectors and so

$$[I_+, U_-] = 0$$

The further significance of this notation will become apparent in the next section.

G, $I_\pm$, $U_\pm$ and $V_\pm$ form an independent set of eight generators. However, other combinations are sometimes useful. If we define

$$U_3 = -\tfrac{1}{2}I_3 + \tfrac{1}{2}\cdot 3^{1/2}M \tag{10.15}$$

then U_+, U_- and U_3 are found to obey the commutation relations of isospin. They are called U-spin operators and have the following physical significance. The electric charge operator is a combination of $SU(3)$ generators, for

$$Q = I_3 + \tfrac{1}{2}Y = I_3 + 3^{-1/2}M \tag{10.16}$$

and all three U-spin operators commute with Q,

$$[U_\pm, Q] = 0, \quad [U_3, Q] = 0$$

as is easily verified. We shall later exploit the analogy between U-spin and isospin.

If we similarly define

$$V_3 = -\tfrac{1}{2}I_3 - \tfrac{1}{2}\cdot 3^{1/2}M \tag{10.17}$$

then V_+, V_- and V_3 form operators of V-spin, but they do not have a similar importance.

To end this section we caution the reader that other conventions for labelling the generators exist.

10.3 Supermultiplets of $SU(3)$

Our first task is to determine the irreducible supermultiplets allowed by $SU(3)$ symmetry. Irreducibility means that all the states of the supermultiplet can be obtained by repeated application of the generators to any one state. In future supermultiplet will be taken to mean *irreducible supermultiplet* unless the contrary is stated.

For some purposes it is convenient to consider finite transformations of the $SU(3)$ group. In this case an irreducible supermultiplet of states forms a basis for a unitary irreducible representation of $SU(3)$. These developments are described by for example, de Swart (1963).

The tensor approach to $SU(3)$ also emphasises the finite transformations of $SU(3)$, and is described by Coleman (1965, 1966) and by Low (1966). We have preferred the shift operator approach whose principles are familiar from the case of angular momentum.

10.3.1 *Weights and lattices of weights*

We saw that out of the eight operators only two, I_3 and $M(Y)$, commute and hence can be simultaneously diagonalised. Thus the states will be labelled by eigenvalues of these two operators. Thus in the notation

$$G = (I_3, M)$$

we have for any state

$$G\,|\alpha, g\rangle = g\,|\alpha, g\rangle \tag{10.18}$$

where the eigenvalues of I_3 and M are written in the vector notation, and α denotes all other quantum numbers required to specify the state. $\boldsymbol{g}$ is called a *weight* and we represent the state $|\alpha, g\rangle$ by a point with position vector $\boldsymbol{g}$ in a *weight diagram*.

We shall assume that it is sufficient to consider supermultiplets consisting of a finite number of states. Mathematically it can be shown that all unitary irreducible representations of *SU*(3) are finite dimensional.

A supermultiplet is defined by giving all the weights of the states comprising it, and the *multiplicity* of each weight, that is the number of distinct states with that weight. All this information is contained in the weight diagram of the supermultiplet.

The significance of the shift operators $I_\pm$, $U_\pm$ and $V_\pm$ is expressed with the aid of their auxiliary vectors $\boldsymbol{i}$, $\boldsymbol{u}$ and $\boldsymbol{v}$ (which can be taken to lie in the weight diagram) by the following theorem.

Shift theorem. Given a state $|\alpha, g\rangle$ of weight $\boldsymbol{g}$, the state $I_+|\alpha, g\rangle$ is either zero or is a state of weight $\boldsymbol{g}+\boldsymbol{i}$ in the same supermultiplet. Similarly,

$I_-|\alpha, g\rangle$ if non-zero, has weight $\boldsymbol{g}-\boldsymbol{i}$

$U_\pm|\alpha, g\rangle$ if non-zero, has weight $\boldsymbol{g}\pm\boldsymbol{u}$

$V_\pm|\alpha, g\rangle$ if non-zero, has weight $\boldsymbol{g}\pm\boldsymbol{v}$

All these relations are proved in the same way.

Proof. From (10.14*c*) we have

$$GU_+ = U_+(G+u)$$

We apply this operator equation to the state $|\alpha, g\rangle$, and use (10.18) to obtain

$$\begin{aligned} GU_+|\alpha, g\rangle &= U_+(G+u)|\alpha, g\rangle \\ &= (g+u)\,U_+|\alpha, g\rangle \end{aligned}$$

Hence, assuming $U_+|\alpha, \boldsymbol{g}\rangle$ is not zero, it is a state with weight $\boldsymbol{g}+\boldsymbol{u}$, and by definition of irreducibility, it lies in the same supermultiplet as $|\alpha, \boldsymbol{g}\rangle$.

It follows that the state $U_+|\alpha, \boldsymbol{g}\rangle$ has eigenvalues $I_3 - \frac{1}{2}$ and $M + \frac{1}{2}\cdot 3^{1/2}$. Thus the shift operator U_+ takes a state with weight $\boldsymbol{g}$ into a state with weight $\boldsymbol{g}+\boldsymbol{u}$, and this is the origin of the term *shift operator*.

Quite generally the effect of a string of shift operators is easily found: the state

$$U_-V_-U_+I_-|\alpha, \boldsymbol{g}\rangle$$

has weight

$$\boldsymbol{g}-\boldsymbol{i}+\boldsymbol{u}-\boldsymbol{v}-\boldsymbol{u}$$

(provided it is non-zero).

The shift theorem together with the hexagonal symmetry of $\boldsymbol{i}$, $\boldsymbol{u}$, and $\boldsymbol{v}$ shows that the weights of the states of a supermultiplet lie in a hexagonal lattice.

Next we have:

Reflection theorem. The weight diagram of a supermultiplet is symmetrical with respect to reflections in the three lines through the origin perpendicular to $\boldsymbol{i}$, to $\boldsymbol{u}$ and to $\boldsymbol{v}$.

Proof (for $\boldsymbol{i}$). The operator

$$P_i = e^{-i\pi I_2}$$

has the properties

$$P_i^{-1}I_3P_i = -I_3$$

$$P_i^{-1}MP_i = M$$

and is unitary

$$P_i^\dagger P_i = 1$$

The first property is proved in appendix C on G-parity, and the second follows from the fact that I_2 commutes with M.

Now given a state $|\alpha, \boldsymbol{g}\rangle$, the state $P_i|\alpha, \boldsymbol{g}\rangle$ belongs to the same supermultiplet and has eigenvalues $(-I_3, M)$, for†

$$\begin{aligned} I_3P_i|\alpha, \boldsymbol{g}\rangle &= -P_iI_3|\alpha, \boldsymbol{g}\rangle \\ &= -i_3P_i|\alpha, \boldsymbol{g}\rangle \end{aligned}$$

and

$$\begin{aligned} MP_i|\alpha, \boldsymbol{g}\rangle &= P_iM|\alpha, \boldsymbol{g}\rangle \\ &= mP_i|\alpha, \boldsymbol{g}\rangle \end{aligned}$$

Finally $P_i|\alpha, \boldsymbol{g}\rangle$ cannot be zero because the unitary operator P_i conserves the normalisation of a state.

† Where necessary for clarity the eigenvalues of I_3 and M are written i_3 and m.

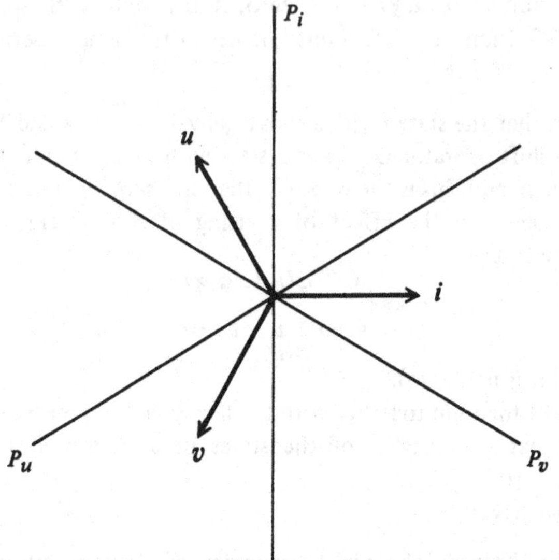

Fig. 10.2. Lines of reflection symmetry associated with P_i, P_u and P_v

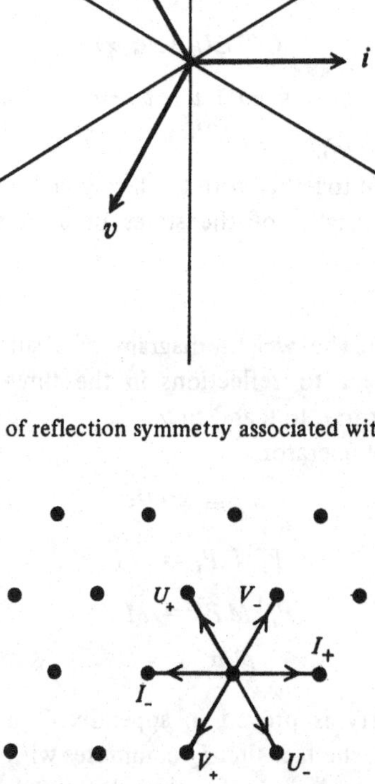

Fig. 10.3. Action of shift operators on the lattice of weights.

Now the weight of the state $P_i\,|\alpha, g\rangle$ can be obtained from that of $|\alpha, g\rangle$ geometrically by reflection in a line perpendicular to the I_3-axis (i.e. the M-axis).

Similarly we can define

$$P_u = e^{-i\pi U_2}$$

which changes the sign of U_3-eigenvalues, but leaves Q unchanged, because Q commutes with the U-spin operators. Geometrically this corresponds to reflection in a line perpendicular to u as shown in fig. 10.2. Finally

$$P_v = e^{-i\pi V_2}$$

generates reflections in a line making an angle of 120° with the previous two.

The resulting lattice of possible weights with hexagonal symmetry with the action of the shift operators is shown in fig. 10.3.

10.3.2 *Highest weight of a supermultiplet*

Since the supermultiplet is assumed to be finite, it is possible to define a *highest* weight (and a *highest state*) as the one with the greatest value of M, and if there are several such weights, the one among them with the greatest value of I_3.

We might have expected several distinct states with the highest weight, g_{max}, but this is forbidden by the final theorem:

Highest weight theorem. The highest weight of a supermultiplet has unit multiplicity.

The proof of this theorem is straightforward but requires several preliminary definitions not otherwise needed here, so we shall omit it and refer the interested reader to Racah (1965, p. 48, theorem 1).

The regularities expressed in the three theorems stated above are almost sufficient to determine the structure of the weight diagram of the general supermultiplet, and hence of the I_3- and M-eigenvalues of the states which compose it.

Consider the weight diagram shown in fig. 10.4. Application of the reflection operator P_i to the highest state $|\alpha, g_{max}\rangle$ gives the state with weight p. Since P_i is unitary it conserves the number of states and hence the multiplicity of the weight p is equal to that of g_{max}, i.e. one.

Similarly P_u and P_v imply weights at q and r; P_i in turn implies weights s and t, all of unit multiplicity.

We now ask: can there be a weight at k corresponding to the state $I_-|\alpha, p\rangle$? The answer is no, because if such a weight existed, the P_i-operator would imply the weight at l contrary to the hypothesis that g_{max} is the highest weight.

Similarly q, r, s and t are extreme weights in the sense implied by this reasoning.

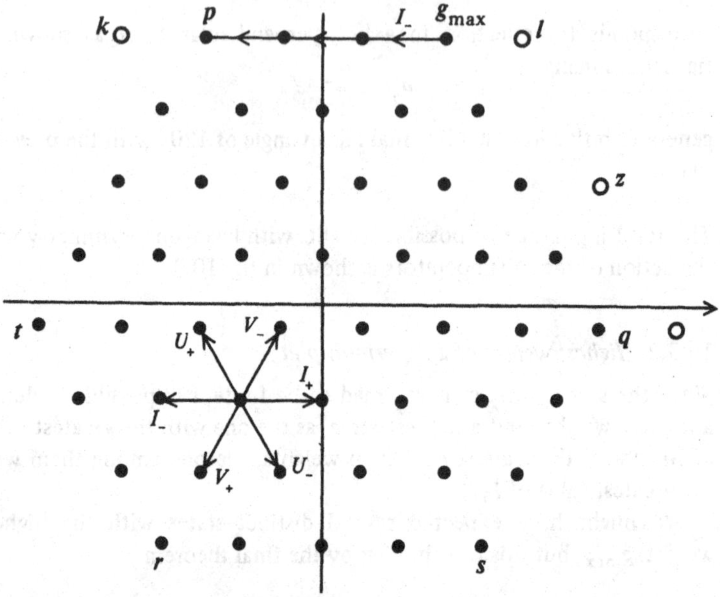

Fig. 10.4. A typical weight diagram. Open circles indicate possible weights which are excluded as described in the text.

This can be expressed in terms of the shift operators, by noting that for $g_{\max}$

$$I_+|\alpha, g_{\max}\rangle = U_+|\alpha, g_{\max}\rangle = V_-|\alpha, g_{\max}\rangle = 0 \quad (10.19a)$$

and similarly for p and q

$$I_-|\alpha, p\rangle = U_+|\alpha, p\rangle = V_-|\alpha, p\rangle = 0 \quad (10.19b)$$

$$I_+|\alpha, q\rangle = U_-|\alpha, q\rangle = V_-|\alpha, q\rangle = 0 \quad (10.19c)$$

Now repeated application of I_- to $|\alpha, g_{\max}\rangle$ generates a string of states with weights as shown. This process must bring us to the weight p when, by (10.19b), it stops. If this were not the case a combination of shifting and reflection would generate infinitely many states. Similar considerations for the strings generated by U_- and V_+ lead to the result that the lattice must be located relative to the origin so that the lines of reflection symmetry pass through weights or half-way between weights.

The arguments so far do not exclude the weight z. We shall now show that the boundary between $g_{\max}$ and q is a straight line.

Consider the situation shown in fig. 10.5, where $g_{\max}$ denotes the the highest weight. We shall show that if weight b has multiplicity 1 and

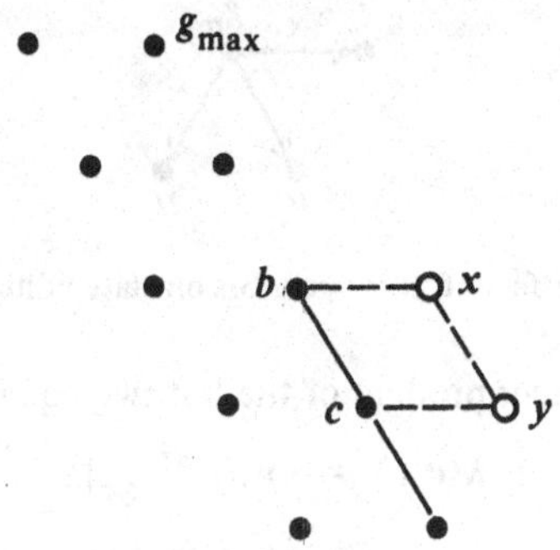

Fig. 10.5. Proof that the boundaries of weight diagrams are straight lines between the extreme weights.

the site x is empty, then the weight c has multiplicity 1 and site y is empty. This result can be used to sweep down the diagonal starting with the case when b is the extreme weight g_{max}, for which the premise is true, and ending when q is reached.

To show that the weight c has only one state corresponding to it, we observe that states with weight c can be generated as follows

$$\left.\begin{aligned} |c\rangle_1 &= U_-|b\rangle \\ |c\rangle_2 &= I_+V_+|b\rangle \\ |c\rangle_3 &= V_+I_+|b\rangle \end{aligned}\right\} \tag{10.20}$$

but from the commutation relations (10.14) these are not independent

$$V_+I_+|b\rangle - I_+V_+|b\rangle = U_-|b\rangle$$

and from the assumption that site x is empty

$$I_+|b\rangle = 0$$

so

$$|c\rangle = |c\rangle_2 = |c\rangle_1 = U_-|b\rangle \tag{10.21}$$

It is not difficult to show that any other string of shift operators applied to a suitable state to generate a state with weight c is reducible to one of the forms (10.20) by use of the commutation relations (10.14). Thus we have shown that c has multiplicity 1.

Next suppose that site y is occupied, then by the uniqueness just proved that state $|c\rangle$ defined by (10.21) can also be reached by

$$I_-|y\rangle = \lambda|c\rangle \tag{10.22}$$

where λ is a possible normalisation.

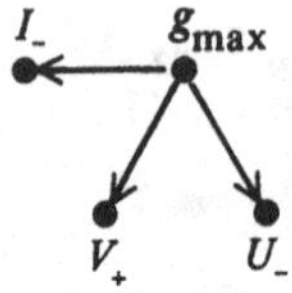

Fig. 10.6. Effect of shift operators on state of highest weight.

On taking the inner product of the last two equations we have

$$\begin{aligned}\lambda\langle c|c\rangle &= \langle y|(I_-)^\dagger U_-|b\rangle \\ &= \langle y|I_+U_-|b\rangle \\ &= \langle y|U_-I_+|b\rangle \\ &= 0\end{aligned}$$

because x is empty.

Since $\langle c|c\rangle > 0$ we have shown that $\lambda = 0$ and thus, by (10.22), our original assumption that site y is occupied was wrong.

Thus the boundary of the weight diagram between g_{max} and q is a straight line and each of the weights on it has unit multiplicity. A similar argument holds for other boundary lines and the result is as shown in fig. 10.4. In certain cases the boundary degenerates into a triangle.

10.3.3 *Multiplicities and labelling of states*

We showed in the preceding section that the boundary weights have unit multiplicity, so it remains to determine the multiplicities of interior weights. Again we start from the highest weight. Looking at fig. 10.6, we note that the states

$$I_-|\alpha, g_{max}\rangle, \quad V_+|\alpha, g_{max}\rangle, \quad U_-|\alpha, g_{max}\rangle \tag{10.23}$$

cannot all be zero, unless g_{max} is the only weight of the supermultiplet.

States of weight $g_{max} + v$ can be constructed as follows

$$V_+|\alpha, g_{max}\rangle, \quad I_-U_-|\alpha, g_{max}\rangle, \quad U_-I_-|\alpha, g_{max}\rangle$$

These are not linearly independent since, by (10.14),

$$I_-U_-|g_{max}\rangle - U_-I_-|g_{max}\rangle = -V_+|g_{max}\rangle$$

and so there are at most two independent states with weight $g + v$. There are exactly two unless either $I_-|g_{max}\rangle$ or $U_-|g_{max}\rangle$ is equal to zero, in which case there is only one.

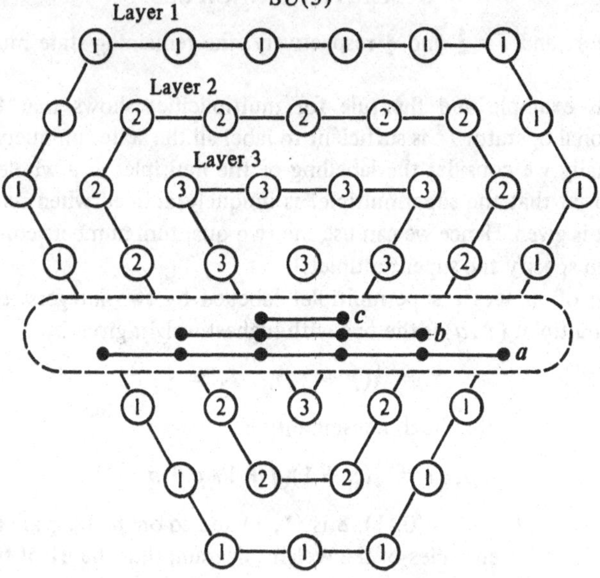

Fig. 10.7. The supermultiplet **81** with the layers indicated. Within the dashed line all of the states with a particular M are displayed and grouped into isospin multiplets with $I = \frac{1}{2}, \frac{3}{2}$ and $\frac{5}{2}$.

By further application of the kind of techniques used here and in the preceding section, one obtains the following results.

In the notation of fig. 10.7, the boundary (layer 1) consists of weights of degeneracy 1; in layer 2 the degeneracy of each weight is 2, and so on, until a triangular layer is reached. The degeneracy is then constant within and on the triangle. Thus the supermultiplet shown contains 81 states. In the special case in which two of (10.23) are zero, the boundary (layer 1) is triangular and *all* the weights have multiplicity 1.

Because in general there are several states with the same weight, i.e. the same pair of eigenvalues of I_3 and $M(Y)$, we need a further state label to distinguish them. A suitable label is the square of the total isospin,

$$I^2 = \tfrac{1}{2}(I_+I_- + I_-I_+) + I_3^2$$

which commutes with both I_3 and $M(Y)$. In effect we classify the states with a fixed value of $M(Y)$ into isospin multiplets. This is illustrated symbolically in fig. 10.7. The state furthest to the right a has $I_3 = +\frac{5}{2}$ and belongs to an $I = \frac{5}{2}$ multiplet. $I_-|a\rangle$ also has $I = \frac{5}{2}$, while the linearly independent state at this site has $I = \frac{3}{2}$. Application of I_- to these two states gives two states at c with

$I_3 = +\frac{1}{2}$, and $I = \frac{5}{2}$ and $\frac{3}{2}$ respectively; the remaining state must have $I = \frac{1}{2}$.

This example and the rule for multiplicities shows that the one additional operator I^2 is sufficient to label all the states uniquely.

Finally we consider the labelling of the multiplet as a whole. It can be proved that the supermultiplet is uniquely defined when its highest weight is given. Hence we can use the two quantum numbers comprising g_{max} to specify the supermultiplet.†

One often sees a supermultiplet labelled by two integers: thus the supermultiplet (p, q) is the one with highest weight given by

$$Y = \tfrac{1}{3}(p+2q), \quad I_3 = \tfrac{1}{2}p$$

It can be shown that its dimensionality is

$$n(p,q) = \tfrac{1}{2}(p+1)(q+1)(p+q+2)$$

Thus **3** is (1, 0), **3*** is (0, 1), **8** is (1, 1) and so on. p and q are also the lengths of adjacent sides of the weight diagram; thus the **81** of fig. 10.7 is (5, 2).

The important point is that two labels are required to specify an *SU*(3) supermultiplet uniquely. However, in practice the total number of states or dimensionality is used because it is more informative. The ambiguities which can occur are easily dealt with.

10.3.4 *The supermultiplets of low dimension*

Let us now consider some of the simplest supermultiplets permitted by the rules which we have derived.

We shall replace the operator M by the *hypercharge* related to it by

$$M = \tfrac{1}{2}\cdot 3^{1/2}Y \qquad (10.12)$$

M was used because it revealed the natural symmetry of the weight diagrams. In order to maintain this symmetry it is convenient to use a scale for the Y-axis different from that on the I_3-axis.

1: The simplest supermultiplet consists of a single state $|0\rangle$ with zero weight. So for this state,

$$Y = I_3 = 0$$

Because this state lies in all three reflection lines, no additional states

† This is analogous to the role of the total isospin quantum number I as the greatest value of I_3 in an isospin multiplet.

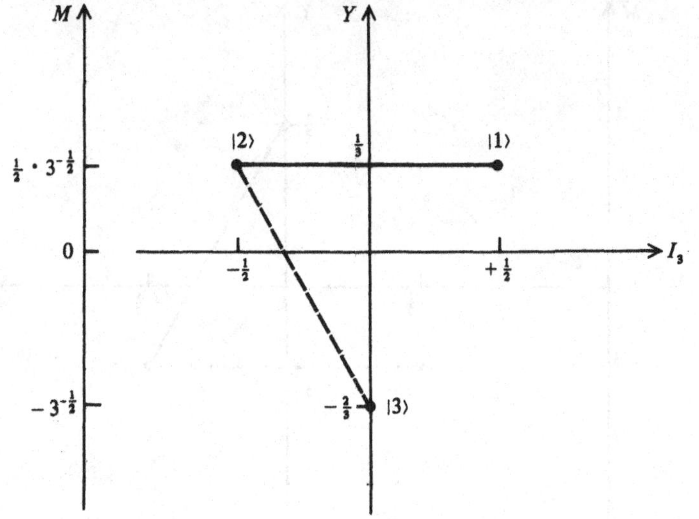

Fig. 10.8. Weight diagram for **3**.

are implied by the reflection theorem. All six shift operators applied to $|0\rangle$ give zero,

$$I_\pm|0\rangle = U_\pm|0\rangle = V_\pm|0\rangle = 0$$

so no further states are implied.

The state $|0\rangle$ is clearly an $I = 0$ state.

3: The smallest non-trivial supermultiplet consists of three states whose weights lie on the reflection lines: see fig. 10.8.

The reflection theorem is satisfied. The highest state is $|1\rangle$: it satisfies

$$I_+|1\rangle = U_+|1\rangle = V_-|1\rangle = 0 \qquad (10.24)$$

and also

$$U_-|1\rangle = 0 \qquad (10.25)$$

We define $|2\rangle$ by

$$I_-|1\rangle = |2\rangle \qquad (10.26)$$

Since $|2\rangle$ can also be reached by reflection by P_i, from the highest state, there is no state further to the left, and so

$$I_-|2\rangle = 0$$

$|1\rangle$ and $|2\rangle$ form an isospin doublet, and we have

$$I_+|2\rangle = |1\rangle \qquad (10.27)$$

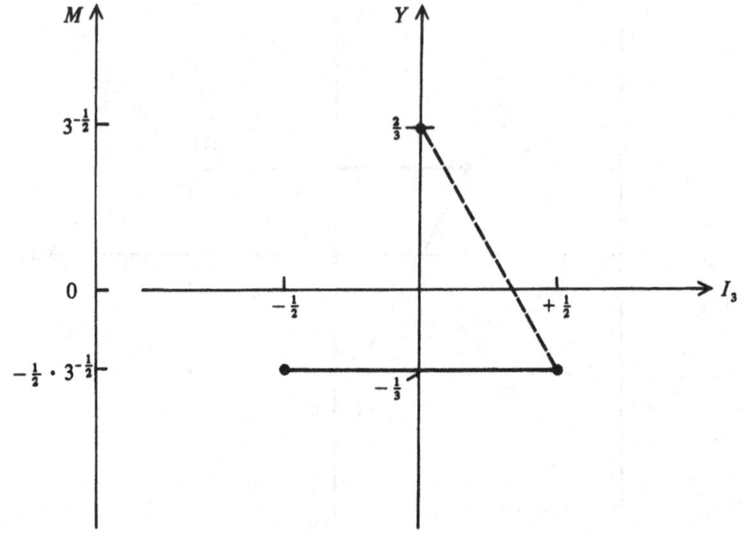

Fig. 10.9. Weight diagram for $\mathbf{3}^*$.

The third state may be defined by

$$U_-|2\rangle = |3\rangle \tag{10.28}$$

In the equations (10.26) to (10.28) we are adopting a generalisation of the Condon and Shortley phase convention as follows:†

Phase convention: The matrix elements of the shift operators $I_\pm$ and $U_\pm$ with respect to the states of a supermultiplet shall be taken real and positive.

With the aid of the commutation relations one finds:

$$U_+|3\rangle = |2\rangle, \quad V_+|1\rangle = |3\rangle, \quad V_-|3\rangle = |1\rangle$$

For example, by (10.25),

$$\begin{aligned} U_-|2\rangle &= U_-I_-|1\rangle \\ &= I_-U_-|1\rangle + V_+|1\rangle \\ &= V_+|1\rangle \end{aligned}$$

† This convention differs from that of de Swart (1963) who requires positive matrix elements for $I_\pm$ and for $V_\pm$ (i.e. $K_\mp$ in his notation). The minus signs in the commutation relations (10.14) preclude simultaneously positive matrix elements for all six shift operators. The reason for our choice is to facilitate the use of U-spin. See appendix C.

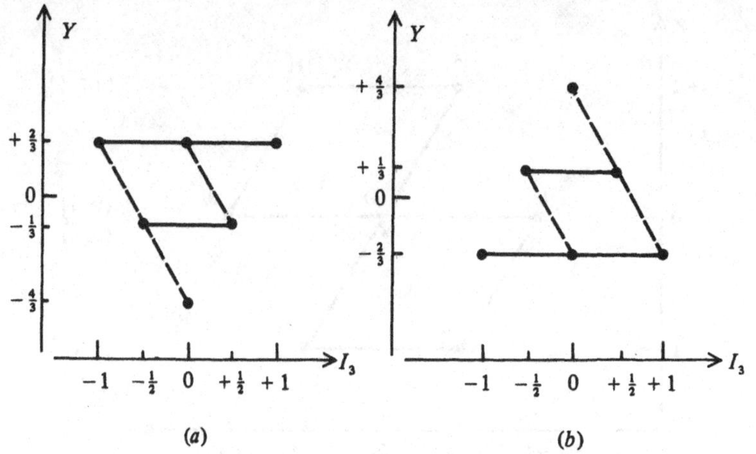

Fig. 10.10. Weight diagram for (*a*) **6** and (*b*) **6***.

Although the eigenvalues of I_3 $(0, \pm\frac{1}{2})$ are appropriate for ordinary particles, those of Y $(\frac{1}{3}, -\frac{2}{3})$ are not. This is due to the identification (10.12), but this does not worry us because the empirical evidence is for octets rather than for triplets. We shall return to the supermultiplets with fractional Y in the next chapter.

We noted in the introduction that the $SU(3)$ algebra developed here also underlies the Sakata model. In that theory the identification of M was

$$M = \tfrac{1}{2}\cdot 3^{1/2}(S + \tfrac{1}{3}B) \quad \text{(Sakata)} \tag{10.29}$$

instead of (10.12). B is the baryon number and S is the strangeness. So the strangeness of the isodoublet and isosinglet become 0 and -1 respectively, so that (p, n) and Λ could be accommodated.

3*: There exists a second supermultiplet of dimension 3 with weight diagram shown in fig. 10.9. This is unlike the case of isospin where there is an essentially unique multiplet of a given dimension.

From the diagram it can be seen that the eigenvalues of Y are different from those in the **3** supermultiplet. This shows that it is not possible by a unitary transformation in the basic states of the form

$$|3^*, n\rangle = \sum_{m=1}^{3} U_{nm}\,|3, m\rangle \tag{10.30}$$

to turn one set into the other: under a transformation (10.30) the operators undergo a similarity transformation

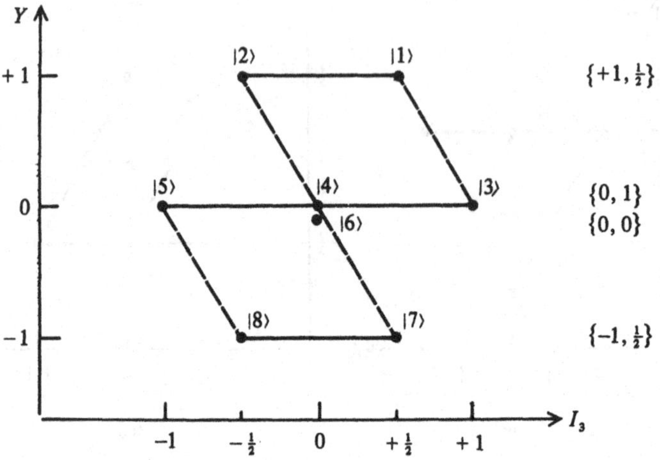

Fig. 10.11. Weight diagram for the octet **8**. The hypercharge–isospin content $\{Y, I\}$ is indicated on the right.

$$Y \to Y' = UYU^{-1}$$

and this, as is well known, preserves the eigenvalues.

The action of the shift operators is

$$U_-|1\rangle = |2\rangle, \quad U_+|2\rangle = |1\rangle$$

$$I_-|2\rangle = |3\rangle, \quad I_+|3\rangle = |2\rangle$$

$$V_-|3\rangle = |1\rangle, \quad V_+|1\rangle = |3\rangle$$

In the Sakata model with (10.29), the antiparticles $\bar{\Lambda}$, $\bar{n}$ and $\bar{p}$ can be be accommodated in the **3***.

6 and **6***: There exist two distinct supermultiplets with six members, and again with fractional hypercharge. Their weight diagrams are shown in fig. 10.10.

8: The weight diagram is shown in fig.10.11. Here Y has integral eigenvalues as a result of the identification (10.12).

$|1\rangle$ is the highest state and $|2\rangle$ defined by

$$I_-|1\rangle = |2\rangle$$

forms an $I = \frac{1}{2}$ doublet with $|1\rangle$.

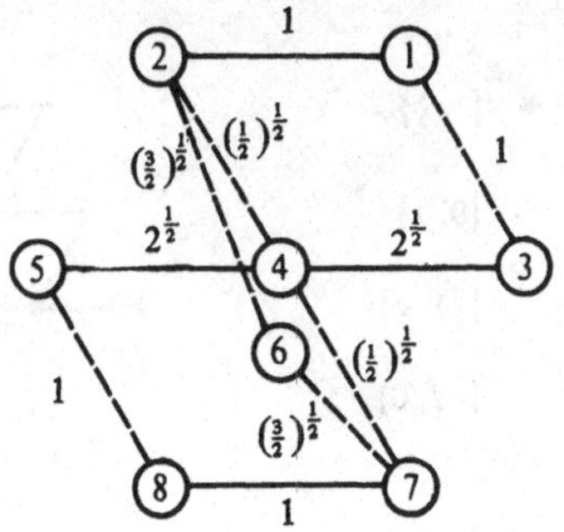

Fig. 10.12. Matrix elements of $I_\pm$ and $U_\pm$ between states of the **8**.

$|3\rangle$ is defined by

$$U_-|1\rangle = |3\rangle$$

The weight $I_3 = Y = 0$ has multiplicity 2: one of the states can be defined as the central member of the $I = 1$ triplet by putting

$$I_-|3\rangle = 2^{1/2}|4\rangle$$

The factor on the right is the standard isospin matrix element $\rho_-(I = 1, I_3 = 1)$ of (3.35). The linearly independent state with $g = 0$ is $|6\rangle$, an isosinglet state.

$|5\rangle$ is obtained by

$$I_-|4\rangle = 2^{1/2}|5\rangle$$

We shall show later in developing the U-spin technique, that

$$U_-|2\rangle = (\tfrac{1}{2})^{1/2}|4\rangle + (\tfrac{3}{2})^{1/2}|6\rangle$$

but we may note here that since U_- does not commute with I-spin we obtain a superposition of $I = 1$ and $I = 0$ states on the right.

Similarly, it can be shown that

$$U_-|4\rangle = (\tfrac{1}{2})^{1/2}|7\rangle, \quad U_-|6\rangle = (\tfrac{3}{2})^{1/2}|7\rangle$$

Finally we define

$$I_-|7\rangle = |8\rangle$$

$$U_-|5\rangle = |8\rangle$$

The matrix elements of U_- and I_- listed are summarised in fig. 10.12. Those of U_+ and I_+ are obtained from them by

$$\langle m|I_+|n\rangle = \langle n|I_-|m\rangle, \quad \langle m|U_+|n\rangle = \langle n|U_-|m\rangle$$

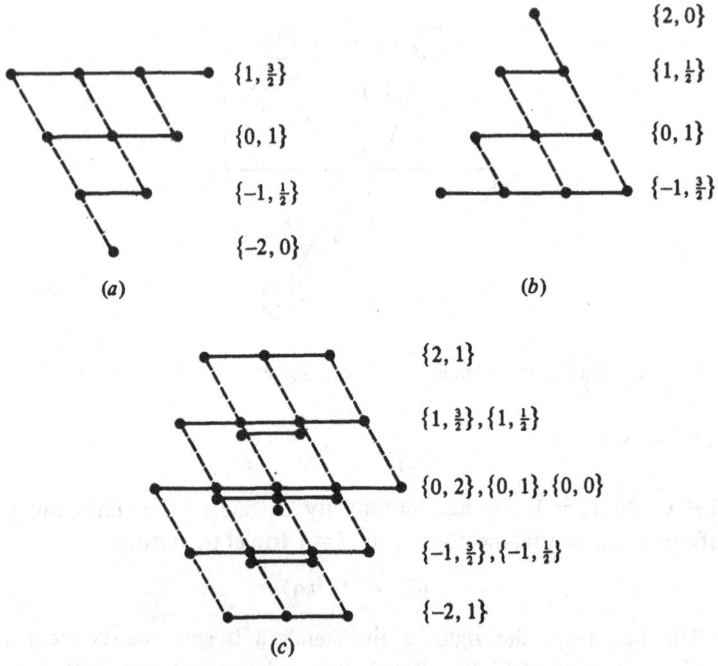

Fig. 10.13. Weight diagrams for (*a*) **10**, (*b*) **10*** and (*c*) **27**.

by Hermiticity and reality. All are positive according to the convention adopted here.

The matrix elements of shift operators in an arbitrary supermultiplet can be obtained by similar techniques, and general formulae are available (Carruthers, 1966).

10, 10* and **27**: The weight diagrams for these supermultiplets, all of which have physical I_3- and Y-values, are shown in fig. 10.13. The notation $\{Y, I\}$ to denote an isospin I multiplet of hypercharge Y is used.

10.4 Assignment of particles and resonances to *SU*(3) supermultiplets

In the remainder of this chapter we shall describe some of the consequences of $SU(3)$ invariance.

It follows from the assumption of $SU(3)$ invariance that all the states of a supermultiplet should be degenerate in energy (mass) with the same spin and parity. The lack of perfect symmetry means that there is only

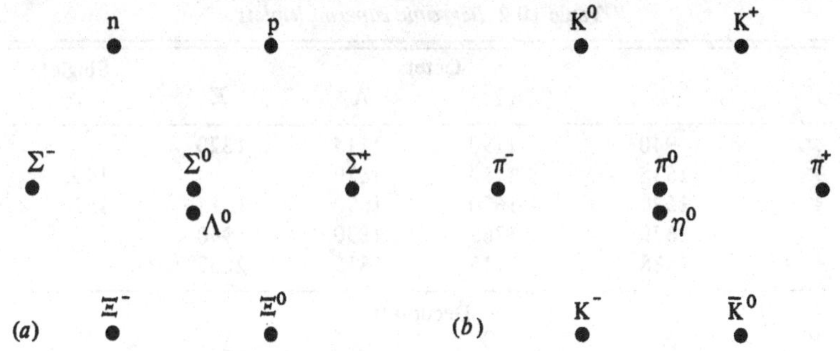

Fig. 10.14. (*a*) $\frac{1}{2}^+$ baryon and (*b*) 0^- meson octets.

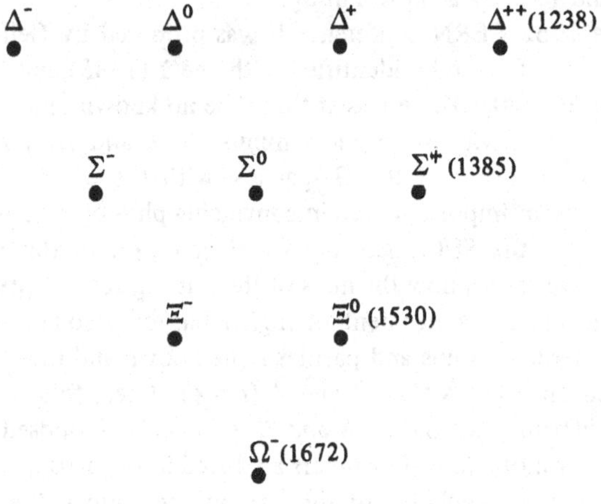

Fig. 10.15. $\frac{3}{2}^+$ decuplet of resonant states.

approximate equality of energy but, guided by the other two properties, it is possible to identify supermultiplets of hadron states. We shall return later to the question of deviations from perfect symmetry.

The *SU*(3) scheme was designed to accommodate octets of $\frac{1}{2}^+$ baryons and 0^- mesons. The assignments are shown in fig. 10.14.

Hadronic resonances which are the products of approximately *SU*(3) symmetric interactions should also form supermultiplets of *SU*(3). The well-known Δ(1236) with spin parity $\frac{3}{2}^+$ has internal quantum numbers $Y = 1$, $I = \frac{3}{2}$, and cannot be fitted into an octet. The decuplet is the smallest supermultiplet which can accommodate such a particle. Its partners must then be baryonic resonances with $\{Y, I\}$-values $\{0, 1\}$,

Table 10.2. *Baryonic supermultiplets*

J^P	Octets N	Σ	Λ	Ξ	Singlets Λ′
$\frac{1}{2}^+$	940	1190	1115	1320	
$\frac{1}{2}^-$	1535	1750	1670		1405
$\frac{3}{2}^-$	1520	1670	1690	1815	1520
$\frac{5}{2}^-$	1670	1765	1830	1940?	
$\frac{5}{2}^+$	1688	1915	1815	2037?	

J^P	Decuplets Δ	Σ	Ξ	Ω^-
$\frac{3}{2}^+$	1236	1385	1530	1675
$\frac{7}{2}^+$	1950	2030		

$\{-1, \frac{1}{2}\}$ and $\{-2, 0\}$, as shown in fig. 10.15.

At the 1962 CERN conference it was proposed by Gell-Mann that the first two of these be identified with the Σ(1385) and the cascade resonance Ξ(1530). There was at that time no known candidate for the remaining state, with strangeness minus three and isospin zero. The subsequent discovery of the Ω^--particle with the predicted quantum numbers was an important step in convincing physicists of the essential correctness of the *SU*(3) scheme. We shall return to this point again when we have shown how the mass of the missing Ω^- was predicted.

The satisfactory assignment of higher baryon resonances to supermultiplets requires spins and parities to be known and these are usually established first for N ($I = \frac{1}{2}$) and Δ ($I = \frac{3}{2}$) states, followed with increasing difficulty by the Σ, Λ and Ξ resonances. Proposed multiplets with three or more known members are listed in table 10.2.

An important regularity of the baryon spectrum is the absence of any $Y = +2$ resonances in KN scattering. This suggests that there are no **10*** or **27** (or higher) supermultiplets, which if true, means that every nucleonic resonance can be assigned to an **8** or a **10** according as it has $I = \frac{1}{2}$ or $I = \frac{3}{2}$. One can then examine the list of hyperon resonances for $Y = 0$, $I = 0$ or 1 partners of these nucleonic states. Further *SU*(3) predictions such as branching ratios for decays, are useful criteria at this stage, and we shall describe this technique in the case of the **10** baryon decuplet in §10.6. The problems of baryon resonance assignments have been reviewed by Levi-Setti (1969) and Harari (1968).

Mesonic resonances firmly assigned to *SU*(3) supermultiplets are listed in table 10.3. Only octets and singlets are found to occur.

It is convenient to introduce here the concept of *conjugate supermultiplet*. Given an *SU*(3) supermultiplet *N*, there is always a conjugate

Table 10.3. *Meson supermultiplets*

J^P	Octets			Singlets
0^-	π	η	K	$\eta'(958)$
1^-	$\rho(770)$	$\phi(1019)$	K(892)	$\omega(784)$
2^+	$A_2(1310)$	f'(1514)	K*(1420)	f(1270)

supermultiplet denoted by N^*. The weight diagram of N is obtained from that of N^* by reflecting it in the origin, so that for each state $|N, Y, I, I_3\rangle$ in N there is a state $|N^*, -Y, I, -I_3\rangle$ in N^*. It can happen that N and N^* are the same (self-conjugate) as with the **8**. The general case is illustrated by the pairs **3**, **3***, **6**, **6*** and **10**, **10*** already encountered.

If a set of particles belong to an N, their antiparticles belong to an N^*.

In the case of baryons distinguished from their antiparticles by baryon number, the antiparticles of an **8**, e.g. the $\frac{1}{2}^+$ baryons, belong to an **8**, while the antiparticles of a **10** form a **10***. For mesons ($B = 0$) a self-conjugate supermultiplet such as **8** or **27** contains both particle and its antiparticle (e.g. π^+ and π^-, K^+ and K^-, etc.). The existence of a meson **10** would imply the existence of a **10***, the two being degenerate in mass as a consequence of the *CPT* theorem, but there is no experimental evidence for this case.

10.5 Broken symmetry: mass formulae and particle mixing

The large deviations from mass equality within the supermultiplets show that invariance under *SU*(3) is not exact. Faced with this situation we try to describe and quantify the deviations from perfect symmetry.

10.5.1 *Gell-Mann's symmetry-breaking hypothesis*

As a conceptual model it is useful to think of the strong interaction Hamiltonian as made up of two parts: a 'very strong' interaction which is invariant under *SU*(3), and a 'medium strong' interaction responsible for the deviations from exact symmetry.

Thus if the medium strong interaction could be switched off, *SU*(3) would become an exact invariance of the hadrons, and supermultiplets would be degenerate in mass, in the same way as the members of an isospin multiplet are assumed to be degenerate in the absence of the electromagnetic interaction.

We express this by writing

$$H_{st} = H_{vs} + H_{ms} \tag{10.31}$$

where H_{vs} commutes with all the $SU(3)$ generators:

$$[H_{vs}, F_i] = 0$$

but

$$[H_{ms}, F_i] \neq 0$$

for some F_i, expressing the fact that H_{ms} is not invariant under all $SU(3)$ transformations. Gell-Mann turned this negative statement (of non-invariance) into a positive statement that H_{ms} has simple transformation properties under $SU(3)$.

To show that this can lead to experimentally testable consequences, we consider an analogous situation in atomic physics.

Consider an atom, with neglect of spin for simplicity. As a consequence of spherical symmetry an energy level E_l with orbital angular momentum l is $(2l+1)$-fold degenerate, the individual states being distinguished by the eigenvalue m_l of L_z.

If a weak magnetic field $\mathfrak{B}$ along the z-axis is imposed, there is an additional energy of interaction

$$H' = \frac{e\hbar}{2mc}\mathfrak{B}L_z = k\,L_z$$

between the field and the effective orbital dipole moment $e\hbar L_z/2mc$. The system is no longer rotationally invariant because H does not commute with L_x and L_y. However, because the perturbation is proportional to a generator L_z of the invariance group of the original system, we can obtain directly the new energy levels. In first-order perturbation theory the energy correction is given by the expectation value

$$\delta E_{lm_l} = (\psi_{lm_l}, H'\psi_{lm_l}) = k\,m_l$$

The perturbation is diagonal in the basis ψ_{lm_l}, so degenerate perturbation theory is not required.

This is the theory of the normal Zeeman effect.

We now return to $SU(3)$ and ask what hypothesis can be made about the transformation properties of H_{ms}? Although full $SU(3)$ symmetry is violated, isospin and hypercharge remain as valid symmetries, so that

$$[Y, H_{ms}] = 0$$

$$[I_3, H_{ms}] = [I_{\mp}, H_{ms}] = 0$$

Gell-Mann proposed that H_{ms} should transform like one of the generators of $SU(3)$ (cf. the example just described). Examination of the $SU(3)$ algebra, (10.14*a*–*i*), shows that the only generator which commutes with Y and $I_\pm$ is Y itself. Thus we are led to formulate the following:

Symmetry-breaking hypothesis: H_{ms} transforms like the hypercharge Y.

It will become apparent in the following discussion that in $SU(3)$, unlike the example of the Zeeman effect, there is a difference between H_{ms} transforming *like* Y and H_{ms} being proportional to Y.

10.5.2 *U-spin*

In deriving the consequences of the symmetry-breaking hypothesis, we shall use the U-spin technique developed by Lipkin and others.†

In the case of exact $SU(3)$ symmetry with the medium strong and electromagnetic interactions switched off, all eight $\frac{1}{2}^+$ baryons would be degenerate in mass, and the subdivision of the octet into isospin multiplets of definite hypercharge would not obviously suggest itself. In mathematical terms: instead of labelling the states by I_3 and Y we could use instead the pair

$$U_3 = -\tfrac{1}{2}I_3 + \tfrac{3}{4}Y \tag{10.32}$$

$$Q = I_3 + \tfrac{1}{2}Y \tag{10.33}$$

As was noted above, the U-spin operators U_3, $U_\pm$ obey the same commutation relations as $I_3, I_\pm$ or $J_3, J_\pm$. Hence all the apparatus of angular momentum can be carried over to U-spin.

In the case of exact $SU(3)$ then, it would be equally valid to classify states of an $SU(3)$ supermultiplet by Q, U_3, and the eigenvalue $U(U+1)$ of the total U-spin

$$U^2 = U_1^2 + U_2^2 + U_3^2$$

where U_1 and U_2 are defined by

$$U_\pm = U_1 \pm \mathrm{i}U_2$$

Then since Q commutes with U-spin, the states of a U-spin multiplet have U_3 running from $-U$ to $+U$ and all have the same charge Q.

† For the reader familiar with group theory, a more rapid derivation can be made by means of tensor operators and the Wigner—Eckart theorem.

We need to be able to relate the two labelling schemes, (U, U_3, Q) and (I, I_3, Y). This is usually quite easy. If a weight has multiplicity 1, (10.32) and (10.33) give the values of U_3 and Q, and the value of the total U-spin can often be deduced. Thus in the baryon octet p and Σ^+ have $(Q, U_3) = (+1, +\frac{1}{2})$ and $(+1, -\frac{1}{2})$ respectively, forming a $U = \frac{1}{2}$ multiplet. n and Ξ^0 correspond to $(0, +1)$ and $(0, -1)$, and both Σ^0 and Λ^0 to $(0, 0)$. However, the states at the origin which have definite total U are superpositions of the states Σ^0 and Λ^0 of definite I. These superpositions are denoted by Σ_U^0 and Λ_U^0 according as $U = 1$ or 0, and may be determined as follows. Suppose that the $U = 1, U_3 = 0$ state is

$$
\begin{aligned}
|\Sigma_U^0\rangle &= |U = 1, U_3 = 0\rangle \\
&= \alpha|\Sigma^0\rangle + \beta|\Lambda^0\rangle
\end{aligned}
\tag{10.34}
$$

where $\alpha^2 + \beta^2 = 1$ since all the symbols $|\rangle$ are normalised. We shall see that α and β can be chosen to be real.

The neutron has $U_3 = 1$ and must belong to a $U = 1$ triplet, so applying the U-spin lowering operator U_- we have

$$U_-|\mathrm{n}\rangle = 2^{1/2}|\Sigma_U^0\rangle$$

Similarly within the $I = 1$ triplet

$$I_-|\Sigma^+\rangle = 2^{1/2}|\Sigma^0\rangle$$

In these equations $2^{1/2}$ is the matrix element of U_- (or I_-) obtained from (3.41) and is positive in our convention.

On taking the inner product of the last two equations we have

$$
\begin{aligned}
2\langle\Sigma^0|\Sigma_U^0\rangle &= \langle\Sigma^+|I_-^\dagger U_-|\mathrm{n}\rangle \\
&= \langle\Sigma^+|I_+ U_-|\mathrm{n}\rangle \\
&= \langle\Sigma^+|U_- I_+|\mathrm{n}\rangle \\
&= \langle\mathrm{p}|\mathrm{p}\rangle \\
&= 1
\end{aligned}
$$

where we used the fact that I_+ and U_- commute, and that

$$U_+|\Sigma^+\rangle = |\mathrm{p}\rangle$$

Since $\langle\Sigma^0|\Sigma_U^0\rangle$ is just α we have $\alpha = \frac{1}{2}$ and thus

$$|\Sigma_U^0\rangle = \tfrac{1}{2}|\Sigma^0\rangle + \tfrac{1}{2}\cdot 3^{1/2}|\Lambda^0\rangle \tag{10.35}$$

The orthogonal linear combination gives the Λ_U^0 state

$$|\Lambda_U^0\rangle = \tfrac{1}{2}\cdot 3^{1/2}|\Sigma^0\rangle - \tfrac{1}{2}|\Lambda^0\rangle \qquad (10.36)$$

Here the position of the minus sign is arbitrary. These results are of course valid for any octet.

In the case of a triangular supermultiplet such as the **10**, see fig. 10.13, there is no multiple occupancy of weights and the states can be regrouped into U-spin multiplets directly. In figs. 10.8 to 10.13 the U-spin multiplets have been indicated by dashed lines.

10.5.3 *The mass formulae*

The usefulness of U-spin in symmetry-breaking considerations lies in the fact that a U-spin multiplet contains states of different (Y, I) and hence, unlike the I-spin multiplet, it contains effects of the symmetry-breaking interaction.

We have postulated that H_{ms} transforms like Y and since

$$Y = U_3 + \tfrac{1}{2}Q$$

and

$$[Q, U] = 0$$

it follows that H_{ms} transforms like a superposition of U_3, a component of a $U = 1$ vector in U-spin space, and of Q, a scalar in U-spin space.

We represent this by writing

$$H_{\text{ms}} = H_{\text{ms}}^{(\text{S})} + H_{\text{ms}}^{(\text{V})} \qquad (10.37)$$

In the spirit of first-order perturbation theory, the corrections to the energies or masses of the members of an octet will be given by the expectation values of the perturbation. Thus

$$\delta m_{\text{p}} = \langle \text{p}|H_{\text{ms}}|\text{p}\rangle \qquad (10.38)$$

and so on.

We go to U-spin labelling and consider the $U = 1$ triplet

$$(\text{n}, \tfrac{1}{2}\Sigma^0 + \tfrac{1}{2}\cdot 3^{1/2}\Lambda^0, \Xi^0)$$

So we require to evaluate

$$\langle U = 1, U_3|H_{\text{ms}}|U = 1, U_3\rangle$$
$$= \langle 1, U_3|H_{\text{ms}}^{(\text{S})}|1, U_3\rangle + \langle 1, U_3|H_{\text{ms}}^{(\text{V})}|1, U_3\rangle$$

The expectation value of a U-spin scalar $H_{\text{ms}}^{(\text{S})}$ cannot depend on U_3, and so

$$\langle 1, U_3 | H_{\mathbf{ms}}^{(\mathrm{S})} | 1, U_3 \rangle = a$$

while the expectation value of the U-spin *vector* part $H_{\mathbf{ms}}^{(\mathrm{V})}$ which transforms like U_3, is proportional to U_3, and thus

$$\langle 1, U_3 | H_{\mathbf{ms}}^{(\mathrm{V})} | 1, U_3 \rangle = bU_3$$

Here a and b are constants characteristic of the supermultiplet considered.

On taking the matrix elements of $H_{\mathbf{ms}}$ between the states of the $U = 1$ triplet in turn we obtain

$$\langle \mathrm{n} | H_{\mathbf{ms}} | \mathrm{n} \rangle = a + b$$

$$\langle \tfrac{1}{2}\Sigma^0 + \tfrac{1}{2}\cdot 3^{1/2}\Lambda^0 | H_{\mathbf{ms}} | \tfrac{1}{2}\Sigma^0 + \tfrac{1}{2}\cdot 3^{1/2}\Lambda^0 \rangle = a$$

$$\langle \Xi^0 | H_{\mathbf{ms}} | \Xi^0 \rangle = a - b$$

Since $H_{\mathbf{ms}}$ conserves I-spin, it has no matrix elements between the states Σ^0 and Λ^0, and so the middle line becomes

$$\tfrac{1}{4}\langle \Sigma^0 | H_{\mathbf{ms}} | \Sigma^0 \rangle + \tfrac{3}{4}\langle \Lambda^0 | H_{\mathbf{ms}} | \Lambda^0 \rangle = a$$

Upon eliminating a and b, we obtain one relation between the mass shifts

$$\tfrac{1}{2}(\delta \mathrm{n} + \delta \Xi^0) = \tfrac{1}{4}(\delta \Sigma^0 + 3\delta \Lambda^0)$$

Since the unperturbed masses are assumed to be equal, we can rewrite this as a relation between the total (observed) masses of the particles

$$\tfrac{1}{2}(\mathrm{N} + \Xi) = \tfrac{1}{4}(\Sigma + 3\Lambda) \tag{10.39}$$

The charge labels have been omitted since, with the neglect of electromagnetic effects, the masses within an isospin multiplet are equal.

Equation (10.39) is the *Gell-Mann–Okubo mass formula*, and should hold for any octet.

For the case of the $\tfrac{1}{2}^+$ baryon octet, the values of the two sides are 1128 MeV and 1135 MeV. If we take the difference over the mean mass in this octet as a measure of the accuracy of the mass formula we find a figure of 1 per cent. This agreement is astonishing bearing in mind the use of first-order perturbation theory.

In the case of the 0^- mesons the agreement is less good. Feynman noted that it can be improved if for mesons we rewrite the formula in terms of the squared masses. Then we have

$$m_{\mathrm{K}}^2 = \tfrac{1}{4}(m_\pi^2 + 3m_\eta^2) \tag{10.40}$$

since $m_{\bar{\mathrm{K}}} = m_{\mathrm{K}}$ by *CPT*-invariance.

The justification for using squared masses may lie in the fact that in field theories for bosons, as opposed to fermions, the squared masses rather than the masses themselves enter naturally, cf. the Klein–Gordon equation and the Dirac equation.

The mass formula for a decuplet is obtained more easily since, as noted above, the members of the U-spin multiplets can be read off directly.

Consider the expectation value of H_{ms} with respect to the members of the $U = \frac{3}{2}$ multiplet

$$|U = \tfrac{3}{2}, U_3\rangle = (\Delta^-, \Sigma^-, \Xi^-, \Omega^-)$$

We find

$$\langle \tfrac{3}{2} U_3 | H_{\text{ms}} | \tfrac{3}{2} U_3 \rangle = a' + b' U_3$$

where a' and b' are constants. Thus the mass shifts in the multiplet vary linearly with U_3 and hence, since Q is constant, with Y. The same is true of the observed masses since the unperturbed masses are equal.

Thus we have an *equal spacing rule* for decuplets

$$\Delta - \Sigma^* = \Sigma^* - \Xi^* = \Xi^* - \Omega \tag{10.41}$$

When this result was first obtained the existence of the Ω^- was not known. The masses of the Δ, Σ^* and Ξ^* resonant states were in good agreement with this formula. The current values are (in MeV)

$$\Delta = 1236, \quad \Sigma^* = 1385, \quad \Xi^* = 1531$$

with widths of about 10 MeV.

This enabled a mass of about 1680 MeV to be predicted for the missing $I = 0$, $Y = -2$ state. It follows that this state must decay weakly since the lowest energy available to the $\Xi\bar{K}$-state is 1812 MeV. The subsequent observation of a strangeness minus three hyperon by Barnes *et al.* (1964) with a mass of 1686 ± 12 MeV was convincing evidence for the soundness of the broken $SU(3)$ scheme, that is, an underlying $SU(3)$ symmetry together with a symmetry-breaking interaction which transforms like the hypercharge.

To end this section we note that with a symmetry-breaking interaction of the kind postulated, it can be shown that the masses in a supermultiplet are given (Okubo, 1962) by

$$m = m_0 + m_1 Y + m_2 [\tfrac{1}{4} Y^2 - I(I+1)] \tag{10.42}$$

where the values of the constants m_0, m_1 and m_2 depend on the supermultiplet considered.

10.5.4 *Mass mixing*

In seeking to assign the 1^- meson resonances in the mass range 700 to 900 MeV to an octet one finds the $\rho(770)$ $(I=1, Y=0)$ $K^*\bar{K}^*(I=\frac{1}{2}, Y=\pm 1)$, but two candidates $\phi(1019)$ and $\omega(784)$ for the $I=Y=0$ state. There are no other 1^- mesons in this mass range, so the simplest hypothesis is that we have an octet *and* a singlet. To see which of ϕ or ω is the octet member one can use the Gell-Mann–Okubo formula to predict the mass of the $I=Y=0$ state. One finds a value of 930 MeV which does not agree well with either ϕ or ω.

An explanation for this discrepancy can be given within the broken $SU(3)$ scheme. Suppose that in the limit of exact $SU(3)$ it happens that an octet and a singlet of 1^- mesons are degenerate or nearly degenerate in mass. Now H_{ms} transforming like $U_3+\frac{1}{2}Q$ will in general have non-zero matrix elements between the singlet state $|1, I_3=Y=0\rangle$ and the octet isosinglet $|8, I=I_3=Y=0\rangle$ because these states are $U=0$ and a superposition of $U=0$ and $U=1$ respectively. (There is no matrix element with $|8, I=1, I_3=Y=0\rangle$ because H_{ms} conserves I-spin.)

These matrix elements will be enhanced by the equality of the unperturbed energies, and degenerate perturbation theory must be used on these two states, with the following results.

The matrix of the perturbation H_{ms} between the unperturbed states denoted by 1 and 8 has the form

$$\begin{pmatrix} \langle 8|H_{\mathrm{ms}}|8\rangle & \langle 8|H_{\mathrm{ms}}|1\rangle \\ \langle 1|H_{\mathrm{ms}}|8\rangle & \langle 1|H_{\mathrm{ms}}|1\rangle \end{pmatrix}$$

however, for reasons indicated above we shall work with the squared masses:

$$\begin{pmatrix} M_{88}^2 & M_{81}^2 \\ M_{18}^2 & M_{11}^2 \end{pmatrix} \tag{10.43}$$

This matrix is Hermitian, and by choice of phases of the states can be made real, and hence symmetric, $M_{18}^2 = M_{81}^2$.

The perturbed energies are given by the eigenvalues of (10.43) which are found to be

$$\left.\begin{matrix} m_\phi^2 \\ m_\omega^2 \end{matrix}\right\} = \tfrac{1}{2}\{(m_{88}^2+m_{11}^2) \pm [(m_{88}^2-m_{11}^2)^2+4m_{81}^4]^{1/2}\} \tag{10.44}$$

The corresponding normalised eigenvectors may be written

$$\begin{pmatrix}\cos\theta\\ \sin\theta\end{pmatrix} \text{ and } \begin{pmatrix}-\sin\theta\\ \cos\theta\end{pmatrix}$$

where θ is given by

$$\tan\theta = \frac{m_\phi^2 - m_{88}^2}{m_{81}^2} = \frac{m_{81}^2}{m_{88}^2 - m_\omega^2} \tag{10.45}$$

and the perturbed (physical) states are then

$$\left.\begin{aligned} |\phi\rangle &= \cos\theta\,|8\rangle + \sin\theta\,|1\rangle \\ |\omega\rangle &= -\sin\theta\,|8\rangle + \cos\theta\,|1\rangle \end{aligned}\right\} \tag{10.46}$$

θ is called the *mixing angle*.

To apply these results to the 1^- mesons we suppose that the mass splitting in the octet is taken into account first, so that m_{88}^2 is related to m_ρ^2 and m_K^2 by the analogue of (10.40).

$$m_{88}^2 = \tfrac{1}{3}(4m_K^2 - m_\rho^2)$$

giving, as noted above,

$$m_{88} = 930 \text{ MeV}$$

Then m_{11}^2 and m_{81}^2 (or θ) are chosen so as to give the observed masses for the ϕ^0- and ω^0-mesons.

The formulae

$$m_{81}^2 = (m_{88}^2 m_{11}^2 - m_\phi^2 m_\omega^2)^{1/2}$$

$$m_\phi^2 + m_\omega^2 = m_{88}^2 + m_{11}^2$$

$$\sin^2\theta = \frac{m_\phi^2 - m_{88}^2}{m_\phi^2 - m_\omega^2}, \quad \cos^2\theta = \frac{m_{88}^2 - m_\omega^2}{m_\phi^2 - m_\omega^2}$$

obtained from (10.44) and (10.45) enable one conveniently to deduce

$$m_{11} = 888 \text{ MeV}, \quad m_{81} = 456 \text{ MeV}, \quad \cos\theta = 0.7685, \quad \theta = 40.0°$$

using as input

$$m_\phi = 1019 \text{ MeV}, \quad m_\omega = 784 \text{ MeV}$$

Thus we may say that the ϕ-meson has 60 per cent probability of being found as an octet member and 40 per cent as a singlet member.

In addition to explaining an apparent deviation from the octet mass formula this picture of particle mixing leads to further predictions (§10.7.3).

Particle mixing is to be expected whenever $SU(3)$ supermultiplets with the same spin and parity and approximately equal masses are found. In the case of the nine 2^+ mesons f(1260), $A_2(1310)$, K*(1420) and f′(1514) the mixing angle is found to be $\theta \simeq 34°$, the f′ being mostly octet member. For the 0^- mesons, mixing of $\eta'(958)$ with $\eta(550)$ is small, $\theta \simeq 10°$.

10.6 Decuplet decays

The assignment of hadron states to $SU(3)$ supermultiplets may be tested by comparing the $SU(3)$ predictions for their decay rates (partial widths) into various modes with the observed values. We shall illustrate this by considering the decays of the $\frac{3}{2}^+$ baryon decuplet into $\frac{1}{2}^+$ baryons and 0^- mesons. We suppose $SU(3)$ symmetry to be exact, then both I-spin and U-spin are conserved in the decay process. The decays allowed by I-spin and Y-conservation are

$$\Delta \to N\pi, \Sigma K; \quad \Sigma^* \to N\bar{K}, \Lambda\pi, \Sigma\pi, \Sigma\eta, \Xi K$$
$$\Xi^* \to \Lambda\bar{K}, \Sigma\bar{K}, \Xi\pi, \Xi\eta; \quad \Omega \to \Xi\bar{K}$$

When the actual masses of the particles are inserted, we find that only four of the decays are energetically allowed.

It will be shown that the amplitudes for all the processes listed can in principle be expressed in terms of a single parameter.

The possible decays of the $U = \frac{3}{2}$ multiplet $(\Delta^-, \Sigma^{*-}, \Xi^{*-}, \Omega^-)$ are

$$D(U = \tfrac{3}{2}) \to B(U = \tfrac{1}{2}) + M(U = 1) \tag{10.47a}$$

$$D(U = \tfrac{3}{2}) \to B(U = 1) + M(U = \tfrac{1}{2}) \tag{10.47b}$$

where

$$B(U = \tfrac{1}{2}) = (\Sigma^-, \Xi^-) \qquad M(U = 1) = (K^0, \pi^0_U, \bar{K}^0)$$

$$B(U = 1) = (n, \Sigma^0_U, \Xi^0) \quad M(U = \tfrac{1}{2}) = (\pi^-, K^-)$$

In each of these processes the various transition matrix elements are all given in terms of a single parameter and a U-spin Clebsch–Gordan coefficient, which can be read off from table 3.2 (cf. the remarks following (10.33)).

Thus for (10.47a) we find

$$\langle \Sigma^- K^0 | \mathcal{T} | \Delta^- \rangle = C^{3/2+3/2}_{1/2+1/2,1+1}\, g = g$$

$$\langle \Xi^- K^0 | \mathcal{T} | \Sigma^{*-} \rangle = C^{3/2+1/2}_{1/2-1/2,1+1}\, g = (\tfrac{1}{3})^{1/2} g$$

$$\langle \Sigma^- \pi^0_U | \mathcal{T} | \Sigma^{*-} \rangle = C^{3/2+1/2}_{1/2+1/2,1\,0}\, g = (\tfrac{2}{3})^{1/2} g$$

Table 10.4. *Comparison of experiment with SU(3) predictions for widths of decuplet decays*

Decay	Width predicted (MeV)	Width measured (MeV)
$\Delta(1236) \to N\pi$	Input	120
$\Sigma(1385) \to \Lambda\pi$	46	35 ± 8
$\to \Sigma\pi$	6	5.2 ± 1.5
$\Xi(1530) \to \Xi\pi$	16	7.5 ± 3

$$\langle \Xi^- \pi_U^0 | \mathcal{T} | \Xi^{*-} \rangle = (\tfrac{2}{3})^{1/2} g$$

$$\langle \Sigma^- \mathrm{K}^0 | \mathcal{T} | \Xi^{*-} \rangle = (\tfrac{1}{3})^{1/2} g$$

$$\langle \Xi^- \mathrm{K}^0 | \mathcal{T} | \Omega^- \rangle = g$$

Now the $U_3 = 0$ states $|\pi_U^0\rangle$ and $|\eta_U^0\rangle$ are expressed in terms of physical meson states by equations analogous to (10.35) and (10.36) which when inverted give

$$|\pi^0\rangle = \tfrac{1}{2}|\pi_U^0\rangle + \tfrac{1}{2}\cdot 3^{1/2}|\eta_U^0\rangle$$

$$|\eta^0\rangle = \tfrac{1}{2}\cdot 3^{1/2}|\pi_U^0\rangle - \tfrac{1}{2}|\eta_U^0\rangle$$

Since the decay

$$\mathrm{D}(U = \tfrac{3}{2}) \to \mathrm{B}(U = \tfrac{1}{2}) + \eta_U(U = 0)$$

is forbidden by *U*-spin, it follows that

$$\langle \Sigma^- \pi^0 | \mathcal{T} | \Sigma^{*-} \rangle = \tfrac{1}{2}\langle \Sigma^- \pi_U^0 | \mathcal{T} | \Sigma^{*-} \rangle = (\tfrac{1}{6})^{1/2} g$$

and

$$\langle \Sigma^- \eta^0 | \mathcal{T} | \Sigma^{*-} \rangle = \tfrac{1}{2}\cdot 3^{1/2}\langle \Sigma^- \pi_U^0 | \mathcal{T} | \Sigma^{*-} \rangle = (\tfrac{1}{2})^{1/2} g$$

and similarly for the Ξ^*-decay

$$\langle \Xi^- \pi^0 | \mathcal{T} | \Xi^{*-} \rangle = (\tfrac{1}{6})^{1/2} g$$

$$\langle \Xi^- \eta^0 | \mathcal{T} | \Xi^{*-} \rangle = (\tfrac{1}{2})^{1/2} g$$

A parallel calculation for the decay (10.47*b*) gives

$$\langle \mathrm{n}\pi^- | \mathcal{T} | \Delta^- \rangle = f, \qquad \langle \Xi^0 \pi^- | \mathcal{T} | \Xi^{*-} \rangle = (\tfrac{1}{3})^{1/2} f$$

$$\langle \mathrm{nK}^- | \mathcal{T} | \Sigma^{*-} \rangle = (\tfrac{1}{3})^{1/2} f, \qquad \langle \Sigma^0 \mathrm{K}^- | \mathcal{T} | \Xi^{*-} \rangle = (\tfrac{1}{6})^{1/2} f$$

$$\langle \Sigma^0 \pi^- | \mathcal{T} | \Sigma^{*-} \rangle = (\tfrac{1}{6})^{1/2} f, \qquad \langle \Lambda^0 \mathrm{K}^- | \mathcal{T} | \Xi^{*-} \rangle = (\tfrac{1}{2})^{1/2} f$$

$$\langle \Lambda^0\pi^- | \mathcal{T} | \Sigma^{*-}\rangle = (\tfrac{1}{2})^{1/2} f, \quad \langle \Xi^0 K^- | \mathcal{T} | \Omega^- \rangle = f$$

f and g are not independent. They are related by I-spin considerations, applied for example to the Ω^--decay. The Ω^- must go to the $I = 0$ state of $\Xi\bar{K}$,

$$[\Xi\bar{K}]_{I=0} = (\tfrac{1}{2})^{1/2}\{|\Xi^0 K^-\rangle - |\Xi^- \bar{K}^0\rangle\}$$

and hence

$$\langle \Xi^0 K^- | \mathcal{T} | \Omega^-\rangle = -\langle \Xi^- \bar{K}^0 | \mathcal{T} | \Omega^- \rangle$$

so that

$$g = -f$$

The remaining charge states can be related to those listed by use of I-spin conservation. Thus in exact $SU(3)$ all the decuplet decay amplitudes are given in terms of a single parameter g.

Experimentally, what is observed is not the lifetime τ of a sample of the decuplet particles, but rather the width $\Gamma = 1/\tau$ of the corresponding resonance in meson–baryon scattering. The extraction of Γ from the data is made by means of models. It is usual to fit the scattering amplitude to a Breit–Wigner form

$$\mathcal{T}(E) = \frac{1}{E_0 - E - \frac{1}{2}\mathrm{i}\Gamma}$$

in which the width Γ is energy dependent.

A variety of forms has been used. A simple energy dependence for decay into a state of orbital angular momentum l, is

$$\Gamma = C^2 g^2 \frac{m_B}{m_D} q^{2l+1} \tag{10.48}$$

where q is the relative momentum of the decay products, m_D and m_B are the decuplet and final baryon masses and C is the Clebsch–Gordan coefficient which we have calculated above. The q^{2l} is intended to represent an angular momentum barrier penetration factor. By using the actual physical particle masses in this expression some account is taken of $SU(3)$ symmetry breaking. g^2 represents an effective decay matrix element squared which is assumed to be the same for all the decuplet decays. It can be determined by fitting to the $\Delta(1236)$ decay width. The other three decay widths may then be predicted and the results are given in table 10.4. The fit is not an outstandingly good one, but displays the right trends.

A similar analysis can be made for other two body decays of supermultiplets of resonances of baryons or of mesons. Several detailed and critical fits have been made, investigating for example the effect of using different energy dependences for Γ. A recent one is that of Samios, Goldberg and Meadows (1974) and further references may be found in the 'Review of particle properties' (1974). The general conclusion seems to be that there is good agreement with the predictions of exact *SU*(3) symmetry in the way it has been used here, for all the supermultiplets so far identified.

10.7 Clebsch–Gordan coefficients for *SU*(3)

In considering the decuplet decays it was not obvious *a priori* that all the decay amplitudes could be expressed in terms of a *single* parameter. The reason for this becomes clear only when we enquire about the *SU*(3) transformation properties of states of two or more particles.

10.7.1 *Clebsch–Gordan series*

Consider two octets of particles which we shall refer to as a baryon octet (B) and a meson octet (M), although the method is quite general. There are 64 possible product states $|BM\rangle$. They do not form a single irreducible supermultiplet of *SU*(3). Instead sets of linear combinations of these states can be found such that under *SU*(3) operations, the states of each set transform only among themselves. Thus each of these sets must correspond to one of the irreducible supermultiplets allowed by the rules of §10.3. Nowhere was it required by those rules that the states of a supermultiplet be *single* particle states.

The first step in the analysis is to determine to which supermultiplets these irreducible sets correspond. We shall show that the 64 $|BM\rangle$ states decompose into a **27**, **10***, **10**, **8**, **8** and **1**. This result is written

$$\mathbf{8}\times\mathbf{8} = \mathbf{27}+\mathbf{10}^{*}+\mathbf{10}+\mathbf{8}+\mathbf{8}+\mathbf{1} \tag{10.49}$$

and is called the *Clebsch–Gordan series* for the product of two octets.

The product states $|BM\rangle$ can be labelled by the (Y, I_3) values of B and M, or by the total Y and total I_3

$$Y = Y_{\mathrm{B}}+Y_{\mathrm{M}}$$

$$I_3 = I_{3\mathrm{B}}+I_{3\mathrm{M}}$$

In the weight language the state $|BM\rangle$ has total weight

$$g = g_{\mathrm{B}}+g_{\mathrm{M}}$$

The 64 possible weights of the product states are obtained by simply adding vectorially all possible pairs of a g_{B} and a g_{M}. A syste-

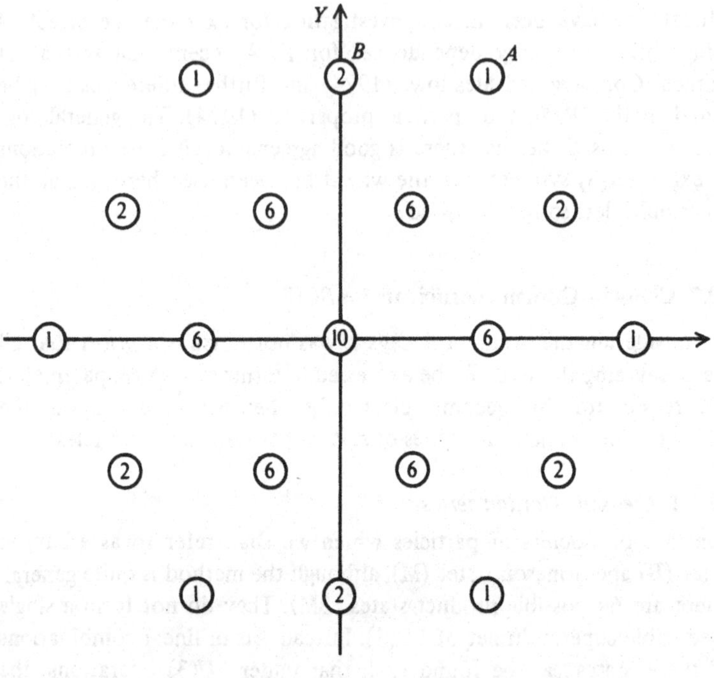

Fig. 10.16. Reducible weight diagram for **8** × **8**. The numbers indicate the multiplicities.

matic way to do this is as follows. We take the weight diagram for octet M and place its origin (0, 0) on each weight of the diagram of octet B in turn, at each step marking the positions where the weights of M lie. The marked points furnish the vectors $\boldsymbol{g}$. The weight diagram formed in this way is shown in fig. 10.16.

To find the irreducible supermultiplets contained in it, we seek the highest weight (A in the diagram). It is the highest weight of a **27**, fig. 10.13(c), and unambiguously signals the presence of a **27** supermultiplet. We therefore remove the corresponding 27 weights leaving the weight diagram in fig. 10.17. We again identify the highest weight: that of a **10*** and so remove the corresponding 10 weights. At the next state two **8**s can be removed followed by a **10** leaving finally a **1** at the origin. In this way we arrive at (10.49).

The technique we have described is perfectly general and can be used to reduce the product of any two (or even more) supermultiplets into its irreducible constituents. The reader may show for himself that the following reductions occur:

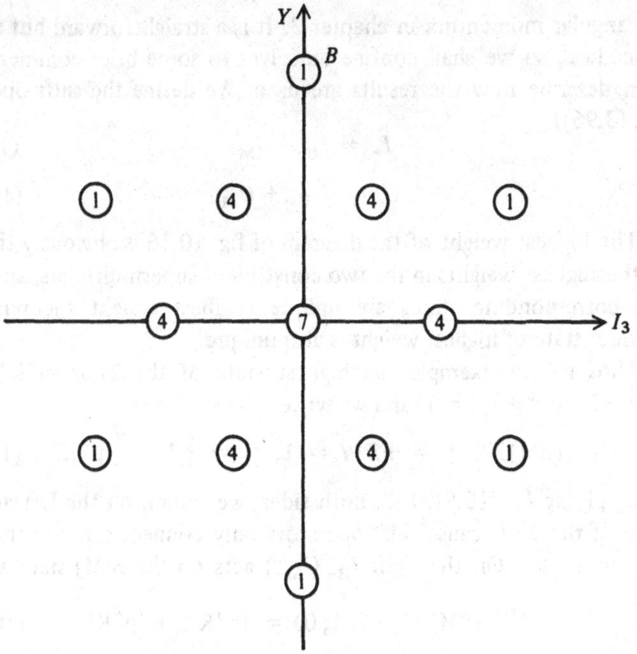

Fig. 10.17. Intermediate weight diagram in the reduction of **8** × **8**. States of the **27** have been removed. B is the highest state of the residue.

$$\mathbf{10} \times \mathbf{8} = \mathbf{35} + \mathbf{27} + \mathbf{10} + \mathbf{8} \tag{10.50a}$$

$$\mathbf{3} \times \mathbf{3}^* = \mathbf{8} + \mathbf{1} \tag{10.50b}$$

$$\mathbf{3} \times \mathbf{3} = \mathbf{6} + \mathbf{3}^* \tag{10.50c}$$

$$\begin{aligned} \mathbf{3} \times \mathbf{3} \times \mathbf{3} &= (\mathbf{6} + \mathbf{3}^*) \times \mathbf{3} \\ &= \mathbf{10} + \mathbf{8} + \mathbf{8} + \mathbf{1} \end{aligned} \tag{10.50d}$$

The **35** occurring in (10.50*a*) has the highest weight $Y = +2, I_3 = +2$, and the following I-spin multiplet structure: $\{Y, I\} = \{+2, 2\}, \{+1, \frac{5}{2}\}, \{+1, \frac{3}{2}\}, \{0, 2\}, \{0, 1\}, \{-1, \frac{3}{2}\}, \{-1, \frac{1}{2}\}, \{-2, 1\}, \{-2, 0\}$ and $\{-3, \frac{1}{2}\}$.

10.7.2 *Clebsch–Gordan coefficients*

The next step is to determine the Clebsch–Gordan coefficients. The method is exactly analogous to that used to find the C–G. coefficients for angular momentum in chapter 3. It is a straightforward but tedious procedure, so we shall confine ourselves to some brief comments and then describe how the results are used. We define the shift operators (cf. (3.96))

$$I_- = I_{B-} + I_{M-} \tag{10.51a}$$

$$U_- = U_{B-} + U_{M-} \tag{10.51b}$$

The highest weight of the diagram of fig. 10.16 is obviously the sum of the highest weights in the two constituent supermultiplets, and since the corresponding states are unique (highest weight theorem) the product state of highest weight is also unique.

Thus in our example the highest state of the 27 is $|p^+K^+\rangle$ with $Y = +2, I_3 = +1, I = 1$, and we write

$$|(BM)27,\ Y = +2,\ I = 1,\ I_3 = +1\rangle = |p^+K^+\rangle \tag{10.52a}$$

On applying I_-, (10.51*a*), to both sides, we obtain on the left another state of the 27 because shift operators only connect states in the same supermultiplet. On the right I_{B-} (I_{M-}) acts on the B(M) state and we find

$$2^{1/2}|(BM)27, +2, 1, 0\rangle = |n^0K^+\rangle + |p^+K^0\rangle \tag{10.52b}$$

The $2^{1/2}$ on the left is just the matrix element (3.47) of I_- between the $I = 1$ states. It is clear that here we are just calculating the NK states with $I = 1$, $Y = +2$, and the second application of I_- gives

$$|(BM)27, +2, 1, -1\rangle = |n^0K^0\rangle \tag{10.52c}$$

Equations (10.52*a–c*) may be summarised as follows

$$|(BM)27, +2, 1, I_3\rangle = [NK]_{1I_3}$$

where the symbol on the right represents the nucleon–kaon state of total isospin $I = 1$ and third component I_3.

To reach other Y-values, we must use U_- of (10.51*b*). It is necessary to know the matrix elements of U_- with respect to the states of the supermultiplets involved, and general formulae are available (e.g. Behrends, 1968). Here we shall quote results without proof.

U_- applied to (10.52*a*) gives

$$2^{1/2}|(BM)27, +1, \tfrac{3}{2}, +\tfrac{3}{2}\rangle = |\Sigma^+K^+\rangle + |p^+\pi^+\rangle \tag{10.53}$$

where we used

$$U_{B-}|p^+\rangle = |\Sigma^+\rangle, \quad U_{M-}|K^+\rangle = |\pi^+\rangle$$

Application of I_- to (10.53) simply generates the other isospin substates of ΣK and $N\pi$ with $I = \frac{3}{2}$, and we can anticipate the result

$$|(BM)27, +1, \tfrac{3}{2}, I_3\rangle = (\tfrac{1}{2})^{1/2}[\Sigma K]_{3/2 I_3} + (\tfrac{1}{2})^{1/2}[N\pi]_{3/2 I_3} \tag{10.54}$$

The numerical coefficients on the right are called *isoscalar factors*, because they are independent of the quantum numbers I_3. When multiplied by the isospin C–G. coefficients implicit in the bracket symbols on the right, they furnish the C–G. coefficients for *SU*(3).

The isoscalar factors are real and equal to the square roots of ratios of integers. The sum of the squares of the isoscalar factors in an expression of the form (10.54) must equal 1.

By repeated application of these techniques we may find the rest of the states of the **27**. The results are tabulated in appendix D.

Passing to the **10***, we see that its highest state can be identified as the $Y = +2, I_3 = 0$ state orthogonal to (10.52*b*); thus

$$|(BM)\mathbf{10}^*, +2, 0, 0\rangle = [NK]_{00}$$

Application of U_- leads to

$$|(BM)\mathbf{10}^*, +1, \tfrac{1}{2}, I_3\rangle$$
$$= \tfrac{1}{2}[N\pi]_{1/2 I_3} + \tfrac{1}{2}[\Sigma K]_{1/2 I_3} + \tfrac{1}{2}[N\eta]_{1/2 I_3} - \tfrac{1}{2}[\Lambda K]_{1/2 I_3}$$

and so on.

We have seen that there are two octets contained in **8** × **8**. These may be chosen to be symmetrical, $\mathbf{8}_S$, and antisymmetrical $\mathbf{8}_A$,† under the interchange $B \rightleftharpoons M$ i.e. $p \rightleftharpoons K^+$, $\Sigma^0 \rightleftharpoons \pi^0$, etc. made on the two particle states. This has been done in the tables of appendix D. Thus $[\Sigma\pi]_{00}$, which from the properties of the isospin C–G. coefficient is symmetrical in its factors, occurs in $|\mathbf{8}_S, 0, 0, 0\rangle$ but not in $|\mathbf{8}_A, 0, 0, 0\rangle$.

The **27** and **1** states are already symmetrical under the just mentioned symmetry operation, while the **10** and **10*** are antisymmetrical.

The symmetry of the couplings becomes physically significant when two identical octets are considered. For example, two 0^- meson octets may only couple as follows

$\mathbf{27}, \mathbf{8}_S, \mathbf{1}$ in states of angular momentum l = even

$\mathbf{10}, \mathbf{10}^*, \mathbf{8}_A$ in states of l = odd

† $\mathbf{8}_A$ and $\mathbf{8}_S$ are also referred to as $\mathbf{8}_F$ and $\mathbf{8}_D$ in the literature.

This follows from the generalised Pauli principle applied to the wavefunction of the two mesons considered as the product of an $SU(3)$ part and a spatial part.

We end this section by giving formal definitions of the C–G. coefficients in the general case (de Swart, 1963).

The supermultiplet labels are denoted by μ, μ_a, μ_b while the individual state labels Y, I, I_3 are denoted collectively by ν.

Given the product states $\phi^{\mu_a}_{\nu_a}\phi^{\mu_b}_{\nu_b}$ formed from the states of two supermultiplets μ_a and μ_b, the states with definite $SU(3)$ transformation properties are given by the linear combinations

$$\phi_{\nu}^{(\mu_a\mu_b)\mu_\gamma} = \sum_{\nu_a\nu_b} \left(\begin{array}{cc|c} \mu_a & \mu_b & \mu_\gamma \\ \nu_a & \nu_b & \nu \end{array}\right) \phi^{\mu_a}_{\nu_a}\phi^{\mu_b}_{\nu_b} \tag{10.55}$$

where μ_γ ranges over all supermultiplets contained in the Clebsch–Gordan series of $\mu_a \times \mu_b$. γ distinguishes multiple occurrences of a particular μ as with $\mathbf{8_A}$ and $\mathbf{8_S}$ in $\mathbf{8} \times \mathbf{8}$

$$\left(\begin{array}{cc|c} \mu_a & \mu_b & \mu_\gamma \\ \nu_a & \nu_b & \nu \end{array}\right)$$

is the $SU(3)$ C–G. coefficient. We have seen how the C–G. coefficient factorises

$$\left(\begin{array}{cc|c} \mu_a & \mu_b & \mu_\gamma \\ \nu_a & \nu_b & \nu \end{array}\right) = \left(\begin{array}{cc|c} \mu_a & \mu_b & \mu_\gamma \\ Y_aI_a & Y_bI_b & YI \end{array}\right) C^{II_3}_{I_aI_{3a}I_bI_{3b}} \tag{10.56}$$

into an isoscalar factor and an isospin C–G. coefficient.

The $SU(3)$ C–G. coefficient satisfies orthogonality relations which permit (10.55) to be inverted giving

$$\phi^{\mu_a}_{\nu_a}\phi^{\mu_b}_{\nu_b} = \sum_{\mu_\gamma\nu} \left(\begin{array}{cc|c} \mu_a & \mu_b & \mu_\gamma \\ \nu_a & \nu_b & \nu \end{array}\right) \phi_{\nu}^{(\mu_a\mu_b)\mu_\gamma} \tag{10.57}$$

10.7.3 *Applications*

The most immediate application of the C–G. apparatus is in counting the number of independent amplitudes for decays or scattering in an $SU(3)$ invariant theory.

To illustrate this, let us reconsider the decuplet decays. The amplitudes for the decay of the $\nu = (Y, I, I_3)$ member of the **10** into a baryon ν_B and a meson ν_M is given by the $\mathcal{T}$-matrix element,

$$\mathcal{T}_{\nu_B \nu_M, \nu} = \langle 8\,\nu_B, 8\,\nu_M | \mathcal{T} | 10\,\nu \rangle$$

We may expand the final BM state by means of (10.57) where μ runs over 27, $\mathbf{10}^*$, 10, 8, 8 and 1,

$$\mathcal{T}_{\nu_B \nu_M, \nu} = \sum_{\mu_\gamma \nu'} \begin{pmatrix} 8 & 8 & \mu_\gamma \\ \nu_B & \nu_M & \nu' \end{pmatrix} \langle (\mathrm{BM})\mu_\gamma \nu' | \mathcal{T} | 10\,\nu \rangle \tag{10.58}$$

If the interaction is invariant under *SU*(3) then

$$\langle (\mathrm{BM})\mu_\gamma \nu' | \mathcal{T} | 10\,\nu \rangle = \delta_{\mu_\gamma 10} \delta_{\nu' \nu} \mathcal{T}_{\mu_\gamma}$$

which expresses the fact that transitions can only take place between states belonging to the same supermultiplet type (in this example $\mathbf{10} \to \mathbf{10}$), and that the matrix element cannot depend on the particular substate concerned (i.e. on ν).

Since **10** only occurs once in $\mathbf{8} \times \mathbf{8}$ the γ is redundant and the decay amplitudes for *all* the decuplet members are expressible in terms of a single amplitude $\mathcal{T}_{10}$ and a C–G. coefficient

$$\mathcal{T}_{\nu_B \nu_M, \nu} = \begin{pmatrix} 8 & 8 & 10 \\ \nu_B & \nu_M & \nu \end{pmatrix} \mathcal{T}_{10}$$

In the case of a resonant baryon octet with two body decay

$$B_8^* \to B_8 + M_8$$

such as the $\frac{1}{2}^-$ or $\frac{3}{2}^-$, the decays are given in terms of two amplitudes $\mathcal{T}_{8A}$ and $\mathcal{T}_{8S}$ because of the double occurrence of **8** in $\mathbf{8} \times \mathbf{8}$.

In the case of meson decay the number of amplitudes is restricted by the generalised Pauli principle.

In the decay of a vector into two pseudo scalar mesons the final state must have $l = 1$ and hence must have an antisymmetric *SU*(3) wavefunction.

The $\mathbf{8}_S$ coupling of the final state is thus excluded, and there is only one *SU*(3) invariant amplitude to describe all the vector meson decays. However, if nonet mixing is taken into account another amplitude is introduced. Since two pseudo-scalar octet mesons coupled to a singlet **1** have no symmetric *SU*(3) wavefunction the decay of a *SU*(3) singlet vector meson into two pseudo-scalar mesons is forbidden. The ϕ–ω mixing picture, and in particular (10.46), tells us that the invariant coupling strengths for the physical ϕ- and ω-mesons to two 0^- mesons are respectively $\cos\theta$ and $-\sin\theta$ times that for the other octet members.

Similarly for the 2^+ meson decays into the pseudo-scalar mesons only 8_S coupling is permitted.

The virtual Yukawa process

$$B \rightleftharpoons B + M$$

which generalises $N \rightleftharpoons \pi N$ can be handled in a similar way. Here there are no symmetry considerations and there are two invariant amplitudes, i.e. coupling constants, in terms of which all the individual baryon–baryon–meson couplings are given in an $SU(3)$ invariant theory.

10.8 Electromagnetic effects in *SU*(3)

Since $SU(3)$ supermultiplets contain particles with different electric charges and these interact differently with the electromagnetic field, it follows that the electromagnetic interaction violates $SU(3)$ symmetry. Nevertheless, as with the medium strong interaction, a simple hypothesis about the way the symmetry is violated leads to testable predictions about the electromagnetic properties of hadrons.

10.8.1 *Magnetic moments of baryons*

The interaction of hadrons with the electromagnetic field is assumed to satisfy the principle of minimal electromagnetic coupling. This can be adequately formulated only within quantum field theory, but it means that the fundamental interaction is with the *charge* of the hadron, and that there is no fundamental magnetic moment. The magnetic moment is assumed to arise partly as a natural consequence of relativistic description of a spinning particle as in the Dirac equation, and partly from the hadronic structure, for example from the currents in the virtual charged pion cloud around a nucleon. For a spin-$\frac{1}{2}$ particle the latter contribution changes the magnetic moment from its Dirac value of 1 nuclear magneton.

The principle of minimal coupling means that the electromagnetic interaction H_{em} transforms like the charge operator Q which is one of the $SU(3)$ generators. Since Q commutes with U-spin all the members of a U-spin multiplet are predicted to have the same electromagnetic properties.

For the magnetic moments of the $\frac{1}{2}^+$ baryon octet we have

$$\mu(p) = \mu(\Sigma^+) \tag{10.59a}$$

$$\mu(\mathrm{n}) = \mu(\Sigma_U^0) = \mu(\Xi^0) \tag{10.59b}$$

$$\mu(\Sigma^-) = \mu(\Xi^-) \tag{10.59c}$$

$\mu(\mathrm{B})$ may be regarded as the expectation value in the baryon state $|\mathrm{B}\rangle$ of a magnetic moment operator,

$$\mu(\mathrm{B}) = \langle \mathrm{B}|\mu|\mathrm{B}\rangle$$

Hence we may transform the unphysical quantity $\mu(\Sigma_U^0)$ by means of (10.35), as follows

$$\mu(\Sigma_0^U) = \tfrac{1}{4}\mu(\Sigma^0) + \tfrac{3}{4}\mu(\Lambda^0) + \tfrac{1}{2}\cdot 3^{1/2}\mu(\Sigma^0\Lambda^0) \tag{10.59d}$$

where $\mu(\Sigma^0\Lambda^0)$ represents the transition magnetic moment which contributes to the decay $\Sigma \to \Lambda\gamma$.

Since H_{em} conserves U-spin,

$$\langle \Sigma_U^0 | H_{\mathrm{em}} | \Lambda_U^0 \rangle = 0$$

and so by (10.35) and (10.36)

$$\tfrac{1}{4}\cdot 3^{1/2}\mu(\Sigma^0) + \tfrac{1}{2}\mu(\Sigma^0\Lambda^0) - \tfrac{1}{4}\cdot 3^{1/2}\mu(\Lambda^0) = 0 \tag{10.59e}$$

To proceed further we observe that it follows from the equation

$$Q = \tfrac{1}{2}Y + I_3$$

and the relation between H_{em} and Q, that H_{em} is a sum of an isoscalar part and an isovector part,

$$H_{\mathrm{em}} = \tfrac{1}{2}H_{\mathrm{em}}^Y + H_{\mathrm{em}}^{I_3}$$

and the same follows for the magnetic moment operator

$$\mu = \tfrac{1}{2}\mu^Y + \mu^{I_3}$$

On taking the matrix element of this equation with respect to the states of an isospin multiplet, μ^Y contributes a constant term while μ^{I_3} gives a contribution linear in I_3, so that

$$\mu(I_3) = a + bI_3$$

When applied to the Σ-hyperon triplet, this leads to the relation

$$\mu(\Sigma^+) + \mu(\Sigma^-) = 2\mu(\Sigma^0) \tag{10.59f}$$

Finally we shall show that

$$\mu(\Lambda^0) = -\tfrac{1}{2}\mu(\Lambda_U^0) \tag{10.59g}$$

where by (10.36)

$$\mu(\Lambda_U^0) = \tfrac{3}{4}\mu(\Sigma^0) + \tfrac{1}{4}\mu(\Lambda^0) - \tfrac{1}{2}\cdot 3^{1/2}\mu(\Sigma^0\Lambda^0) \qquad (10.59h)$$

To obtain (10.59*g*) we note that only the isoscalar part μ^Y can contribute to the magnetic moment of the $I = 0$ Λ^0,

$$\mu(\Lambda^0) = \langle\Lambda^0|\tfrac{1}{2}\mu^Y|\Lambda^0\rangle \qquad (10.60)$$

Next we exploit the fact that the reflection operator

$$P_v = e^{-i\pi V_2}$$

introduced in § 10.3.1 has the property of interchanging I-spin with U-spin, and Y with Q. Thus

$$P_v^{-1} Y P_v = -Q \qquad (10.61)$$

and

$$P_v|\Lambda_U^0\rangle = |\Lambda^0\rangle \qquad (10.62)$$

This last equation is a special case of the general rule that P_v takes a state $|U = u, U_3 = u_3\rangle$ into $|I = u, I_3 = u_3\rangle$ in the same supermultiplet.

From (10.61) we obtain

$$P_v^{-1}\mu^Y P_v = -\mu^Q = -\mu$$

and then

$$\begin{aligned}\langle\Lambda^0|\tfrac{1}{2}\mu^Y|\Lambda^0\rangle &= \tfrac{1}{2}\langle\Lambda_U^0|P_v^{-1}\mu^Y P_v|\Lambda_U^0\rangle\\ &= -\tfrac{1}{2}\langle\Lambda_U^0|\mu^Q|\Lambda_U^0\rangle\\ &= -\tfrac{1}{2}\mu(\Lambda_U^0)\end{aligned}$$

giving (10.59*g*).

After using (10.59*a*) and (10.59*h*) to eliminate $\mu(\Lambda_U^0)$ and $\mu(\Sigma_U^0)$ we have seven relations between the nine magnetic moments including $\mu(\Sigma^0\Lambda^0)$. All the baryon magnetic moments are thus expressed in terms of those of the proton and neutron. One finds

$$\begin{aligned}\mu(\Sigma^+) &= \mu_p & \mu(\Lambda^0) &= \tfrac{1}{2}\mu_n\\ \mu(\Sigma^0) &= -\tfrac{1}{2}\mu_n & \mu(\Xi^0) &= \mu_n\\ \mu(\Sigma^-) &= -\mu_p - \mu_n & \mu(\Xi^-) &= -\mu_p - \mu_n\\ \mu(\Sigma^0\Lambda^0) &= -\tfrac{1}{2}\cdot 3^{1/2}\mu_n\end{aligned}$$

A comparison of the predictions with experiment is shown in table 10.5. The agreement is moderately good, although there appears to be a genuine disagreement in the case of $\mu(\Lambda^0)$ which is the best determined experimentally.

Table 10.5. *Comparison between observed magnetic moments and the SU(3) predictions*

	Experimental	Predicted
p	2.79	Input
n	−1.91	Input
Σ^+	2.59 ± 0.46	2.79
Σ^0	?	0.95
Σ^-	?	−0.88
Λ^0	−0.67 ± 0.06	−0.95
Ξ^0	?	−1.91
Ξ^-	−1.93 ± 0.75	−0.88

10.8.2 *Other electromagnetic processes*

Since the electromagnetic interaction represents the coupling of photons to hadrons, minimal coupling implies that the photon is a $U=0$ system, because it couples to the charge Q.

With the aid of U-spin one can exploit the consequences of this fact. For example consider photo-production of the decuplet resonances, in particular

$$\gamma + \mathrm{p} \to \Delta^0 + \pi^+$$

$$\gamma + \mathrm{p} \to \Sigma^{*0} + \mathrm{K}^+$$

The initial state $|\gamma\mathrm{p}\rangle$ has $U=\frac{1}{2}$, $U_3=\frac{1}{2}$, while Δ^0 and π^+ have $U=1$, $U_3=+1$ and $U=\frac{1}{2}$, $U_3=-\frac{1}{2}$ respectively so $\langle\Delta^0\pi^+|\mathcal{T}|\gamma\mathrm{p}\rangle$ is proportional to a U-spin Clebsch–Gordan coefficient $C^{1/2+1/2}_{1+1,\,1/2-1/2}$. Hence

$$\frac{\langle\Sigma^{*0}\mathrm{K}^+|\mathcal{T}|\gamma\mathrm{p}\rangle}{\langle\Delta^0\pi^+|\mathcal{T}|\gamma\mathrm{p}\rangle} = \frac{C^{1/2+1/2}_{10,\,1/2+1/2}}{C^{1/2+1/2}_{1+1,\,1/2-1/2}} = -2^{-1/2}$$

Another prediction concerns the decays $\pi^0 \to \gamma\gamma$ and $\eta^0 \to \gamma\gamma$. Since

$$|\pi^0_U\rangle = \tfrac{1}{2}|\pi^0\rangle + \tfrac{1}{2}\cdot 3^{1/2}|\eta^0\rangle$$

has $U=1$, the decay $\pi^0_U \to \gamma\gamma$ is forbidden

$$\langle\gamma\gamma|\mathcal{T}|\pi^0_U\rangle = 0$$

and hence for the physical decays

$$\Gamma(\pi^0 \to \gamma\gamma) = 3\Gamma(\eta^0 \to \gamma\gamma)$$

A phase space correction to take account of the $\pi - \eta$ mass difference is necessary before this can be tested.

10.8.3 *Mass formulae including electromagnetic effects*

Electromagnetic perturbation of the masses must be the same within a *U*-spin multiplet. Thus for the baryon octet we have

$$\delta\Sigma^+ = \delta p$$

$$\delta n^0 = \delta\Xi^0$$

$$\delta\Sigma^- = \delta\Xi^-$$

and therefore

$$\delta\Sigma^+ - \delta p + \delta n - \delta\Xi^0 + \delta\Xi^- - \delta\Sigma^- = 0$$

The unshifted masses as well as the contributions from the medium strong interaction H_{ms}, which are the same within an isospin multiplet, cancel out of this last form and hence we obtain

$$(n-p)-(\Sigma^- - \Sigma^+)+(\Xi^- - \Xi^0) = 0 \qquad (10.63)$$

This is the Coleman–Glashow six mass formula. An alternative derivation based on the 'parallelogram law' of Feldman and Matthews (1965) shows that it should be valid to all orders in H_{em} and to all orders in H_{ms} but with neglect of cross-terms starting with $H_{ms}H_{em}$. Since contributions from H_{em} and H_{ms} are of order α (i.e. 1 per cent) and 10 per cent respectively we may expect (10.63) to be valid to better than 1 per cent.

Using the 'Review of particle properties' (1973) we find

$$(n-p)-(\Sigma^- - \Sigma^+) = 1.293 - 7.93 = -6.6 \pm 0.1 \text{ MeV}$$

and

$$\Xi^- - \Xi^0 = 6.39 \pm 0.6 \text{ MeV},$$

a very satisfactory agreement.

This is a good test of $SU(3)$ as it is based not on specific transformation properties of the symmetry-breaking interactions, but only on the assumptions that H_{ms} conserves *I*-spin and that H_{em} conserves *U*-spin.

Space does not permit us to describe Cabibbo's elegant synthesis of strangeness-conserving and strangeness-changing leptonic weak interactions within an $SU(3)$ framework. We refer the reader to Gell-Mann and Ne'eman (1964), Gasiorowicz (1966) or Frazer (1966).

CHAPTER 11

THE QUARK MODEL

We have seen in the preceding chapter that the $SU(3)$ symmetry scheme is successful in ordering the hadron states and in explaining some of the relations between them, for example mass differences and decay rates. However, many questions remain unanswered. As with any pure symmetry scheme, $SU(3)$ gives a catalogue of possibilities: for example it gives an infinite list of supermultiplets but is not able to say which of these will be occupied by the particles of nature. A related puzzle is that the smallest supermultiplets, $\mathbf{3}$, $\mathbf{3}^*$, $\mathbf{6}$ etc., do not appear to correspond to any known particles. Also the origin of the symmetry-breaking interaction is a mystery.

One way to try to answer these questions is to go beyond $SU(3)$ and look at particular physical models which incorporate $SU(3)$. The simplest and most successful of these is the *quark model* originated by Gell-Mann and Zweig and elaborated by many other people (Gell-Mann, 1964; Zweig, 1964, 1965).

11.1 Quarks

The basic assumptions of the quark model are as follows:

(*a*) There exists a fundamental triplet of strongly interacting particles called quarks, denoted by† u, d and s, with the following properties: u and d form a strangeness zero isospin doublet, $I = \frac{1}{2}$, with isospin up and down respectively, while s is an isospin singlet, $I = 0$, with strangeness -1. All three have baryon number $B = \frac{1}{3}$.

(*b*) The observed hadrons are bound states of two or more quarks and/or antiquarks.

The quantum numbers of the quarks are listed in table 11.1. It can be seen that the quarks fit into a $\mathbf{3}$ of $SU(3)$, and the antiquarks into a $\mathbf{3}^*$, as shown in fig. 11.1.

† The quarks are often denoted by p, n and λ, because of the analogy with the Sakata model in which the proton, neutron and Λ^0 form a triplet. However, lower-case p and n are standard notation for the physical proton and neutron, and N is used for the nucleon, so a different notation seems desirable.

Table 11.1. *Quantum numbers of the quarks*

Quark label	B	Y	I	I_3	$Q = I_3 + \frac{1}{2}Y$	$S = Y - B$
u	$\frac{1}{3}$	$\frac{1}{3}$	$\frac{1}{2}$	$+\frac{1}{2}$	$\frac{2}{3}$	0
d	$\frac{1}{3}$	$\frac{1}{3}$	$\frac{1}{2}$	$-\frac{1}{2}$	$-\frac{1}{3}$	0
s	$\frac{1}{3}$	$-\frac{2}{3}$	0	0	$-\frac{1}{3}$	-1
$\bar{s}$	$-\frac{1}{3}$	$\frac{2}{3}$	0	0	$\frac{1}{3}$	$+1$
$\bar{d}$	$-\frac{1}{3}$	$-\frac{1}{3}$	$\frac{1}{2}$	$+\frac{1}{2}$	$\frac{1}{3}$	0
$\bar{u}$	$-\frac{1}{3}$	$-\frac{1}{3}$	$\frac{1}{2}$	$-\frac{1}{2}$	$-\frac{2}{3}$	0

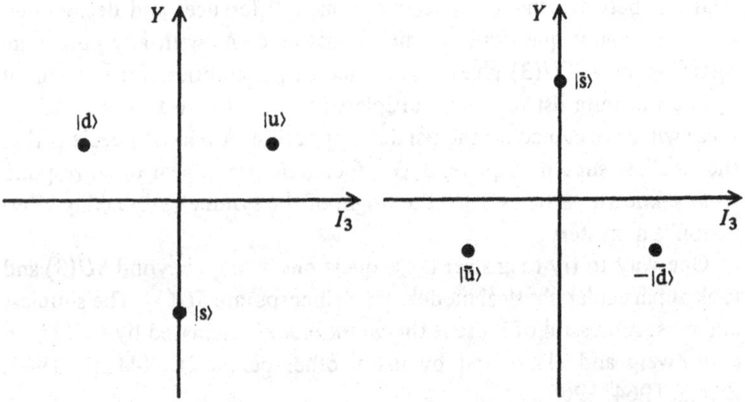

Fig. 11.1 Weight diagrams of the quark 3 and antiquark 3*. The phase convention for the antiquark isodoublet ($\bar{d}$, $\bar{u}$) is in accordance with appendix C.

If the binding forces are symmetrical with respect to the three quarks, then the model is invariant under $SU(3)$ and the bound states must form supermultiplets of $SU(3)$.†

By assuming a particular form of quark–quark interaction we may calculate the energies of the bound states, i.e. the masses of the physical particles, and their spins and parities.

Since the quarks have baryon number $B = \frac{1}{3}$, the observed baryons are constructed out of three quarks, qqq, while mesons with $B = 0$ are formed from equal numbers of quarks and antiquarks. The simplest possibility is a quark–antiquark pair $q\bar{q}$.

We shall start by showing how the observed meson and baryon supermultiplets can be constructed from the quarks. The symmetry of the

† This is analogous to the situation in light nuclei: if the forces are symmetrical with respect to protons and neutrons, the energy levels form isospin multiplets.

wavefunction under permutations of the quarks will be important here. Then we shall make the additional assumption that the quarks have spin $\frac{1}{2}$ and that the quark–quark forces are independent both of the $SU(3)$ character (i.e. u, d or s) of the quarks and of their spin orientation (↑ or ↓). This leads us to the $SU(6)$ symmetry group. Although this can be formulated as a symmetry independently of the quark model, we shall use it as an aid to book-keeping in the classification of the states. It will then be seen how the spins and parities of the baryon and meson states can be successfully accounted for. Other consequences of the quark model such as the electromagnetic properties of baryons are briefly reviewed and we end with some remarks on the question of the existence of free quarks.

11.2 Quark model of mesons

If we assume that mesons have the simplest structure, namely $q\bar{q}$, the nine possible states can be displayed on a weight diagram, fig. 11.2. Clearly there is an $SU(3)$ octet and a singlet.

Assuming that a strongly attractive force exists between q and $\bar{q}$ giving rise to binding in states of the appropriate angular momentum and parity, we have a model of the observed mesons. We shall say more about the space and spin dependence of the forces later. For the present we are concerned with the $SU(3)$ aspects, and the important result is that the $q\bar{q}$ structure allows only octets and singlets of meson states, in accord with observation.

In order to form a higher supermultiplet such as a 27, which the observation of, for example, doubly charged mesons or KK resonances would imply, we should have to consider structures such as $q\bar{q}q\bar{q}$. As yet there is no need to do this.

The $q\bar{q}$ structure of the octet mesons is shown in table 11.2. The only non-trivial calculation involved is the determination of the correct linear combinations of $q\bar{q}$ states with $I_3 = Y = 0$. Bearing in mind that $I_{\pm}$ and $U_{\pm}$ have positive matrix elements between the states $+|\bar{s}\rangle$, $-|\bar{d}\rangle$ and $+|\bar{u}\rangle$ of $\bar{q}$ (see appendix C), the $I = 1, I_3 = 0$ state is obtained from the $I = 1, I_3 = +1$ state $-|u\bar{d}\rangle$ by the shift operator

$$(I_-^q + I_-^{\bar{q}})(-|u\bar{d}\rangle) = -|d\bar{d}\rangle + |u\bar{u}\rangle$$

followed by normalisation to give

$$|I = 1, I_3 = 0\rangle = 2^{-1/2}(|u\bar{u}\rangle - |d\bar{d}\rangle) \tag{11.1}$$

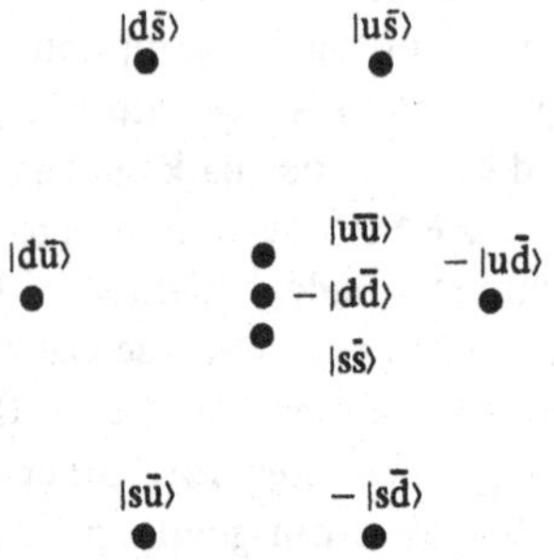

Fig. 11.2. Quark–antiquark states displayed on the reducible weight diagram of 3 × 3*.

Table 11.2. *Quark structure of pseudo-scalar and vector mesons*

Pseudo-scalar	Vector	Quark structure
K^+	K^{*+}	$\|u\bar{s}\rangle$
K^0	K^{*0}	$\|d\bar{s}\rangle$
π^+	ρ^+	$-\|u\bar{d}\rangle$
π^0	ρ^0	$2^{-1/2}(\|u\bar{u}\rangle-\|d\bar{d}\rangle)$
π^-	ρ^-	$\|d\bar{u}\rangle$
η^0	ϕ_8	$6^{-1/2}(-\|u\bar{u}\rangle-\|d\bar{d}\rangle+2\|s\bar{s}\rangle)$
$\bar{K}^0$	$\bar{K}^{*0}$	$-\|s\bar{d}\rangle$
K^-	K^{*-}	$\|s\bar{u}\rangle$
$\eta^{0'}$	ω_1	$3^{-1/2}(\|u\bar{u}\rangle+\|d\bar{d}\rangle+\|s\bar{s}\rangle)$

The $U=1$, $U_3=0$ state at the origin obtained by the U-spin lowering operator applied to $|d\bar{s}\rangle$ is

$$|U=1, U_3=0\rangle = 2^{-1/2}(-|d\bar{d}\rangle+|s\bar{s}\rangle)$$

Then, from (10.35) relating I-spin and U-spin eigenstates in an octet, we find

$$|I=0, I_3=0\rangle = 6^{-1/2}(-|u\bar{u}\rangle-|d\bar{d}\rangle+2|s\bar{s}\rangle) \quad (11.2)$$

Finally the remaining state at the origin orthogonal to (11.1) and (11.2) must be the $SU(3)$ singlet

$$|1, Y=I=I_3=0\rangle = 3^{-1/2}(|u\bar{u}\rangle+|d\bar{d}\rangle+|s\bar{s}\rangle) \quad (11.3)$$

If the $q\bar{q}$ interaction is completely symmetrical between the three kinds of quarks, all nine bound $q\bar{q}$ states must be degenerate in energy. However, we may expect the $q\bar{q}$ interaction to depend on the $SU(3)$ supermultiplet character of the system, **8** or **1**, and in this case the masses of the octet and singlet bound states will differ.

Let us now turn to violation of $SU(3)$ symmetry in this model. A simple and natural hypothesis is that the mass splittings among the observed mesons are due to a difference in mass between the singlet quark s and the isodoublet (u, d). We wish to retain isospin symmetry and so we keep m_u equal to m_d, but set

$$m_s = M + m$$

where

$$M = m_u = m_d$$

It is assumed that the bound state wavefunctions have been calculated in the symmetry limit $m = 0$ and that they are insensitive to the change of mass. On taking the matrix elements of the perturbed Hamiltonian with respect to the states given in table 11.2, we find for the masses of the mesons,

$$\begin{aligned} m(\mathrm{K}^*) &= m(\mathrm{K}) = 2M + m - E_{\mathrm{B}}(8) \\ m(\rho) &= 2M - E_{\mathrm{B}}(8) \\ m(\phi_8) &= (\tfrac{1}{6}+\tfrac{1}{6})2M + \tfrac{4}{6}(2M+2m) - E_{\mathrm{B}}(8) \\ &= 2M + \tfrac{4}{3}m - E_{\mathrm{B}}(8) \\ m(\omega_1) &= (\tfrac{1}{3}+\tfrac{1}{3})2M + \tfrac{1}{3}(2M+2m) - E_{\mathrm{B}}(1) \\ &= 2M + \tfrac{2}{3}m - E_{\mathrm{B}}(1) \end{aligned}$$

The notation here is that $2M - E_{\mathrm{B}}(\mathbf{n})$ is the binding energy in the symmetry limit in the supermultiplet **n**.

We see that the mass splitting is given by

$$m = m(\mathrm{K}^*) - m(\rho)$$

so, using the experimental data,

$$m = 110\,\mathrm{MeV}$$

We also find

$$4m(\mathrm{K}^*) = m(\rho) + 3m(\phi_8) \qquad (11.4)$$

which is just the Gell-Mann–Okubo formula. However, there is now a non-zero matrix element of the mass term in the $q\bar{q}$ Hamiltonian between the ϕ_8 and ω_1 states

$$\langle \phi_8 | H | \omega_1 \rangle = 18^{-1/2}\{-2M - 2M + 2(2M+2m)\}\mathscr{I} = -2\cdot 2^{1/2}\mathscr{I}m/3$$

where $\mathscr{I}$ is an overlap integral of $q\bar{q}$ space–spin wavefunctions. This can only be obtained from a more detailed model. We shall not pursue this further, but if $\mathscr{I}$ is known the mixing angle can be calculated using the formulae of §10.5.4.

If the $q\bar{q}$ potential has the maximal symmetry referred to above, then

$$E_B(8) = E_B(1) \tag{11.5}$$

In this case it is expected that the physical states are not ϕ_8 and ω_1 of the table, but the ones in which the quarks of different mass are separated. Thus we are led to

$$|\phi\rangle = |s\bar{s}\rangle \tag{11.6a}$$

$$|\omega\rangle = 2^{-1/2}(|u\bar{u}\rangle + |d\bar{d}\rangle) \tag{11.6b}$$

with masses

$$m(\phi) = 2M + 2m - E_B$$

$$m(\omega) = 2M - E_B$$

Hence in this particular case it is predicted that

$$m(\omega) = m(\rho) \tag{11.7a}$$

$$m(\omega) + m(\phi) = 2m(K^*) \tag{11.7b}$$

These are satisfied experimentally to about 1 per cent, which gives a limit on the accuracy of (11.5).

The states (11.6) when written in terms of the states $|\phi_8\rangle$ and $|\omega_1\rangle$ correspond to a mixing angle (in this case called *ideal mixing*) of

$$\tan\theta = 2^{-1/2} \quad \text{or} \quad \theta = 35.3^\circ$$

which is very close to the value obtained in the phenomenological analysis of vector meson mixing.

In the case of the pseudo-scalar mesons the equalities (11.7) are not satisfied and the mixing of the $\eta'(958)$ with $\eta(550)$ is small, $\theta \simeq 10^\circ$. This suggests that in this case it is not a good approximation to equate the binding energies as in (11.5).

11.3 Quark model of baryons

The low-lying baryon states ($B = 1$) can be constructed out of three quarks, qqq. It has to be assumed that the properties of the qq force are such that three quarks bind in a stable configuration but that two or four do not. Again we start by considering the $SU(3)$ properties of the bound system.

11.3.1 *Classification of three quark states*

Consider first the two quark states. On combining the weight diagrams

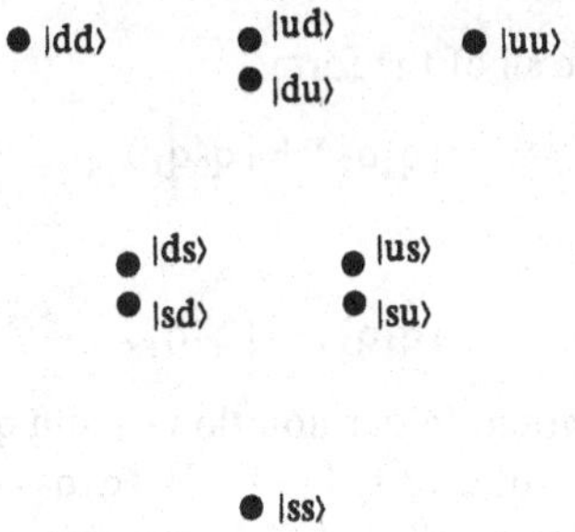

Fig. 11.3. Two quark states displayed on the reducible weight diagram of 3 × 3.

of two **3**s by the method of § 10.7 we find the result shown in fig. 11.3. $|uu\rangle$ is the highest state of a **6**, fig. 10.10(*a*). The states of the **6** are

$$|dd\rangle \qquad 2^{-1/2}\{|ud\rangle + |du\rangle\} \qquad |uu\rangle$$

$$2^{-1/2}\{|ds\rangle + |sd\rangle\} \quad 2^{-1/2}\{|us\rangle + |su\rangle\}$$

$$|ss\rangle$$

Here the three extreme states were read off directly, and in the case of the doubly occupied weights the correct linear combinations are obtained by use of shift operators

$$I_-|uu\rangle = |du\rangle + |ud\rangle$$

$$U_-|dd\rangle = |sd\rangle + |ds\rangle$$

When the weights of the **6** are removed from fig. 11.3, we see that the weights of a **3*** remain. Since the corresponding states must be orthogonal to those of the **6**, they are

$$2^{-1/2}\{|ud\rangle - |du\rangle\}$$

$$2^{-1/2}\{|ds\rangle - |sd\rangle\} \quad 2^{-1/2}\{|us\rangle - |su\rangle\}$$

This completes the reduction

$$\mathbf{3} \times \mathbf{3} = \mathbf{6} + \mathbf{3}^*$$

We see here another instance of the rule that when combining two identical supermultiplets, all the states of an irreducible supermultiplet formed from them have the same definite symmetry type under interchange of their constituents. Thus, denoting the general two quark state by

$$|q_1q_2\rangle$$

the states of the **6** are all of the form

$$|q_1q_2\rangle + |q_2q_1\rangle \tag{11.8}$$

and those of the $\mathbf{3}^*$ are

$$|q_1q_2\rangle - |q_2q_1\rangle \tag{11.9}$$

apart from normalisation. In our notation q_1 and q_2 are variable indices which take the values u, d or s. In (11.8), (q_1q_2) can be uu, dd, ss, ud, ds or su, giving six states; while (11.9) is non-zero only if $q_1 \neq q_2$, allowing three states.

The next step is to combine the two quark states with a third quark. Compounding the symmetrical qq states of the **6** with a one quark **3** we find the weight diagram of fig. 11.4. The highest state $|uuu\rangle$ with $Y=1$, $I_3 = +\frac{3}{2}$ is that of a **10**: when the weights of a **10** are removed those of an **8** remain. Therefore

$$\mathbf{6} \times \mathbf{3} = \mathbf{10} + \mathbf{8} \tag{11.10}$$

Compounding the antisymmetrical qq states of $\mathbf{3}^*$ with the third quark in the same way we find

$$\mathbf{3}^* \times \mathbf{3} = \mathbf{8} + \mathbf{1} \tag{11.11}$$

Thus the 3^3 states of three quarks decompose into the $SU(3)$ supermultiplets **10**, **8** (twice) and **1**. This four-fold classification is in exact correspondence with the classification of states of three particles according to their symmetry type under permutations of the particles.

We shall summarise without proof the classification of three particle states by symmetry type. This is generalisation of the well known classification of two particle states into symmetric and antisymmetric parts.

In general terms we have a system of three identical objects each of which can be in any one of a number n of states, and we denote by $|q_1q_2q_3\rangle$ a state of the system in which the first object is in state q_1, the second is in q_2 and the third in q_3. For the case of quarks $n = 3$ and q_1, q_2 and q_3 can each take the values u, d or s.

Then we can always choose the states of the system to be one of the four symmetry types:

1. *Totally symmetric*, S: This state is unchanged under any permutation of the particle labels

$$|(q_1q_2q_3)S\rangle = |q_1q_2q_3\rangle + |q_2q_3q_1\rangle + |q_3q_1q_2\rangle + |q_2q_1q_3\rangle + |q_3q_2q_1\rangle + |q_1q_3q_2\rangle \tag{11.12}$$

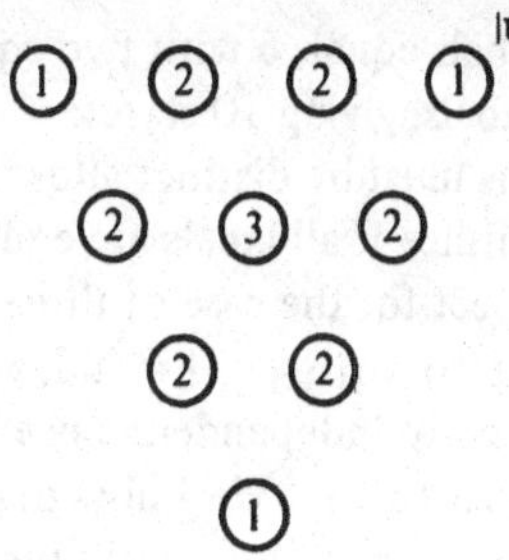

Fig. 11.4. Reducible weight diagram of **6** × **3**.

Of course if two or more labels become equal this expression simplifies.

2. *Totally antisymmetric*, A: Under an even (odd) permutation this state is unchanged (multiplied by -1):

$$|(q_1q_2q_3)A\rangle = |q_1q_2q_3\rangle - |q_2q_1q_3\rangle + |q_2q_3q_1\rangle - |q_3q_2q_1\rangle + |q_3q_1q_2\rangle - |q_1q_3q_2\rangle \qquad (11.13)$$

Here the three labels must all be different.

3 and 4. *Mixed symmetry*: Here there are two states for each choice of quark labels. The effect of an arbitrary permutation is to transform each state into a certain linear combination of the same two states. These two states can be chosen in many different ways. A convenient choice is

$$|(q_1q_2q_3)M_1\rangle = 2|q_1q_2q_3\rangle + 2|q_2q_1q_3\rangle - |q_2q_3q_1\rangle - |q_3q_2q_1\rangle - |q_3q_1q_2\rangle - |q_1q_3q_2\rangle \qquad (11.14)$$

$$|(q_1q_2q_3)M_2\rangle = 2|q_1q_2q_3\rangle - 2|q_2q_1q_3\rangle - |q_2q_3q_1\rangle + |q_3q_2q_1\rangle - |q_3q_1q_2\rangle + |q_1q_3q_2\rangle \qquad (11.15)$$

With the aid of the orthogonality relations

$$\langle q_1'q_2'q_3'|q_1q_2q_3\rangle = \delta_{q_1'q_1}\delta_{q_2'q_2}\delta_{q_3'q_3}$$

it can be shown that the four states are mutually orthogonal. One may also verify that

$$|q_1q_2q_3\rangle = \tfrac{1}{6}\{|S\rangle + |A\rangle + |M_1\rangle + |M_2\rangle\}$$

from which it follows that any three quark state may be expressed in terms of the four standard states.

Next let us count the number of distinct three quark states of each symmetry type.

Symmetric: 3 with all labels equal, 6 with two equal and one distinct, and 1 with three distinct labels, giving 10 states.

Antisymmetric: all labels must be distinct allowing 1 state.

Mixed state, M: M_1 vanishes if all labels are equal. It is symmetric in q_1 and q_2 so we might expect for the case of three labels distinct, three states $|M_1\rangle$ corresponding to $(q_1q_2, q_3) = (ud, s)$, (su, d) or (ds, u). However, these are not linearly independent since their sum is a totally symmetric state proportional to $|S\rangle$. We know that $|M_1\rangle$ is orthogonal to $|S\rangle$; hence it has only two states with three different quarks. Similarly for the case of two labels equal and one distinct, the states obtained by setting $(q_1q_2, q_3) = (uu, d)$ and (ud, u) are proportional, allowing only six states of this kind. The conclusion is that there are eight states of symmetry type M_1. Similarly one finds eight states of type M_2.

We have now exposed the connection between symmetry types and $SU(3)$ supermultiplets for three quark states. In the Clebsch–Gordan decomposition

$$\begin{aligned} 3 \times 3 \times 3 &= (6 + 3^*) \times 3 \\ &= \mathbf{10} + \mathbf{8} + \mathbf{8}' + \mathbf{1} \end{aligned} \tag{11.16}$$

the states of the **10** are totally symmetric, the **1** is totally antisymmetric, while the states of **8** and **8**$'$ are of mixed symmetry.

The three quark states normalised and with overall phase factors to agree with our previous conventions are listed in table 11.3.

A note on the definition of the mixed symmetry states M_1 and M_2. M_1 was obtained from $|q_1q_2q_3\rangle$ by the following sequence:

(*a*) symmetrise in positions 1 and 2;

(*b*) antisymmetrise in positions 1 and 3;

(*c*) symmetrise in positions 1 and 2.

M_2 was obtained by an analogous procedure but with antisymmetrisation in (*a*) and (*c*) and symmetrisation in step (*b*).

The proper tool for the symmetry analysis of n particle states is the theory of the permutation group on n objects and its representations, as given for example in Hamermesh (1962). The states S, A, M_1 and M_2 correspond to the Young tableaux

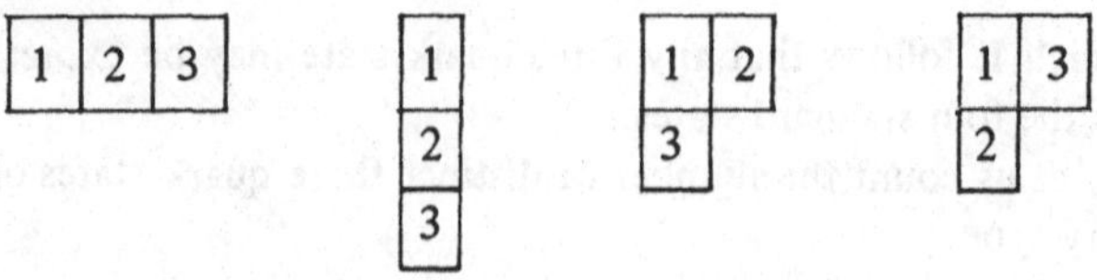

Table 11.3. *Quark structure of baryon states* $|\mathbf{n}, Y, I, I_3\rangle$

$$|10, +1, \tfrac{3}{2}, +\tfrac{3}{2}\rangle = |uuu\rangle$$
$$|10, +1, \tfrac{3}{2}, +\tfrac{1}{2}\rangle = 3^{-1/2}\{|udu\rangle + |duu\rangle + |uud\rangle\}$$
$$|10, +1, \tfrac{3}{2}, -\tfrac{1}{2}\rangle = 3^{-1/2}\{|ddu\rangle + |udd\rangle + |dud\rangle\}$$
$$|10, +1, \tfrac{3}{2}, -\tfrac{3}{2}\rangle = |ddd\rangle$$
$$|10, 0, 1, +1\rangle = 3^{-1/2}\{|usu\rangle + |suu\rangle + |uus\rangle\}$$
$$|10, 0, 1, 0\rangle = 6^{-1/2}\{|uds\rangle + |dsu\rangle + |sud\rangle + |dus\rangle + |sdu\rangle + |usd\rangle\}$$
$$|10, 0, 1, -1\rangle = 3^{-1/2}\{|dsd\rangle + |sdd\rangle + |dds\rangle\}$$
$$|10, -1, \tfrac{1}{2}, +\tfrac{1}{2}\rangle = 3^{-1/2}\{|ssu\rangle + |uss\rangle + |sus\rangle\}$$
$$|10, -1, \tfrac{1}{2}, -\tfrac{1}{2}\rangle = 3^{-1/2}\{|ssd\rangle + |dss\rangle + |sds\rangle\}$$
$$|10, -2, 0, 0\rangle = |sss\rangle$$

$$|8, +1, \tfrac{1}{2}, +\tfrac{1}{2}\rangle = 6^{-1/2}\{2|uud\rangle - |udu\rangle - |duu\rangle\}$$
$$|8, +1, \tfrac{1}{2}, -\tfrac{1}{2}\rangle = 6^{-1/2}\{|dud\rangle + |udd\rangle - 2|ddu\rangle\}$$
$$|8, 0, 1, +1\rangle = 6^{-1/2}\{2|uus\rangle - |usu\rangle - |suu\rangle\}$$
$$|8, 0, 1, 0\rangle = 12^{-1/2}\{2|uds\rangle - |dsu\rangle - |sud\rangle + 2|dus\rangle - |sdu\rangle - |usd\rangle\}$$
$$|8, 0, 1, -1\rangle = 6^{-1/2}\{2|dds\rangle - |dsd\rangle - |sdd\rangle\}$$
$$|8, 0, 0, 0\rangle = \tfrac{1}{2}\{|usd\rangle + |sud\rangle - |sdu\rangle - |dsu\rangle\}$$
$$|8, -1, \tfrac{1}{2}, +\tfrac{1}{2}\rangle = 6^{-1/2}\{|sus\rangle + |uss\rangle - 2|ssu\rangle\}$$
$$|8, -1, \tfrac{1}{2}, -\tfrac{1}{2}\rangle = 6^{-1/2}\{|sds\rangle + |dss\rangle - 2|ssd\rangle\}$$

$$|8', +1, \tfrac{1}{2}, +\tfrac{1}{2}\rangle = 2^{-1/2}\{|udu\rangle - |duu\rangle\}$$
$$|8', +1, \tfrac{1}{2}, -\tfrac{1}{2}\rangle = 2^{-1/2}\{|udd\rangle - |dud\rangle\}$$
$$|8', 0, 1, +1\rangle = 2^{-1/2}\{|usu\rangle - |suu\rangle\}$$
$$|8', 0, 1, 0\rangle = \tfrac{1}{2}\{|usd\rangle + |dsu\rangle - |sdu\rangle - |sud\rangle\}$$
$$|8', 0, 1, -1\rangle = 2^{-1/2}\{|dsd\rangle - |sdd\rangle\}$$
$$|8', 0, 0, 0\rangle = 12^{-1/2}\{2|uds\rangle - |dsu\rangle - |sud\rangle - 2|dus\rangle + |sdu\rangle + |usd\rangle\}$$
$$|8', -1, \tfrac{1}{2}, +\tfrac{1}{2}\rangle = 2^{-1/2}\{|uss\rangle - |sus\rangle\}$$
$$|8', -1, \tfrac{1}{2}, -\tfrac{1}{2}\rangle = 2^{-1/2}\{|dss\rangle - |sds\rangle\}$$

$$|1, 0, 0, 0\rangle = 6^{-1/2}\{|uds\rangle + |dsu\rangle + |sud\rangle - |dus\rangle - |sdu\rangle - |usd\rangle\}$$

The recipe given in the text corresponds to using Thrall's (1941) modification of the Young symmetrisers.

11.3.2 *Physics of three quark states*

We have seen that the three quarks can form only the supermultiplets **10**, **8** and **1**. The totally symmetric decuplet **10** is identified with the $SU(3)$ decuplet of excited baryons with spin $\frac{3}{2}$, namely Δ^{++}, Δ^{+}, Δ^{0}, Δ^{-}, Σ^{*+}, Σ^{*0}, Σ^{*-}, Ξ^{*0}, Ξ^{*-}, Ω^{-} as indicated in table 11.3.

The octet of ordinary baryons of spin $\frac{1}{2}$, p, n, Σ^{+}, Σ^{0}, Σ^{-}, Λ^{0}, Ξ^{0} and Ξ^{-}, is allocated to a mixed symmetry octet. Each state of this octet will in general correspond to a superposition of the corresponding states of the mixed symmetry octets M_1 and M_2, but without discussing the details of the space–spin part of the wavefunction, it is not possible to say what this combination is. The correlation between $SU(3)$ multiplet and ordinary spin is discussed in $SU(6)$.

No particle corresponding to the unitary singlet three quark state has hitherto been observed.

Thus assuming that the qq forces lead to binding in a favourable way we see that the quark model can explain why only decuplets and octets of $B = 1$ states are observed.

It is an important empirical regularity of the baryon spectrum that no $Y = +2$ resonances have been firmly established. Such states would require a **27** or **10*** of $SU(3)$, and in the quark model these could only be made out of (qqq)(q$\bar{q}$), or, more generally, three quarks combined with several q$\bar{q}$ pairs.

Any states which cannot be formed as qqq bound states (if baryons) or as q$\bar{q}$ bound states (if mesons) are called *exotic*. The absence of exotic states in Nature gives strong support for the hypotheses of the simple quark model.

Symmetry breaking can be introduced for the baryons in the same way as for the mesons. For a decuplet, the states with $Y = +1, 0, -1$ and -2 have 0, 1, 2 and 3 s-type quarks respectively, giving immediately the equal spacing rule

$$\Sigma - \Delta = \Xi - \Sigma = \Omega - \Xi = m$$

From the observed masses we obtain for the quark mass difference

$$m = 145\,\text{MeV}$$

which is rather higher than the estimate of 110 MeV made using the vector meson data.

If the same argument is applied to an octet, an equal spacing rule is predicted there too,

$$\Sigma - \text{N} = \Xi - \Sigma$$

$$\Lambda = \Sigma$$

and the first of these is in poor agreement with observation: $\Sigma - \text{N} = 254\,\text{MeV}$; $\Xi - \Sigma = 125\,\text{MeV}$. This discrepancy must be attributed to deviations from $SU(3)$ symmetry in the qq interaction (Federman, Rubinstein and Talmi, 1966).

11.4 Introduction of spin into the quark model: $SU(6)$

The successes of the $SU(3)$ aspects of the quark picture suggest that it may be worth while taking a more realistic view of quarks. Since we must account for the half-integral spin of the baryons it is necessary to ascribe a half-integral spin to the quarks.

Let us assume that the three quarks each carry an intrinsic spin of $\frac{1}{2}\hbar$. Since no particles with the properties of these quarks have been conclusively identified experimentally, it must be supposed that the quark mass is large ($\gtrsim$ 10 GeV) in order to bring the cross-section for production of $q\bar{q}$ pairs down below the limits imposed by accelerator searches.

One could start by assuming a force law between quarks and between quark and antiquark, and proceed to calculate the properties of the baryon (qqq) and meson ($q\bar{q}$) bound states. In carrying out this programme one is hampered by the difficulty of doing relativistic bound state calculations, especially in the case of *three* particles. Even the $q\bar{q}$ bound state appears to be an extreme relativistic case for the total quark mass of say 20 GeV is almost completely cancelled by the binding energy to give mesons with masses of about 1 GeV.

However, Morpurgo (1965) has argued that non-relativistic concepts may be applicable. If we assume the range of the strong interaction between quark and antiquark to be typical of the other hadrons, $R \sim \hbar/m_v c$ where m_v is the mass of a vector meson.†

The typical momenta of quarks in a well of width R is $\hbar/R \sim m_v c$, so the quark velocities are

$$v \sim \frac{m_v c}{M_q} \quad \text{or} \quad v/c \sim m_v/M_q \sim 1/10$$

and a non-relativistic calculation should be applicable.

The reason for examining qq and $q\bar{q}$ forces in more detail is to calculate the energies, spins and parities of the bound states; that is, their spatial properties. A possible way to side-step detailed calculations with particular force laws is to postulate additional invariances of the qq interaction and use symmetry arguments to deduce the consequences. Since we are interested in space–time properties and their interplay with $SU(3)$ properties it is natural to seek an invariance which combines the two.

It was suggested by Gursey and Radicati (1964), by Pais (1964) and

† A neutral vector meson ($J = 1$) or 'gluon' which couples to the baryon number B of a hadron gives an attractive force between quark ($B = +1$) and antiquark ($B = -1$) in direct analogy to the attractive interaction between particles of opposite electric charge due to photon exchange.

by Sakita (1964) that it might be a useful approximation to assume that the qq interaction is approximately spin independent as well as $SU(3)$ independent. Mathematically this corresponds to treating the six states of a quark (u, d or s with spin up, ↑, or down, ↓) as equivalent, and leads us to the invariance group $SU(6)$. This is the set of all unitary transformations on the six basic states of a spin-$\frac{1}{2}$ quark (or antiquark).

$$\mathrm{u}\uparrow, \mathrm{u}\downarrow, \mathrm{d}\uparrow, \mathrm{d}\downarrow, \mathrm{s}\uparrow, \mathrm{s}\downarrow \tag{11.17}$$

subject only to the condition that the determinant of the transformation is + 1.

In order that we can speak meaningfully of $SU(6)$ transformations affecting spin but not orbital angular momentum as invariances, it must be assumed that the spin and orbital angular momentum are to a good approximation, separately conserved. This in turn requires the spin–orbit interaction between quarks to be small. Now it is well known that the spin–orbit interaction is a relativistic effect, so that the assumption of $SU(6)$ invariance is a non-relativistic approximation.

Many ways have been proposed for relativistic generalisations of $SU(6)$. They are reviewed by, for example, Pais (1966). We shall not discuss such attempts here, except to remark that the resulting groups, $U(6, 6), SL(6, C)$ etc., cannot simply be interpreted as invariance groups of a Hamiltonian in the same way as $SU(3)$ or $SU(6)$.

11.4.1 *The SU(6) transformations*

We shall denote an oriented quark state by $|\mathrm{Q}\rangle$, where Q can take one of the six values u↑, u↓, d↑, d↓, s↑ or s↓. $|\bar{\mathrm{Q}}\rangle$ represents an antiquark state, $\bar{\mathrm{s}}$↑, $\bar{\mathrm{s}}$↓, $\bar{\mathrm{d}}$↑, $\bar{\mathrm{d}}$↓, $\bar{\mathrm{u}}$↑ or $\bar{\mathrm{u}}$↓.

The general $SU(6)$ transformation is a linear unitary transformation with unit determinant on the six states $|\mathrm{Q}\rangle$. As usual the generators of the group defined by infinitesimal transformations are of more immediate physical significance. By analogy with $SU(3)$ it may be seen without detailed derivation that there are 2 × 15 shift operators which transform any state of the six into any other. This is illustrated in fig. 11.5. The bold line corresponds to the operator transforming $|\mathrm{u}\uparrow\rangle$ into $|\mathrm{d}\downarrow\rangle$ and its Hermitian conjugate with the reverse action. There are also five diagonal operators whose eigenvalues may be chosen as follows: the z-component of spin of each kind of quark, s_z^{u}, s_z^{d} and s_z^{s}, and the resultant Y and I_3.

By taking linear combinations of these operators we may form three operators which act on the spin variables but which are indifferent to

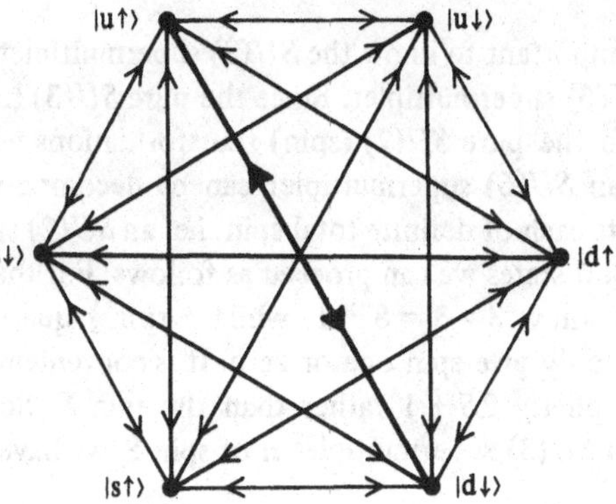

Fig. 11.5. Action of the 30 shift operators of $SU(6)$ on the oriented quark states.

the kind of quark, and similarly eight operators which act on the quark degree of freedom irrespective of the spin orientation. This shows how the transformations of spin $SU(2)$ and $SU(3)$ are contained within $SU(6)$. The details need not detain us.

The technique of weight diagrams which was useful in the case of $SU(3)$ is less useful here. It would correspond to a classification of states by the five diagonal operators already referred to. However, the physical particle states are not only eigenstates of the total Y and I_3 and S_z of their constituent quarks, but also of the total isospin I and total spin S. Now I and S do not commute with S_z^{u}, S_z^{d} and S_z^{s} so, except for special cases, the physical states would not correspond to single points in the weight diagram, which in this case is five dimensional and rather difficult to visualise. Instead we can set up the physical states by more intuitive considerations.

11.4.2 *Meson states in SU(6)*

With the inclusion of spin there are 36 quark–antiquark states $|Q\bar{Q}\rangle$. By an analysis similar to that for the $|q\bar{q}\rangle$ states in §11.1 we can show that the 36 states separate into 35 and 1, the latter being unconnected with the 35 by the $SU(6)$ shift operators. It follows that the 35 and 1 are supermultiplets of $SU(6)$, and thus

$$6 \times 6 = 35 + 1$$

SU(6) supermultiplets are denoted by their dimensionality in italic numerals.

It is very important to know the *SU*(3) supermultiplet and spin content of an *SU*(6) supermultiplet. Since the pure *SU*(3) transformations commute with the pure *SU*(2) (spin) transformations within *SU*(6) it follows that an *SU*(6) supermultiplet can be decomposed into *SU*(3) supermultiplets each of definite total spin, i.e. an *SU*(2) supermultiplet.

For the *6* × *6* states we can proceed as follows. For the *SU*(3) supermultiplets we know $\mathbf{3} \times \mathbf{3} = \mathbf{8} + \mathbf{1}$, while a spin-$\frac{1}{2}$ quark and a spin-$\frac{1}{2}$ antiquark can only give spin one or zero. It is convenient to work with the spin multiplicity $2S + 1$ rather than the spin S. So, denoting by $(2S + 1, \mathbf{n})$ an *SU*(3) supermultiplet **n** of spin S, we have the following possibilities

$$(3, 8), (1, 8), (3, 1), (1, 1) \tag{11.18}$$

which may be referred to as 'triplet eight' and so on. Clearly the first three compose the *35* and the last the *1* of *SU*(6).

We can now consider the physics. If *SU*(6) invariance is approximately satisfied in Nature, then the states of the *35* should be roughly degenerate in energy while the *1* can occur at a different energy. According to (11.18) we should therefore expect an octet and a singlet of spin one mesons (3, 8) and (3, 1) and an octet of spin zero mesons (1, 8). Turning to the spatial part of the $Q\bar{Q}$ wavefunction, we might expect the lowest bound state to be an s-state ($l = 0$), then remembering that the relative parity of a fermion–antifermion pair is odd, we expect the $Q\bar{Q}$ bound states to be of negative parity.

To summarise, the quark model with *SU*(6) invariance predicts an octet and a singlet of 1^- mesons and an octet of 0^- mesons. This is in a very satisfactory agreement with the observed low-lying states of the meson spectrum.

It is interesting that the *35* allows nine vector mesons but only eight 0^- mesons. A ninth 0^- meson must go in the *1*. This may account for the results of the phenomenological analysis of mixing where it is found that the mixing of the octet and singlet states is much smaller for the 0^- mesons than for the 1^- mesons.

Table 11.4. *Mesonic resonances in the 35 multiplet with* $L = 1$

$(2s + 1, \mathbf{n})_{J^{PC}}$	$I = \frac{1}{2}, Y = \pm 1$	$I = 1, Y = 0$	$I = 0, Y = 0$	$I = 0, Y = 0$
$(3, 8)_{2^{++}}(3, 1)_{2^{++}}$	K(1420)	$A_2(1310)$	f(1270)	f′(1514)
$(3, 8)_{1^{++}}(3, 1)_{1^{++}}$	?Q	$A_1(1100)$	D(1285)	E(1420)
$(3, 8)_{0^{++}}(3, 1)_{0^{++}}$	?K	?δ(960)	S^*	
$(1, 8)_{1^{+-}}(1, 1)_{1^{+-}}$		B(1235)		

Mesons of spin greater than one can be understood as states of the $q\bar{q}$ system with non-zero orbital angular momentum L. Even in this case, only $SU(3)$ octets and singlets can be formed.

Let us consider the $q\bar{q}$ *35* with $L = 1$. Vector addition of the total spin $S = 0$ or 1 with $L = 1$ leads to the following multiplet structure. The notation is $(2S + 1, n)_J^P$ where J is the total angular momentum of the state, i.e. the observed spin of the resonance, and P is its parity.

$$
\begin{aligned}
[35, L = 1]: &(3,8)_2^+ \quad (3,8)_1^+ \quad (3,8)_0^+ \\
&(3,1)_2^+ \quad (3,1)_1^+ \quad (3,1)_0^+ \\
&\qquad\quad (1,8)_1^+
\end{aligned}
$$

All these states are of positive parity. In general, meson states formed from $q\bar{q}$ with orbital angular momentum L have parity $(-1)^{L+1}$.

The only other firmly established meson supermultiplet is the $J^P = 2^+$ nonet. This can be accommodated in the $[35, L = 1]$ leaving several spin zero and spin one positive parity states to be filled. For some of these states there are experimentally observed candidates. In some cases, for example, $A_1(1100)$, it is not clear whether one has observed a resonance or several resonances or a kinematical enhancement. This must be borne in mind when referring to the tentative assignments made in table 11.4. The C-parity $(-1)^{L+S}$ of the neutral states is opposite for the spin triplet and spin singlet states.

If $SU(6)$ symmetry were exact, all the states of the *35* would be degenerate in mass. The lack of degeneracy is attributed to deviations from perfect symmetry in the $q\bar{q}$ forces. The following kinds of $q\bar{q}$ interaction can be distinguished.

(*a*) Interactions for which all six kinds of quark are equivalent. This in fact leads to a symmetry higher than $SU(6)$: all 36 $q\bar{q}$ states would be degenerate.

(*b*) General $SU(6)$ invariant interaction, separating *35* and *1*.

(*c*) $SU(3)$ invariant central force, separating different $SU(3)$ supermultiplets.

(*d*) Spin–spin interaction, splitting multiplets with different S.

(*e*) $SU(3)$ invariant non-central forces such as a spin–orbit force, which separates states of the same S and L but different J.

(*f*) $SU(3)$ symmetry-breaking interaction, H_{ms}.

All these kinds of force are required to produce the spectrum of levels observed in the $L = 1$ states. It is difficult to disentangle the various effects because of their number and because of uncertainties in the $L = 1$ particle assignments. For a discussion of the progress which has been made in discerning the trends, the reader is referred to the specialist literature.

11.4.3 *Baryon states in SU*(6)

With the inclusion of spin there are $6 \times 6 \times 6 = 216$ three quark states $|Q_1Q_2Q_3\rangle$. In order to classify these into $SU(6)$ supermultiplets, we start with two quarks and use the techniques of symmetrisation. The 36 two quark states may be chosen to be symmetric or antisymmetric under permutation of their labels. Thus we form

$$|Q_1Q_2\rangle + |Q_2Q_1\rangle$$

$$|Q_1Q_2\rangle - |Q_2Q_1\rangle$$

when Q_1 and Q_2 vary over the six possible values. There are $6 + (\frac{1}{2} \times 6 \times 5) = 21$ of the former and $\frac{1}{2}(6 \times 5) = 15$ of the latter. They therefore correspond to the supermultiplets *21* and *15* of $SU(6)$.

On bringing up the third quark to the two quark *21*, we can form a set of totally symmetric states, $(Q_1Q_2Q_3)S\rangle$ in the notation of §11.3.1. Counting the possibilities of all labels equal, two equal and one distinct, and all three distinct, we have

$$6 + (6 \times 5) + (6 \times 5 \times 4/2 \times 3) = 56$$

The remaining $(6 \times 21) - 56 = 70$ states correspond to state vectors of mixed symmetry.

Similarly on bringing up the third quark to the antisymmetric 15 we can obviously form a set of totally antisymmetric states $|(Q_1Q_2Q_3)A\rangle$ of which there are $6 \times 5 \times 4/2 \times 3 = 20$. The remaining $(6 \times 15) - 20 = 70$ again correspond to states of mixed symmetry. Thus

$$6 \times 6 \times 6 = 56 + 70 + 70' + 20 \qquad (11.19)$$

Again we have the four-fold decomposition of a state of three objects according to their permutation symmetry.

Next we require the $SU(3)$ and spin content of these $SU(6)$ supermultiplets. According to the considerations of §11.2, we can expect three quarks to form supermultiplets **10**, **8** or **1** of $SU(3)$, while from the addition of three spins of $\frac{1}{2}$ we obtain total spin values $\frac{3}{2}$ or $\frac{1}{2}$, or in terms of spin multiplets,

$$2 \times 2 \times 2 = 4 + 2 + 2$$

The quartet $(S = \frac{3}{2})$ corresponds to the totally symmetrical addition

Table 11.5. *Overall symmetry of a three particle state which is a product of spin and SU(3) states, each of a given symmetry*

		Symmetry of $SU(3)$ part		
		S	M	A
Symmetry of spin part	S	S	M	A
	M	M	S, M and A	M
	A	A	M	S

of three spins of $\frac{1}{2}$, while the two doublets correspond to the two states of mixed symmetry. Since there are only two possible values of S_z, there can be no totally antisymmetrical combination of three spins of $\frac{1}{2}$.

To find how the $SU(3)$ and $SU(2)$ multiplets combine into $SU(6)$ supermultiplets we introduce the rules for multiplication of symmetry types for three objects. This is a generalisation of the simple rule for product wavefunctions of two particles, e.g. a symmetrical $SU(3)$ state multiplied by an antisymmetrical spin state is antisymmetrical under interchange of both $SU(3)$ and spin labels.

The rules for three particles are given in table 11.5. The last row has to be ignored in the present application to three quark states because as mentioned already there is no totally antisymmetric spin state.

From the table we see that a totally antisymmetric $SU(6)$ state can be obtained in two ways: spin state S (spin $\frac{3}{2}$) multiplied by $SU(3)$ state A (i.e. **1**) or by spin state M (spin $\frac{1}{2}$) multiplied by $SU(3)$ state M (i.e. **8**). We denote these two possibilities by (4, **1**) and (2, **8**) respectively. Hence the totally antisymmetric *20* of $SU(6)$ has the following spin–$SU(3)$ multiplet structure,

$$\mathit{20} = (4, \mathbf{1}) + (2, \mathbf{8}) \tag{11.20}$$

Similarly we find the composition of the mixed symmetry $SU(6)$ state by reading off where M occurs in the body of the table 11.5. We find

$$\mathit{70} = (2, \mathbf{10}) + (4, \mathbf{8}) + (2, \mathbf{8}) + (2, \mathbf{1}) \tag{11.21}$$

Lastly the totally symmetric 56 can be formed as $S \times S$ or $M \times M$, and thus

$$\mathit{56} = (4, \mathbf{10}) + (2, \mathbf{8}) \tag{11.22}$$

We see from these results that the 56 of $SU(6)$ can accommodate an octet of spin-$\frac{1}{2}$ baryons and a decuplet of spin-$\frac{3}{2}$ baryons, which are exactly the $SU(3)$–spin combinations of the low-lying baryon states. It is natural therefore to assign the lowest-lying baryons to the *56* of $SU(6)$.

There is however a problem in doing this, if we take the quark model seriously. Since the $SU(3)$–spin part of these states is totally symmetrical it must be multiplied by a totally antisymmetrical *spatial* wavefunction to give a totally antisymmetrical overall state vector as required for spin-$\frac{1}{2}$ quarks.

Now unless the qq interaction is very peculiar, it is difficult to understand how a totally antisymmetrical spatial wavefunction can correspond to the state of lowest energy. It must have nodal surfaces where any two of its coordinates become equal, and the curvature of the wavefunction between these nodal surfaces corresponds to kinetic energy. It would be much more likely that the lowest energy state has a totally symmetrical wavefunction which varies in a smoother fashion and thereby lowers its energy. Fermi statistics would then require the $SU(6)$ part of the wavefunction to be totally antisymmetric, i.e. the *20*. This cannot accommodate the **10** which has to go in the *56*. The attractive feature of $SU(6)$ in accounting for the approximate degeneracy of the $\frac{1}{2}^+$ octet and $\frac{3}{2}^+$ decuplet is thereby lost. We must return again to the problem of symmetry when we discuss excited baryon states.

11.4.4 *Baryon wavefunctions in SU(6)*

The calculation of explicit wavefunctions for the baryon states of the *56* is facilitated by the requirement of total symmetry with respect to interchange of quark labels. One starts by writing down states of three orientated quarks with the correct total Y, I_3 and S_z. It is then necessary to find combinations with the required total I and total S. Finally total symmetry must be imposed.

We give some examples. As usual in dealing with symmetrisation the overall normalisation is best left until the end of the calculation. $\approx$ indicates that the right-hand side is not normalised.

We have immediately

$$|\Delta^{++}, S_z = +\tfrac{3}{2}\rangle = |u{\uparrow}u{\uparrow}u{\uparrow}\rangle$$

For Δ^+, $S_z = +\frac{3}{2}$, we take two u and one d quarks all with spin up, e.g. $|u{\uparrow}u{\uparrow}d{\uparrow}\rangle$. Symmetrisation then gives

$$|\Delta^+, S_z = +\tfrac{3}{2}\rangle \approx |u{\uparrow}u{\uparrow}d{\uparrow}\rangle + |u{\uparrow}d{\uparrow}u{\uparrow}\rangle + |d{\uparrow}u{\uparrow}u{\uparrow}\rangle \quad (11.23)$$

All the states of the **10** with $S_z = +\frac{3}{2}$ can simply be read off from table 11.3. by inserting the spin orientation, $\uparrow\uparrow\uparrow$.

Application of the spin-lowering operator to (11.23) gives

$$
\begin{aligned}
|\Delta^+, S_z = +\tfrac{1}{2}\rangle \approx & |u{\downarrow}u{\uparrow}d{\uparrow}\rangle + |u{\uparrow}u{\downarrow}d{\uparrow}\rangle + |u{\uparrow}u{\uparrow}d{\downarrow}\rangle \\
& + |u{\downarrow}d{\uparrow}u{\uparrow}\rangle + |u{\uparrow}d{\downarrow}u{\uparrow}\rangle + |u{\uparrow}d{\uparrow}u{\downarrow}\rangle \\
& + |d{\downarrow}u{\uparrow}u{\uparrow}\rangle + |d{\uparrow}u{\downarrow}u{\uparrow}\rangle + |d{\uparrow}u{\uparrow}u{\downarrow}\rangle \quad (11.24)
\end{aligned}
$$

To obtain the proton state with $S_z = +\frac{1}{2}$ we need quarks uud with spin orientations ↑↑↓, that is, u↑u↑d↓ and u↑u↓d↑ and their permutations.

We have to impose the condition that the total I and total S are each $\frac{1}{2}$. This can be done as follows. The two quark state

$$|u{\uparrow}d{\downarrow}\rangle + |d{\downarrow}u{\uparrow}\rangle - |u{\downarrow}d{\uparrow}\rangle - |d{\uparrow}u{\downarrow}\rangle$$

has $I = 0$ and $S = 0$. It is just

$$(ud - du) \times ({\uparrow}{\downarrow} - {\downarrow}{\uparrow})$$

We now bring up the third quark which must be a u↑. The resultant state

$$|u{\uparrow}d{\downarrow}u{\uparrow}\rangle + |d{\downarrow}u{\uparrow}u{\uparrow}\rangle - |u{\downarrow}d{\uparrow}u{\uparrow}\rangle - |d{\uparrow}u{\downarrow}u{\uparrow}\rangle$$

is not totally symmetrical but it is symmetrical under interchange in the first two places. We therefore add on the states obtained by (*a*) exchanging the first and third labels and (*b*) exchanging the second and third labels. The result is

$$
\begin{aligned}
|p, S_z = +\tfrac{1}{2}\rangle \approx & 2|u{\uparrow}d{\downarrow}u{\uparrow}\rangle + 2|u{\uparrow}u{\uparrow}d{\downarrow}\rangle + 2|d{\downarrow}u{\uparrow}u{\uparrow}\rangle \\
& - |u{\uparrow}u{\downarrow}d{\uparrow}\rangle - |u{\uparrow}d{\uparrow}u{\downarrow}\rangle - |u{\downarrow}d{\uparrow}u{\uparrow}\rangle \\
& - |d{\uparrow}u{\downarrow}u{\uparrow}\rangle - |d{\uparrow}u{\uparrow}u{\downarrow}\rangle - |u{\downarrow}u{\uparrow}d{\uparrow}\rangle \quad (11.25)
\end{aligned}
$$

The normalisation factor is $18^{-1/2}$. Application of the isospin lowering operator gives the neutron state, and continued use of $I_\pm$ and $U_\pm$,† together with the orthogonality relations in the case of Λ^0, allows us to complete the octet of baryons with $S = \frac{1}{2}$. The wavefunctions for the $S_z = +\frac{1}{2}$ states for the baryon octet are listed in table 11.6.

11.4.5 *Excited states of baryons*

A model for the baryons which can be extended to excited states with considerable success is the symmetric oscillator model of Greenberg (1964), which we shall briefly describe here. For more details we refer the reader to Faiman and Hendry (1968, 1969) where proofs of some of the assertions made here can be found.

† Not $V_\pm$ which would give states violating the phase conventions specified in §10.3.4.

Table 11.6. *Wavefunctions for the $S_z = +\frac{1}{2}$ states of the baryon octet in the quark model*

$$|p, S_z=+\tfrac{1}{2}\rangle = 18^{-1/2}\{2|u\uparrow u\uparrow d\downarrow\rangle + 2|u\uparrow d\downarrow u\uparrow\rangle + 2|d\downarrow u\uparrow u\uparrow\rangle - |u\uparrow d\uparrow u\downarrow\rangle - |u\uparrow u\downarrow d\uparrow\rangle - |u\downarrow d\uparrow u\uparrow\rangle - |d\uparrow u\uparrow u\downarrow\rangle - |d\uparrow u\downarrow u\uparrow\rangle - |u\downarrow u\uparrow d\uparrow\rangle\}$$

$$|n, S_z=+\tfrac{1}{2}\rangle = 18^{-1/2}\{-2|d\uparrow d\uparrow u\downarrow\rangle - 2|d\uparrow u\downarrow d\uparrow\rangle - 2|u\downarrow d\uparrow d\uparrow\rangle + |u\uparrow d\uparrow d\downarrow\rangle + |u\uparrow d\downarrow d\uparrow\rangle + |d\downarrow u\uparrow d\uparrow\rangle + |d\uparrow u\uparrow d\downarrow\rangle + |d\uparrow d\downarrow u\uparrow\rangle + |d\downarrow d\uparrow u\uparrow\rangle\}$$

$$-|\Sigma^+, S_z=+\tfrac{1}{2}\rangle = 18^{-1/2}\{2|u\uparrow u\uparrow s\downarrow\rangle + 2|u\uparrow s\downarrow u\uparrow\rangle + 2|s\downarrow u\uparrow u\uparrow\rangle - |u\uparrow s\uparrow u\downarrow\rangle - |u\uparrow u\downarrow s\uparrow\rangle - |u\downarrow u\uparrow s\uparrow\rangle - |s\uparrow u\uparrow u\downarrow\rangle - |s\uparrow u\downarrow u\uparrow\rangle - |u\downarrow s\uparrow u\uparrow\rangle\}$$

$$\begin{aligned}|\Sigma^0, S_z=+\tfrac{1}{2}\rangle = 36^{-1/2}\{&2|u\uparrow d\uparrow s\downarrow\rangle + 2|u\uparrow s\downarrow d\uparrow\rangle + 2|s\downarrow u\uparrow d\uparrow\rangle + 2|d\uparrow u\uparrow s\downarrow\rangle + 2|d\uparrow s\downarrow u\uparrow\rangle + 2|s\downarrow d\uparrow u\uparrow\rangle \\ -&|u\uparrow s\uparrow d\downarrow\rangle - |u\uparrow d\downarrow s\uparrow\rangle - |d\downarrow u\uparrow s\uparrow\rangle - |d\uparrow s\uparrow u\downarrow\rangle - |d\uparrow u\downarrow s\uparrow\rangle - |u\downarrow d\uparrow s\uparrow\rangle \\ -&|s\uparrow u\uparrow d\downarrow\rangle - |s\uparrow d\downarrow u\uparrow\rangle - |d\downarrow s\uparrow u\uparrow\rangle - |s\uparrow d\uparrow u\downarrow\rangle - |s\uparrow u\downarrow d\uparrow\rangle - |u\downarrow s\uparrow d\uparrow\rangle\}\end{aligned}$$

$$|\Sigma^-, S_z=+\tfrac{1}{2}\rangle = 18^{-1/2}\{2|d\uparrow d\uparrow s\downarrow\rangle + 2|d\uparrow s\downarrow d\uparrow\rangle + 2|s\downarrow d\uparrow d\uparrow\rangle - |d\uparrow s\uparrow d\downarrow\rangle - |d\uparrow d\downarrow s\uparrow\rangle - |d\downarrow d\uparrow s\uparrow\rangle - |s\uparrow d\uparrow d\downarrow\rangle - |s\uparrow d\downarrow d\uparrow\rangle - |d\downarrow s\uparrow d\uparrow\rangle\}$$

$$\begin{aligned}|\Lambda^0, S_z=+\tfrac{1}{2}\rangle = 12^{-1/2}\{&|u\uparrow s\uparrow d\downarrow\rangle + |u\uparrow d\downarrow s\uparrow\rangle + |d\downarrow s\uparrow u\uparrow\rangle - |d\uparrow s\uparrow u\downarrow\rangle - |d\uparrow u\downarrow s\uparrow\rangle - |u\downarrow s\uparrow d\uparrow\rangle \\ +&|s\uparrow u\uparrow d\downarrow\rangle + |s\uparrow d\downarrow u\uparrow\rangle + |d\downarrow u\uparrow s\uparrow\rangle - |s\uparrow d\uparrow u\downarrow\rangle - |s\uparrow u\downarrow d\uparrow\rangle - |u\downarrow d\uparrow s\uparrow\rangle\}\end{aligned}$$

$$|\Xi^0, S_z=+\tfrac{1}{2}\rangle = 18^{-1/2}\{-2|s\uparrow s\uparrow u\downarrow\rangle - 2|s\uparrow u\downarrow s\uparrow\rangle - 2|u\downarrow s\uparrow s\uparrow\rangle + |u\uparrow s\uparrow s\downarrow\rangle + |u\uparrow s\downarrow s\uparrow\rangle + |s\downarrow u\uparrow s\uparrow\rangle + |s\uparrow u\uparrow s\downarrow\rangle + |s\uparrow s\downarrow u\uparrow\rangle + |s\downarrow s\uparrow u\uparrow\rangle\}$$

$$|\Xi^-, S_z=+\tfrac{1}{2}\rangle = 18^{-1/2}\{-2|s\uparrow s\uparrow d\downarrow\rangle - 2|s\uparrow d\downarrow s\uparrow\rangle - 2|d\downarrow s\uparrow s\uparrow\rangle + |d\uparrow s\uparrow s\downarrow\rangle + |d\uparrow s\downarrow s\uparrow\rangle + |s\downarrow d\uparrow s\uparrow\rangle + |s\uparrow d\uparrow s\downarrow\rangle + |s\uparrow s\downarrow d\uparrow\rangle + |s\downarrow s\uparrow d\uparrow\rangle\}$$

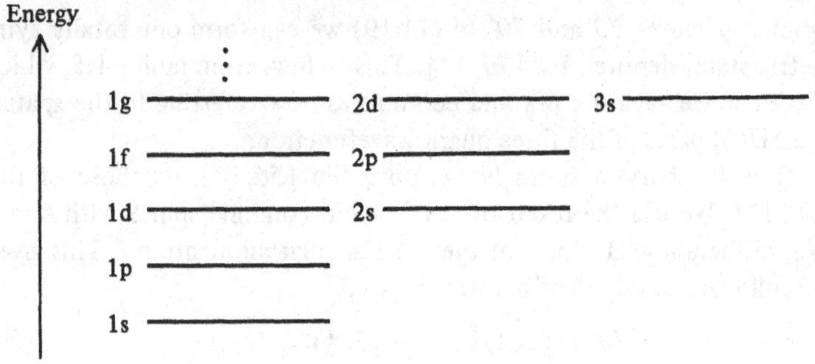

Fig. 11.6. One particle levels in a three-dimensional harmonic oscillator potential.

It is supposed that the quarks interact in pairs via a non-relativistic harmonic oscillator potential. A system of three such quarks can be analysed by means of an equivalent shell model in which three non-interacting quarks move in a common harmonic oscillator well. The shell model states are easily enumerated except that certain states are 'spurious' because they correspond in the original physical model to motion of the centre of mass. The other assumption of the model is that the overall quark wavefunction

$$\psi_{\text{space}}\psi_{SU(6)}$$

can be totally symmetrical, and furthermore that totally symmetrical states lie lowest in energy. We shall say more about this at the end of this section.

The order of one particle energy levels in the three-dimensional harmonic oscillator is well known, and is shown in fig. 11.6. The lowest shell model state is that in which all three quarks are in the 1s-state, and is written $(1s)^3$. Such a state is totally symmetrical and has total orbital angular momentum $L = 0$, and positive parity. When multiplied by the totally symmetric $SU(6)$ wavefunction *56*, it gives the wavefunction of the $\frac{1}{2}^+$ and $\frac{3}{2}^+$ baryons.

Ignoring for the present the problem of Fermi statistics we go on to the excited states. The first excited state in the shell model corresponds to one particle being in the 1p-state: $(1s)^2\,1p$. We might expect three such states, but the symmetrical combination of these is spurious, because it describes a motion of the centre of mass. The other two independent combinations are allowed: they are of mixed symmetry, have $L = 1$ and negative parity.

Taking these spatial states of mixed symmetry with the $SU(6)$ mixed symmetry states *70* and *70′* of (11.19) we can form one totally symmetric state, denoted by $[70, 1^-]$. This follows from table 11.5, which applies if we regard rows and columns as now referring to the spatial and $SU(6)$ parts of the three quark wavefunction.

Thus the baryon states lying above the $[56, 0^+]$ are those of the $[70, 1^-]$. We use the result of (11.21) and combine spin S with $L = 1$ to give the allowed values of the total angular momentum J. This gives the following states all of negative parity,

$$\begin{aligned}[70, 1^-]: &(2, 1)^-_{1/2}, \quad (2, 1)^-_{3/2}\\ &(2, 8)^-_{1/2}, \quad (2, 8)^-_{3/2}\\ &(4, 8)^-_{1/2}, \quad (4, 8)^-_{3/2}, \quad (4, 8)^-_{5/2}\\ &(2, 10)^-_{1/2}, \quad (2, 10)^-_{3/2}\end{aligned}$$

Table 11.7. *Assignment of baryon resonances to the states of* $[70, 1^-]$ *in the harmonic oscillator quark model*

	Δ	N	Σ	Λ	Ξ
$(2, 1)^-_{1/2}$				Λ(1405)	
$(2, 1)^-_{3/2}$				Λ(1520)	
$(2, 8)^-_{1/2}$		N(1535)	Σ( )	Λ(1670)	Ξ( )
$(2, 8)^-_{3/2}$		N(1520)	Σ(1670)	Λ(1690)	Ξ(1820)?
$(4, 8)^-_{1/2}$		N(1700)	Σ( )	Λ( )	Ξ( )
$(4, 8)^-_{3/2}$		N( )	Σ(1940)	Λ( )	Ξ( )
$(4, 8)^-_{5/2}$		N(1670)	Σ(1765)	Λ(1830)	Ξ(1930)?
$(2, 10)^-_{1/2}$	Δ(1650)		Σ(1750)		Ξ( )
$(2, 10)^-_{3/2}$	Δ(1670)		Σ( )		Ξ( )

Using the 'Review of particle properties' (1973) we find that all the firmly established negative parity baryon resonances in the mass range 1400–2000 MeV can be fitted into these multiplets. Equally important, there are no established resonances which are not allowed by the model. Because of the mass splittings within the $SU(3)$ supermultiplets and the overlapping of different supermultiplets, some of the assignments are ambiguous unless further information is available. The present model can be used to calculate two body decay rates of the resonances. This has been done by Feynman, Kislinger and Ravndal (1971). We have adopted their assignments which are given in table 11.7.

The states next in energy in this model involve two quanta of excitation of the harmonic oscillator. The possible configurations are $(1s)^2 2s$, $(1s)^2 1d$ and $1s(1p)^2$, and the requirement of overall symmetry leads to the following states, all of positive parity:

$$[56, 0^+] \quad [70, 0^+] \quad [20, 1^+]$$

$$[56, 2^+] \quad [70, 2^+]$$

Tentative assignments of the higher energy baryons to these multiplets have been made (Harari, 1968; Dalitz and Horgan 1973) but we shall not discuss them further.

The symmetric quark model has been extended to describe strong and electromagnetic decays of hadrons (Feynman, Kislinger and Ravndal, 1971), and as mentioned above these predictions can be used to test assignments of observed particles to multiplets.

11.4.6 *Quark statistics*

It is assumed in the symmetric oscillator model that the overall wavefunction for the three quarks is totally symmetrical. However, according to the spin-statistics theorem, fermion quarks should have totally antisymmetrical wavefunctions. The empirical evidence seems to favour the assumption of the model. One way out of this dilemma, which was pointed out by Greenberg (1964), is to suppose that quarks obey one of the generalised kinds of statistics, in particular the *para-Fermi statistics* of order 3.

This kind of statistics can be described as follows. The quark state depends on what may be called a 'secret' variable ζ which has three possible values, $\zeta = 1$, 2 or 3; three being the *order* of the statistics. However, no physical process whatsoever can distinguish the three values of ζ for a single quark. Thus in this picture the quark state $|u\rangle$ is really one of three $|u, \zeta\rangle$ but $|u, 1\rangle$, $|u, 2\rangle$ and $|u, 3\rangle$ are physically indistinguishable. Furthermore, for a system of several quarks the overall wavefunction has the form

$$\Psi = \psi_{\text{space}} \phi_{SU(6)} \chi_{\text{secret}}(\zeta)$$

and *para-Fermi* statistics demands that Ψ be totally antisymmetric under simultaneous exchange of space, spin, $SU(3)$ and the secret variable ζ. Specialising to three quarks and using the rules of table 11.5 once again, we have the following possibilities

$$\Psi = \psi_{\text{space}-SU(6)} \chi_{\text{secret}}$$

$$A = S \times A$$

$$A = M \times M$$

$$A = A \times S$$

Thus *para-Fermi* statistics permits the space–$SU(6)$ part of the three quark wavefunction to be symmetrical. Moreover, since only one totally antisymmetrical $\chi_{\text{secret}}(\zeta_1, \zeta_2, \zeta_3)$ can be formed for three quarks, the corresponding $\psi_{\text{space}-SU(6)}$ is unique. This hypothesis does permit the space–$SU(6)$ part of the three quark wavefunction to be '*M*' or '*A*' as well, but whether these occur and at what energies depends on the details of the q–q forces.

The reason for introducing the *para-Fermi* statistics is that by the general principles of relativistic quantum field theory, *para-Fermi* statistics are permitted for spin-$\frac{1}{2}$ particles, while Bose statistics (i.e. the direct postulate that the space–$SU(6)$ part is always symmetrical) are not. It should be mentioned that other ways of evading the statistics problem have been proposed. These include models in which quarks have 'colour' or 'charm' instead of the secret variable ζ. For a further discussion and references see Lipkin (1973).

11.5 Magnetic moments of baryons

The treatment of the electromagnetic interactions of hadrons in the quark model is based on the *additivity hypothesis*. Let us consider magnetic moments. The magnetic moment operator of a hadron is assumed to be given by the sum of the magnetic moments of the constituent quarks. For a baryon

$$\boldsymbol{\mu} = \boldsymbol{\mu}_{q_1} + \boldsymbol{\mu}_{q_2} + \boldsymbol{\mu}_{q_3} \tag{11.26}$$

Furthermore the magnetic moment operator of a quark is assumed to be given by

$$\boldsymbol{\mu}_{q_i} = \mu_0 Q \boldsymbol{\sigma} \tag{11.27}$$

where Q is the charge $(+\frac{2}{3}, -\frac{1}{3}, -\frac{1}{3})$, according as $q_i =$ u, d or s, and $\boldsymbol{\sigma}$ is the Pauli spin operator. μ_0 is a scale magnetic moment. This proportionality of magnetic moment and charge is not the same thing as the statement that the anomalous magnetic moments of quarks are zero. Indeed, as we shall see, the quark moments are anomalously different from $\frac{2}{3}$ or $-\frac{1}{3}$ times $e\hbar/2m_Q c$. Rather the proportionality follows from $SU(3)$ invariance and in particular from the form of the magnetic moment operator,

$$\boldsymbol{\mu} = \tfrac{1}{2}\boldsymbol{\mu}^Y + \boldsymbol{\mu}^{I_3}$$

as discussed in §10.8.1. Proceeding as in the analysis of baryon moments, we have

$$\langle u|\mu|u\rangle = \tfrac{1}{2}\langle u|\mu^Y|u\rangle + \langle u|\mu^{I_3}|u\rangle$$

and

$$\langle s|\mu|s\rangle = \tfrac{1}{2}\langle s|\mu^Y|s\rangle$$

because the isovector part cannot contribute to the $I = 0$ state $|s\rangle$.

The reflection operator P_v has the properties

$$P_v|u\rangle = |s\rangle$$

and

$$P_v^{-1}\mu^Y P_v = -\mu$$

so that

$$\langle s|\mu|s\rangle = \tfrac{1}{2}\langle u|P_v^{-1}\mu^Y P_v|u\rangle$$

$$= -\tfrac{1}{2}\langle u|\mu|u\rangle \qquad (11.28)$$

Then by U-spin invariance,

$$\langle s|\mu|s\rangle = \langle d|\mu|d\rangle \qquad (11.29)$$

Equation (11.27) simply summarises (11.28) and (11.29). From the simple expressions (11.26) and (11.27) for μ and with the baryon wave-functions given in table 11.6, it is a straightforward matter to calculate the magnetic moments of the baryons. We find for the proton with $S_z = +\frac{1}{2}$

$$\langle p, S_z = +\tfrac{1}{2}|\mu_z|p, S_z = +\tfrac{1}{2}\rangle$$

$$= 18^{-1}\mu_0\{3{\cdot}4[2\langle u\uparrow|Q\sigma_z|u\uparrow\rangle + \langle d\downarrow|Q\sigma_z|d\downarrow\rangle]$$

$$+ 6[\langle u\uparrow|Q\sigma_z|u\uparrow\rangle + \langle d\uparrow|Q\sigma_z|d\uparrow\rangle + \langle u\downarrow|Q\sigma_z|u\downarrow\rangle]\}$$

$$= \tfrac{1}{18}\mu_0\{3{\cdot}4(\tfrac{4}{3} + \tfrac{1}{3}) + 6(-\tfrac{1}{3})\}$$

$$= \mu_0$$

Similarly, for the neutron

$$\langle n, S_z = +\tfrac{1}{2}|\mu_z|n, S_z = +\tfrac{1}{2}\rangle = -\tfrac{2}{3}\mu_0$$

and thus the quark model with the additivity hypothesis predicts

$$\mu_n/\mu_p = -2/3 \qquad (11.30)$$

Experimentally the ratio is -0.685. The prediction (11.30) was originally obtained as a pure $SU(6)$ result (Beg, Lee and Pais, 1964).

Equation (11.30) together with pure $SU(3)$ considerations mean that all the baryon moments, including the $3/2^+$ decuplet, are given in terms of the proton moment.

Relations between radiative decays of the baryons, e.g. $\Delta(1236) \to p\gamma$ (and also mesons, $\omega^0 \to \pi^0\gamma$), can be obtained by similar techniques (Kokkedee, 1969).

The additivity hypothesis also underlies the treatment of leptonic decays of baryons in the quark model. It is assumed that the fundamental processes are the leptonic decays of quarks.

The emission of the *charged* lepton pair (neglecting neutral currents) means that there are two possible processes

$$d \to u + l^- + \bar{\nu} \tag{11.31a}$$

$$s \to u + l^- + \bar{\nu} \tag{11.31b}$$

together with the corresponding charge conjugate processes.

The first of these is strangeness conserving, $\Delta S = 0$, and involves an isospin change of unity $|\Delta I| = 1$, while in the second $\Delta Q = \Delta S = 1$ and $|\Delta I| = \frac{1}{2}$. Thus the quark model affords a natural explanation for these empirical rules of weak decays. By combining the amplitudes for processes (11.31*a*) and (11.31*b*) with coefficients $G \cos \theta_c$ and $G \sin \theta_c$ we obtain the Cabbibo formulation of semileptonic decays (van Royen and Weisskopf, 1967).

11.6 Free quarks

In all that has gone before, we have been considering bound systems of quarks and antiquarks. We now move to the question of the possible existence of single quarks as free particles.

If the quarks are capable of existing separately, at least one of the three types ought to be stable. Spontaneous transitions (e.g. by beta decay) between the different types of quark may occur, and one cannot predict with any certainty which will be stable or sufficiently long-lived to be observed as free particles. Searches have therefore been made for particles causing ionisation characteristic of a charge either $\frac{1}{3}e$ or $\frac{2}{3}e$, i.e. ionisation $\frac{1}{9}$ or $\frac{4}{9}$ of normal.

From the failure of early searches with accelerators to reveal any such particles, it was concluded that, if they exist at all, they have a mass of several GeV. This would be reconciled with the known masses of orthodox particles by allowing large binding energy to reduce the total mass of a bound system of several quarks. Later searches have therefore been concentrated on processes in which there is enough energy available to break orthodox particles into separate quarks of individual mass 5 GeV or more. With the continued absence of any unambiguous evidence, experiments have progressively moved towards increasingly rigorous exclusion of spurious effects: to find an object which could be a quark has now become of negligible interest, in comparison with the major task of finding an object which can be nothing else.

Methods of search may be divided into production experiments with high energy accelerators, production experiments with cosmic rays, and searches for trapped quarks in ordinary matter.

Among the accelerator experiments, one at the CERN intersecting storage rings (Bott-Bodenhausen *et al.*, 1972) has set an upper limit of $3 \times 10^{-34}\,\mathrm{cm}^2$ for production of quarks of mass 5–22 GeV and charge $\frac{1}{3}e$ (6×10^{-34} for mass up to 13 GeV, for charge $\frac{2}{3}e$); earlier experiments have set even stricter limits for lower quark masses. Cosmic ray experiments and searches for trapped quarks are less easy to interpret in terms of upper limits to cross-sections for specified processes.

It is clear from the present experimental results that free quarks are either severely inhibited from independent existence, or incapable of it. Possible reasons for this situation are still a field for speculation. It is clear that the discovery of free quarks would pose new problems, e.g. concerning spin and statistics, but meanwhile the status of the quark, as either a particle or a collection of quantum numbers, remains unclear.

11.7 Further developments

In the present chapter we have described the essential features of the quark model of hadrons. In conformity with the theme of this book stress has been laid on the symmetry aspects of the quark picture. Quarks used in this way are sometimes called 'valence quarks' or 'constituent quarks'. There are many other aspects of the quark model such as W-spin, $SU(6)_W$, constituent versus current quarks, identity of quarks and partons, and so on, which have not been touched on. For a survey of these topics with references to the original literature the reader may refer to Lipkin (1973) and Weyers (1973) and reports of recent conferences.

See also the postscript on Recent Developments (p. 366) added in the 1980 reprint.

APPENDIX A

CROSS-SECTIONS AND $\mathcal{T}$-MATRIX ELEMENTS

We derive the relations between $\mathcal{T}$-matrix elements and cross-sections and decay rates, by a method which avoids the customary artifice of enclosing the system in a large space–time box.

A.1 Definition of the cross-section

Consider the process

$$a + b \rightarrow c + d \tag{A.1}$$

in which a beam of particles a, of well-defined momentum, is directed onto a target of particles b, and detectors count the particles c or d or both. If $\mathfrak{N}_c$ particles of type c are counted over a time T and none are missed, then the counting rate is $\mathfrak{N}_c/T$. The counting rate is proportional to the number N_b of particles b in the target (provided the beam bathes the whole target and multiple scattering and shadowing are neglected) and to the incident flux J_a. J_a is the number of particles a crossing unit area perpendicular to the beam per unit time. Thus

$$\mathfrak{N}_c/T = \sigma(ab \rightarrow cd) N_b J_a \tag{A.2}$$

where the constant of proportionality is the *cross-section* for the process (A.1).

If we detect $\delta\mathfrak{N}_c$ particles c emitted into a solid angle $\delta\Omega$, we obtain a partial cross-section $\mathrm{d}\sigma/\mathrm{d}\Omega$ given by

$$\delta\mathfrak{N}_c/T = \frac{\mathrm{d}\sigma}{\mathrm{d}\Omega}\delta\Omega N_b J_a \tag{A.3}$$

Equation (A.2) may be put into a form which is more symmetrical with respect to a and b as follows. The incident flux J_a is the density of particles a in the beam × the velocity of a as seen from the target

$$J_a = \rho_a v \tag{A.4}$$

More generally if the particles b are moving v is the relative velocity

$$v_{ab} = |v_a - v_b| \tag{A.5}$$

In the cases we shall discuss v_a and v_b lie along the same line.

If ρ_b denotes the density of the target of volume V, then

$$N_b = \rho_b V \tag{A.6}$$

Inserting (A.4), (A.5) and (A.6) into (A.2) we have

$$\mathcal{N}_c = \sigma v_{ab} \rho_a \rho_b VT \tag{A.7}$$

If ρ_a and ρ_b vary in time or space the expression on the right should generalise to

$$\mathcal{N}_c = \sigma v_{ab} \int d^3x \int dt \rho_a(x, t) \rho_b(x, t) \tag{A.8}$$

This expression is also valid if the beam does not bathe the whole target.

The quantity

$$\Phi = v_{ab} \int d^3x \int dt \rho_a(x, t) \rho_b(x, t) \tag{A.9}$$

is called the mutual flux of a and b.

Thus we have

$$\sigma(ab \to cd) = \mathcal{N}_c/\Phi \tag{A.10}$$

σ is invariant under Lorentz transformations along the direction of relative motion. Consider (A.7): $\mathcal{N}_c$ a number, and VT a space–time volume are Lorentz invariants. ρ_a and ρ_b are not Lorentz invariant. The number density ρ_a in a frame in which the particles are moving with energy E_a and momentum p_a exceeds that in the rest-frame $\mathring{\rho}_a$ by the Lorentz contraction factor,

$$(1 - v_a^2)^{-1/2} = E_a/m_a$$

So

$$\rho_a = \mathring{\rho}_a E_a/m_a$$

Hence

$$\rho_a \rho_b v_{ab} = (\mathring{\rho}_a \mathring{\rho}_b v_{ab} E_a E_b)/(m_a m_b)$$

This can be expressed in terms of Lorentz invariants, as follows: in the rest-frame of b,

$$\frac{E_a E_b}{m_a m_b} v_{ab} = \frac{E_a}{m_a} v_a = \frac{|p_a|}{m_a}$$

$$= \frac{\{(p_a \cdot p_b)^2 - m_a^2 m_b^2\}^{1/2}}{m_a m_b} \tag{A.11}$$

Thus

$$\mathcal{N}_c = \sigma \mathring{\rho}_a \mathring{\rho}_b \frac{\{(p_a \cdot p_b)^2 - m_a^2 m_b^2\}^{1/2}}{m_a m_b} VT$$

A.2 Relativistic transition probability

With the conventions for relativistic momentum states of §4.3 (in particular (4.45)) the probability of finding particle c with momentum in the range $(\boldsymbol{p}_c, \boldsymbol{p}_c + \mathrm{d}\boldsymbol{p}_c)$ and d in the range $(\boldsymbol{p}_d, \boldsymbol{p}_d + \mathrm{d}\boldsymbol{p}_d)$ after scattering is

$$\mathrm{d}w = |(\phi_{\boldsymbol{p}_c\boldsymbol{p}_d}, \mathcal{T}\phi_{\boldsymbol{p}_a\boldsymbol{p}_b})|^2 \frac{\mathrm{d}^3\boldsymbol{p}_c}{(2\pi)^3 2E_c} \frac{\mathrm{d}^3\boldsymbol{p}_d}{(2\pi)^3 2E_d} \tag{A.12}$$

Space and time displacement invariance imply energy–momentum conservation which is expressed by

$$(\phi_{\boldsymbol{p}_c\boldsymbol{p}_d}, \mathcal{T}\phi_{\boldsymbol{p}_a\boldsymbol{p}_b}) =$$
$$(2\pi)^4\delta^{(3)}(\boldsymbol{p}_c + \boldsymbol{p}_d - \boldsymbol{p}_a - \boldsymbol{p}_b)\delta(E_c + E_d - E_a - E_b)T(\boldsymbol{p}_c\boldsymbol{p}_d, \boldsymbol{p}_a\boldsymbol{p}_b) \tag{A.13}$$

where the $(2\pi)^4$ is introduced for convenience later. Thus the matrix element on the left is a singular quantity. However it is an idealisation to speak of perfectly sharp momenta initially. So we replace the initial momentum eigenstate

$$\phi_{\boldsymbol{p}_a\boldsymbol{p}_b} = \phi^a_{\boldsymbol{p}_a}\phi^b_{\boldsymbol{p}_b}$$

by a superposition

$$\Psi = \int \frac{\mathrm{d}^3\boldsymbol{p}_1}{(2\pi)^{3/2}(2E_1)^{1/2}} a_1(\boldsymbol{p}_1)\phi^a_{\boldsymbol{p}_1} \int \frac{\mathrm{d}^3\boldsymbol{p}_2}{(2\pi)^{3/2}(2E_2)^{1/2}} a_2(\boldsymbol{p}_2)\phi^b_{\boldsymbol{p}_2}$$

where the momentum space amplitudes $a_1(\boldsymbol{p}_1)$ and $a_2(\boldsymbol{p}_2)$ are sharply peaked about $\boldsymbol{p}_a$ and $\boldsymbol{p}_b$ respectively. The coefficients are defined as in §4.3.

According to the principle of superposition the amplitude is now

$$A = \frac{1}{(2\pi)^3}\int \frac{\mathrm{d}^3\boldsymbol{p}_1}{(2E_1)^{1/2}} \int \frac{\mathrm{d}^3\boldsymbol{p}_2}{(2E_2)^{1/2}} (\phi_{\boldsymbol{p}_c\boldsymbol{p}_d}, \mathcal{T}\phi_{\boldsymbol{p}_1\boldsymbol{p}_2}) a_1(\boldsymbol{p}_1) a_2(\boldsymbol{p}_2) \tag{A.14}$$

and (A.12) becomes

$$\mathrm{d}w = |A|^2 \frac{\mathrm{d}^3\boldsymbol{p}_c}{(2\pi)^3 2E_c} \frac{\mathrm{d}^3\boldsymbol{p}_d}{(2\pi)^3 2E_d} \tag{A.15}$$

We substitute (A.13) into (A.14) and use the fact that $a_1(\boldsymbol{p}_1)$ and $a_2(\boldsymbol{p}_2)$ are sharply peaked, to replace the function $T(\boldsymbol{p}_c\boldsymbol{p}_d, \boldsymbol{p}_1\boldsymbol{p}_2)$ by its value at $\boldsymbol{p}_1 = \boldsymbol{p}_a, \boldsymbol{p}_2 = \boldsymbol{p}_b$ outside the integral,

$$A = 2\pi T(p_c p_d, p_a p_b) \times \int \frac{\mathrm{d}^3 p_1}{(2E_1)^{1/2}} \int \frac{\mathrm{d}^3 p_2}{(2E_2)^{1/2}} a_1(p_1) a_2(p_2) \delta^{(4)}(p_c + p_d - p_1 - p_2) \tag{A.16}$$

where $\delta^{(4)}$ denotes the four-dimensional delta function. Taking the modulus squared of (A.16), we have

$$|A|^2 = \frac{(2\pi)^2 |T(p_c p_d, p_a p_b)|^2 \delta^{(4)}(p_c + p_d - p_a - p_b)}{4E_a E_b} \times \int \mathrm{d}^3 p_1' \mathrm{d}^3 p_2' \mathrm{d}^3 p_1 \mathrm{d}^3 p_2 \; a_1^*(p_1') a_2^*(p_2') a_1(p_1) a_2(p_2) \delta^4(p_1' + p_2' - p_1 - p_2) \tag{A.17}$$

where we used

$$\delta^{(4)}(p_c + p_d - p_1' - p_2') \delta^{(4)}(p_c + p_d - p_1 - p_2) = \delta^{(4)}(p_c + p_d - p_1 - p_2) \delta^{(4)}(p_1' + p_2' - p_1 - p_2)$$

and then set $p_1 = p_a$, $p_2 = p_b$ in the first factor. Also, for sharply peaked $a_1(\boldsymbol{p})$ and $a_2(\boldsymbol{p})$, we set $E_1 = E_1' = E_a$ and $E_2 = E_2' = E_b$.

A.3 Mutual flux

Corresponding to a momentum–space wavefunction $a(\boldsymbol{p})$ we define a wavefunction in space–time

$$\psi(x, t) = \int \frac{\mathrm{d}^3 p}{(2\pi)^{3/2} (2E_p)^{1/2}} a(p) \mathrm{e}^{\mathrm{i} p \cdot x - \mathrm{i} E_p t} \tag{A.18}$$

which can easily be shown to satisfy the Klein–Gordon equation

$$\left(\frac{\partial^2}{\partial t^2} - \nabla^2 + m^2 \right) \psi(x, t) = 0$$

We *define* a position probability density and a probability flux in terms of ψ by

$$\left.\begin{aligned} \rho(x, t) &= \mathrm{i} \left\{ \psi^* \frac{\partial \psi}{\partial t} - \frac{\partial \psi^*}{\partial t} \psi \right\} \\ J(x, t) &= -\mathrm{i} \{ \psi^* (\nabla \psi) - (\nabla \psi)^* \psi \} \end{aligned}\right\} \tag{A.19}$$

It can be verified that with these definitions the continuity equation

$$\frac{\partial \rho}{\partial t} + \nabla \cdot J = 0$$

is satisfied.

For a ψ which tends to 0 as $|x| \to \infty$ the continuity equation can be used to show that probability is conserved

$$\frac{\partial}{\partial t} \int d^3x \rho(x, t) = 0$$

On substituting (A.18) into (A.19) we find, for a sharply peaked $a(p)$, that a factor of E_p can be cancelled to give

$$\rho(x, t) = \frac{1}{(2\pi)^3} \int d^3p d^3p' a^*(p') a(p) e^{i(p-p')\cdot x - i(E_p - E_{p'})t} \quad \text{(A.20)}$$

An expression of the form (A.20) gives $\rho_a(x, t)$ in terms of $a_1(p)$, and similarly $\rho_b(x, t)$ in terms of $a_2(p)$. On substituting these into the definition (A.9) for Φ and remembering the integral formula

$$\int d^4x e^{iP\cdot x - iEt} = (2\pi)^4 \delta^{(3)}(P) \delta(E)$$

we find for the mutual flux,

$$\Phi = \frac{v_{ab}}{(2\pi)^2} \times$$

$$\int d^3p'_1 d^3p'_2 d^3p_1 d^3p_2 a_1^*(p'_1) a_2^*(p'_2) a_1(p_1) a_2(p_2) \delta^{(4)}(p'_1 + p'_2 - p_1 - p_2)$$

(A.21)

A.4 Cross-section

On evaluating

$$\sigma = \left(\int dw\right) / \Phi$$

by means of (A.15), (A.17) and (A.21) the integral over the initial momentum space amplitudes drops out leaving the desired relation between the $\mathcal{T}$-matrix element and cross-section

$$\sigma = \frac{(2\pi)^4}{4E_a E_b v_{ab}}$$

$$\times \int \frac{d^3p_c d^3p_d}{(2\pi)^3 2E_c (2\pi)^3 2E_d} \delta^{(4)}(p_c + p_d - p_a - p_b) |T(p_c p_d, p_a p_b)|^2$$

(A.22)

We showed in §A.1 that $E_a E_b v_{ab}$ is a Lorentz invariant. The integral

$$\int \frac{d^3p}{2E}$$

is Lorentz invariant, because the Jacobian of the transformation

$$p \to p' = \Lambda p$$

is simply E/E', as is easily verified for the special case of a boost along the z-axis,

$$\int \frac{d^3p}{2E} = \int \frac{d^3p'}{2E}\left|\frac{\partial(p_x p_y p_z)}{\partial(p'_x p'_y p'_z)}\right| = \int \frac{d^3p'}{2E'}$$

the four-dimensional δ-function is Lorentz invariant; the Lorentz invariance of σ was discussed in §A.1 and that of $|T|^2$ follows from the discussion of §4.10.3.

We have neglected spins. If any of the particles have spin then σ and T will carry helicity labels, but the preceding derivations are unchanged.

Four of the six integrations in (A.22) can be done with the δ-function. Specialising to the CM frame we find

$$\sigma = \frac{1}{64\pi^2 W^2}\frac{p_{cd}}{p_{ab}}\int |T|^2 d\Omega \qquad \text{(A.23)}$$

where p_{ab} and p_{cd} are the particle momenta in the initial and final states and W is the total energy

$$W = E_a + E_b = E_c + E_d$$

If T is expressed in terms of the differently normalised two particle states (4.85), then (A.23) becomes

$$\sigma = \frac{4\pi^2}{p_{ab}^2}\int |T'|^2 d\Omega \qquad \text{(A.24)}$$

A.5 Decay rates

Consider a decay process

$$A \to a + b + c + \ldots \qquad \text{(A.25)}$$

The number δN_a of particles of type a detected in time δt is proportional to the number $N_A(t)$ of A present and to δt,

$$\delta N_a = \Gamma N_A(t)\delta t \tag{A.26}$$

Γ is a constant called the decay width. Since the detection of a heralds the decay of A,

$$\delta N_a = -\delta N_A$$

and

$$\delta N_A = -\Gamma N_A(t)\delta t$$

which is the radioactive decay law and shows that $1/\Gamma$ is the lifetime of A for decay by (A.25).

We specialise to a two body decay

$$A \to a + b$$

for ease of writing.

The total number of a particles detected is

$$N_a = \int \frac{\mathrm{d}N_a}{\mathrm{d}t}\mathrm{d}t = \int \Gamma N_A(t)\,\mathrm{d}t = \Gamma\int \rho_A(x,t)\,\mathrm{d}^3x\mathrm{d}t$$

So

$$\Gamma = N_A \Big/ \left(\int \rho_A \mathrm{d}^3x\mathrm{d}t\right) \tag{A.27}$$

We have introduced the density of A particles ρ_A.

By a calculation analogous to that for the cross-section we obtain

$$\Gamma = \frac{(2\pi)^4}{2E_A}\int \frac{\mathrm{d}^3p_a\mathrm{d}^3p_b}{(2\pi)^3 2E_a (2\pi)^3 2E_b}\delta^{(4)}(p_A - p_a - p_b)\,|T(p_a, p_b; p_A)|^2 \tag{A.28}$$

On specialising to the rest-frame of A (the CM of the decay), we find

$$\Gamma = \frac{p_{ab}}{32\pi^2 m_A^2}\int \mathrm{d}\Omega\,|T|^2 \tag{A.29}$$

If the final state is expressed in terms of the two particle state (4.85), then (A.29) becomes

$$\Gamma = \frac{1}{2m_A}\int \mathrm{d}\Omega\,|T'|^2 \tag{A.30}$$

APPENDIX B

DENSITY MATRIX DESCRIPTION OF POLARISATION

B.1 Definition of the density matrix

A quantum-mechanical system prepared under definite conditions is described by a wavefunction or state vector and conversely the state vector contains all possible information about the system. Sometimes we have to deal with systems of which we have less than complete knowledge. This usually arises when we have a large number N of copies of the same system, as for example in a beam of protons emerging from an accelerator. For any one proton we may only have statistical information about its spin orientation: 50 per cent up and 50 per cent down. The quantum-mechanical formalism can be extended to cover such cases. The ensemble of systems is then described by a density matrix instead of by a state vector.

We are only concerned with the density matrix description of spin states, that is of polarisation phenomena. It is assumed that all the particles in the ensemble are in the same spatial state, for example that they all have the same momentum k; so that any reference to the spatial state is suppressed.

A particle of spin s in a well-defined state is described by a state vector ψ which can be represented as a $(2s+1)$-component column vector, whose elements c_λ are the amplitudes for finding the particle with spin projection λ along some specified quantisation axis.

We have

$$\psi = \sum_\lambda c_\lambda \phi_\lambda \tag{B.1}$$

In matrix notation

$$\psi = \begin{bmatrix} c_s \\ c_{s-1} \\ c_{s-2} \\ \cdot \\ \cdot \\ \cdot \\ c_{-s} \end{bmatrix}, \quad \phi_s = \begin{bmatrix} 1 \\ 0 \\ 0 \\ \cdot \\ \cdot \\ \cdot \\ 0 \end{bmatrix}, \quad \phi_{s-1} = \begin{bmatrix} 0 \\ 1 \\ 0 \\ \cdot \\ \cdot \\ \cdot \\ 0 \end{bmatrix}, \quad \text{etc.}$$

The ϕ_λ will usually be eigenfunctions of the z-component of spin S_z, or the helicity operator,

$$\mathcal{H} = \frac{\boldsymbol{J}\cdot\boldsymbol{P}}{|\boldsymbol{P}|}$$

However, it is only necessary that they be a complete set of spin states.

The expectation value of any observable quantity A associated with the spin of the particle, e.g. x-component of spin, is given by

$$\begin{aligned}(\psi, A\psi) &= \sum_{\mu\lambda} c_\mu^* c_\lambda (\phi_\mu, A\phi_\lambda) \\ &= \sum_{\mu\lambda} c_\mu^* c_\lambda A_{\mu\lambda} \qquad \text{(B.2)}\end{aligned}$$

where we have introduced the expansion (B.1), and $A_{\mu\lambda}$ denotes a matrix element of A in the basis of ϕ_λ.

Now imagine a number N of the particles for which it is known that a fraction N_1 are in the state $\psi^{(1)}$, N_2 in the state $\psi^{(2)}$ and so on. What will be the mean value obtained on measuring A for all the members of the ensemble?

For a fraction $w_1 = N_1/N$ of the cases we obtain

$$(\psi^{(1)}, A\psi^{(1)}) = \sum_{\mu\lambda} c_\mu^{(1)*} c_\lambda^{(1)} A_{\mu\lambda}$$

where $c_\lambda^{(1)}$ is the expansion coefficient of $\psi^{(1)}$ in the basis ϕ_λ. In a fraction $w_2 = N_2/N$, we obtain

$$\sum_{\mu\lambda} c_\mu^{(2)*} c_\lambda^{(2)} A_{\mu\lambda}$$

and so on. Hence the expectation value of A averaged over the whole ensemble is

$$\langle A \rangle = \sum_r w_r \sum_{\mu\lambda} c_\mu^{(r)*} c_\lambda^{(r)} A_{\mu\lambda} \qquad \text{(B.3)}$$

Consider the form of this expression. The matrix elements $A_{\mu\lambda}$ depend only on the quantity A we have chosen to measure and on the basis ϕ_λ in which the states $\psi^{(r)}$ are expanded.

The quantity

$$\rho_{\lambda\mu} = \sum_r w_r c_\lambda^{(r)} c_\mu^{(r)*} \qquad \text{(B.4)}$$

on the other hand contains the information about the state of the ensemble of particles in the beam. It is called the *density matrix* for the beam. In terms of $\rho_{\lambda\mu}$ (B.3) may be written

$$\langle A \rangle = \sum_{\mu\lambda} \rho_{\lambda\mu} A_{\mu\lambda}$$

or

$$\langle A \rangle = \mathrm{Tr}\,(\rho A) \qquad \text{(B.5)}$$

where ρ and A are $(2s+1)\times(2s+1)$ matrices, and Tr denotes the trace

$$\mathrm{Tr}\,(M) = \sum_\lambda M_{\lambda\lambda}$$

The density matrix was introduced into quantum mechanics by von Neumann and Landau. It was applied to polarisation problems by Dalitz (1952) and others.

The density matrix contains the maximum information about the state of polarisation of the beam. From it the result of any measurement can be predicted. For two beams with equal density matrices all measurements give the same result for each.

It follows from the definition (B.4) that ρ is a Hermitian matrix

$$\rho_{\lambda\mu} = \rho^*_{\mu\lambda} \tag{B.6}$$

Thus $(2s+1)^2$ real numbers are required to specify ρ.

The trace is

$$\mathrm{Tr}\,(\rho) = \sum_\lambda \rho_{\lambda\lambda} = \sum_\lambda \sum_r w_r |c_\lambda^{(r)}|^2$$

$\sum_r w_r |c_\lambda^{(r)}|^2$ is the fraction of cases in which the particle is found in the state λ, hence the sum over λ gives unity, and so

$$\mathrm{Tr}\,(\rho) = 1$$

If we had carried the spatial wavefunction in our equations, Tr (ρ) would give the intensity of the beam. We shall assume this to be done and write

$$\mathrm{Tr}\,(\rho) = \mathcal{I} \tag{B.7}$$

The expression (B.5) for the expectation of A must then be written

$$\langle A \rangle = \frac{\mathrm{Tr}\,(\rho A)}{\mathrm{Tr}\,(\rho)} \tag{B.8}$$

The case $A = 1$ serves to check this.

B.2 Density matrix for spin one half

Let us consider the properties of ρ using the case $s = \frac{1}{2}$ as an example. This covers the beams of protons, neutrons and other baryons. In addition the photon has only two helicity states and therefore a 2×2 density matrix is sufficient to describe polarised photons.

For a beam of spin-$\frac{1}{2}$ particles we can write

$$\rho_{\mu\lambda} = \begin{pmatrix} \rho_{11} & \rho_{12} \\ \rho_{21} & \rho_{22} \end{pmatrix}$$

where by (B.6) and (B.7)

$$\left.\begin{aligned} \rho_{11} + \rho_{22} &= \mathscr{I} \\ \rho_{12} = \overset{*}{\rho}_{21}; \quad \rho_{11} &\text{ and } \rho_{22} \text{ are real} \end{aligned}\right\} \tag{B.9}$$

(It is convenient to use the labels 1 and 2 rather than $\pm \frac{1}{2}$.) Now any 2×2 Hermitian matrix may be expanded in terms of the Pauli spin matrices and the unit 2×2 matrix. Thus

$$\rho = a1 + \boldsymbol{b}\cdot\boldsymbol{\sigma}$$

where

$$\boldsymbol{\sigma} = (\sigma_x, \sigma_y, \sigma_z)$$

Since

$$\mathrm{Tr}\,(\sigma_x) = \mathrm{Tr}\,(\sigma_y) = \mathrm{Tr}\,(\sigma_z) = 0, \quad \mathrm{Tr}\,(1) = 2$$

(B.7) requires

$$a = \mathscr{I}/2$$

so we put

$$\rho = \tfrac{1}{2}\mathscr{I}(1 + \boldsymbol{P}\cdot\boldsymbol{\sigma}) \tag{B.10}$$

$\mathscr{I}$ and $\boldsymbol{P}$ are the four numbers required to fix ρ. $\boldsymbol{P}$ is called the *polarisation* vector of the beam. Its physical significance is as follows.

From the properties of the Pauli matrices

$$\sigma_x\sigma_y = -\sigma_y\sigma_x = \mathrm{i}\sigma_z, \text{ etc.} \tag{B.11a}$$

and

$$\sigma_x^2 = \sigma_y^2 = \sigma_z^2 = 1 \tag{B.11b}$$

we calculate

$$\mathrm{Tr}\,(\sigma_i\rho) = P_i\,\mathscr{I} \tag{B.12}$$

Thus by (B.8) $\boldsymbol{P}$ is the expectation value for the measurement of $\boldsymbol{\sigma}$ on the beam described by ρ. Explicitly

$$\begin{aligned} \mathscr{I} &= \rho_{11} + \rho_{22} \\ \mathscr{I}P_x &= \rho_{21} + \rho_{12} \\ \mathscr{I}P_y &= -\mathrm{i}(\rho_{21} - \rho_{12}) \\ \mathscr{I}P_z &= \rho_{11} - \rho_{22} \end{aligned}$$

For example

$$P_z = \frac{\text{number with spin up} - \text{number with spin down}}{\text{total number}}$$

We may imagine performing a Stern–Gerlach type of experiment on the beam.† Its effect is to measure the fraction of cases in which the z-projection of spin is $+\frac{1}{2}$ and $-\frac{1}{2}$, and hence to determine P_z. Similarly P_x and P_y may be determined by reorientating the apparatus, and thus the complete density matrix (B.10) may be determined.

In order to understand further the content of the density matrix we consider some special cases.

(*a*) Suppose all the particles of the beam are in the same state

$$\psi = \begin{pmatrix} c_1 \\ c_2 \end{pmatrix}$$

Then in the general definitions (B.3) and (B.4) there is no sum over r, and (B.4) becomes

$$\rho = \begin{pmatrix} c_1 c_1^* & c_1 c_2^* \\ c_2 c_1^* & c_2 c_2^* \end{pmatrix} \tag{B.13}$$

The intensity is simply

$$\mathscr{I} = \mathrm{Tr}\,(\rho) = |c_1|^2 + |c_2|^2$$

and the components of the polarisation vector are

$$\left.\begin{aligned} P_x &= \frac{2\,\mathrm{Re}\,(c_2 c_1^*)}{|c_1|^2+|c_2|^2} \\ P_y &= \frac{2\,\mathrm{Im}\,(c_2 c_1^*)}{|c_1|^2+|c_2|^2} \\ P_z &= \frac{|c_1|^2-|c_2|^2}{|c_1|^2+|c_2|^2} \end{aligned}\right\} \tag{B.14}$$

On the other hand, since $\boldsymbol{P}$ is the expectation of $\boldsymbol{\sigma}$ for the beam, and hence for each particle in the state ψ, we can write

$$\boldsymbol{P} = (\psi, \boldsymbol{\sigma}\psi)/(\psi, \psi) \tag{B.15}$$

which is the same as (B.14).

From (B.14), it follows that

$$|\boldsymbol{P}|^2 = P_x^2 + P_y^2 + P_z^2 = 1 \tag{B.16}$$

† In fact, uncertainty principle arguments show that the Stern–Gerlach experiment fails for charged particles such as free protons (see for example Mott (1929)), so a scattering experiment which effects the desired measurement indirectly would have to be used.

Hence in this case $\boldsymbol{P}$ is a unit vector whose components are the expectation values of the components of $\boldsymbol{\sigma}$.

Next we recall that for a spin-$\frac{1}{2}$ particle in a state ψ there always exists a direction of quantisation $\boldsymbol{n}$ such that the particle has spin projection $+\frac{1}{2}$ along it, that is

$$\boldsymbol{\sigma}\cdot\boldsymbol{n}\psi = \psi \tag{B.17}$$

($\boldsymbol{\sigma}$ is twice the spin operator). This simply amounts to choosing a new coordinate system in the two component spin space so that $\boldsymbol{\sigma}\cdot\boldsymbol{n}$ and ψ are brought to the form

$$\begin{pmatrix} 1 & 0 \\ 0 & -1 \end{pmatrix} \quad \text{and} \quad \begin{pmatrix} 1 \\ 0 \end{pmatrix}$$

We shall now show that the direction $\boldsymbol{n}$ is given by $\boldsymbol{P}$.

The properties (B.11) may be summarised as

$$\sigma_i\sigma_j + \sigma_j\sigma_i = 2\delta_{ij}, \quad i, j = 1, 2, 3$$

Take the ψ expectation of this identity

$$(\psi, \sigma_i\sigma_j\psi) + (\psi, \sigma_j\sigma_i\psi) = 2\delta_{ij}(\psi, \psi)$$

Hence

$$(\psi, \boldsymbol{n}\cdot\boldsymbol{\sigma}\sigma_j\psi) + (\psi, \sigma_j\boldsymbol{n}\cdot\boldsymbol{\sigma}\psi) = 2n_j(\psi, \psi)$$

Now in the terms on the left side we let the Hermitian operator $\boldsymbol{n}\cdot\boldsymbol{\sigma}$ act to the left and to the right in the respective terms, and use (B.17) to obtain

$$2(\psi, \sigma_j\psi) = 2n_j(\psi, \psi)$$

and thus

$$n_j = (\psi, \sigma_j\psi)/(\psi, \psi) = P_j$$

In summary, a beam for which all the particles are in the same state ψ is called *fully polarised*; the density matrix is

$$\rho = \tfrac{1}{2}\mathscr{I}(1 + \boldsymbol{P}\cdot\boldsymbol{\sigma})$$

where $\boldsymbol{P}$ is along the direction of polarisation and

$$|\boldsymbol{P}| = 1$$

(*b*) Next we consider a beam for which the density matrix is

$$\rho = \tfrac{1}{2}\mathscr{I}1 \tag{B.18}$$

If a Stern–Gerlach type of experiment is performed on such a beam, then for any orientation of the apparatus there will be equal fractions in the two split beams. A beam of particles described by (B.18) is therefore *unpolarised*.

(*c*) Suppose next that

$$\rho = \tfrac{1}{2}\mathcal{I}(1 + P\sigma_z) = \mathcal{I}\begin{pmatrix} \tfrac{1}{2}(1+P) & 0 \\ 0 & \tfrac{1}{2}(1-P) \end{pmatrix} \tag{B.19}$$

If $P = 0$, we have case (*b*) while if $P = +1$ (or -1) we have the case of a fully polarised beam (*a*). For P lying between 0 and 1, (B.19) describes a partially polarised beam in which a fraction $\tfrac{1}{2}(1+P)$ of particles have $s_z = +\tfrac{1}{2}$, and $\tfrac{1}{2}(1-P)$ have $s_z = -\tfrac{1}{2}$. Alternatively, since

$$\rho = \mathcal{I}\left\{\tfrac{1}{2}(1-P)\begin{pmatrix} 1 & 0 \\ 0 & 1 \end{pmatrix} + P\begin{pmatrix} 1 & 0 \\ 0 & 0 \end{pmatrix}\right\}$$

we can say that a fraction $(1-P)$ of the beam is unpolarised and a fraction P is fully polarised in the z-direction. (If P is negative this becomes the $-z$-direction.)

These two descriptions are equally valid since they cannot be distinguished experimentally.

(*d*) The general density matrix

$$\rho = \tfrac{1}{2}\mathcal{I}(1 + \boldsymbol{P}\cdot\boldsymbol{\sigma}) \tag{B.20}$$

can be reduced to the form (B.19) by a rotation of the quantisation axis.

Thus we can say of the beam described by the ρ of (B.20):

(1) If $|\boldsymbol{P}| = 1$, the beam is fully polarised along $\boldsymbol{P}$.

(2) If $0 < |\boldsymbol{P}| < 1$, the beam is partially polarised, having fractions $\tfrac{1}{2}(1 \pm |\boldsymbol{P}|)$ with spin projection $\pm\tfrac{1}{2}$ along $\boldsymbol{P}$.

(3) If $\boldsymbol{P} = 0$, the beam is unpolarised.

B.3 Generalisations

The case of spin 1 serves to illustrate the direction in which the above considerations must be generalised for $s > \tfrac{1}{2}$. $3^2 = 9$ real numbers specify the spin-1 density matrix. The representation (B.10) may be replaced by

$$\rho = \mathcal{I}(\tfrac{1}{3}1 + \sum_i P_i S_i + \ldots)$$

where $S_i = (S_x S_y S_z)$ are the 3×3 spin matrices corresponding to spin 1.

Five more Hermitian operators are required to complete the expansion of the Hermitian matrix. Suitable operators are formed from $S_xS_y + S_yS_z$, etc. and S_xS_x, etc.

The coefficients of these quantities are called the moments of ρ. The presence of these extra terms corresponds to the fact that in the Stern–Gerlach experiment there would be three beams; up, down and undeviated. The quantity

$$\mathrm{Tr}\,(\rho S_z)/\mathrm{Tr}\,(\rho) = \text{fraction up} - \text{fraction down}$$

gives no information about the undeviated beam.

Consider a system of two particles a and b with spins s_a and s_b. If the particles are non-interacting and in definite states they are described by a product wavefunction $\psi^{(a)}\chi^{(b)}$. Correspondingly the two beams of particles a and particles b in a scattering experiment may be described by a product density matrix

$$\rho^{(a)}_{\lambda\lambda'}\rho^{(b)}_{\mu\mu'} = \rho^{(ab)}_{\lambda\mu,\lambda'\mu'} \qquad \text{(B.21)}$$

This may be considered as a matrix whose rows are labelled by pairs of indices $\lambda\mu$, and columns by $\lambda'\mu'$. After the beams have interacted there will be correlations between the spins, and the density matrix will be of the form

$$\rho^{(ab)\mathrm{final}}_{\lambda\mu,\lambda'\mu'} \qquad \text{(B.22)}$$

but may not be separated into the form (B.21).

B.4 Density matrix and scattering

We consider the scattering of a beam of particles of spin ½ off a spinless target, for example πN scattering, and we draw on the notation of §4.8.

If the particles of the beam are prepared in a definite spin state, for example in an eigenstate of helicity λ, the scattering amplitude into the eigenstate of helicity μ may be regarded as a matrix

$$f_{\mu\lambda} = (\chi_\mu, f\chi_\lambda)$$

Hence the amplitude for scattering into state μ for a general initial spin state $\Sigma\, c^{\mathrm{i}}_\lambda \chi_\lambda$ is

$$c^{\mathrm{f}}_\mu = \sum_\lambda f_{\mu\lambda} c^{\mathrm{i}}_\lambda$$

We may form the density matrix for the scattered beam

$$\rho^{\mathrm{f}}_{\mu\mu'} = \sum_\lambda f_{\mu\lambda} c^{\mathrm{i}}_\lambda f^*_{\mu'\lambda'} c^{\mathrm{i}*}_{\lambda'}$$

and rearrange to display the dependence on the density matrix for the initial state (which is fully polarised in this case)

$$\rho^{\mathrm{f}}_{\mu\mu'} = \sum_{\lambda\lambda'} f_{\mu\lambda} c^{\mathrm{i}}_{\lambda} c^{\mathrm{i}*}_{\lambda'} f^{*}_{\mu'\lambda'}$$

or

$$\rho^{\mathrm{f}} = f\rho^{\mathrm{i}} f^{\dagger} \tag{B.23}$$

Quite generally, if the incident beam is in an arbitrary polarisation state described by a density matrix ρ^{i}, the density matrix for the scattered beam is given by (B.23). ρ^{f} depends of course on the angle of scattering through the scattering amplitude $f_{\mu\lambda}$.

It is convenient to normalise ρ^{i} to unit intensity in the incident beam,

$$\mathrm{Tr}\,\rho^{\mathrm{i}} = 1$$

so that for an unpolarised beam

$$\rho^{\mathrm{i}} = \tfrac{1}{2}\begin{pmatrix} 1 & 0 \\ 0 & 1 \end{pmatrix}$$

We find for the trace of ρ^{f},

$$\begin{aligned} \mathrm{Tr}\,\rho^{\mathrm{f}} &= \mathrm{Tr}\,(f\cdot\tfrac{1}{2}\cdot f^{\dagger}) \\ &= \tfrac{1}{2}\sum_{\mu\lambda} f_{\mu\lambda} f^{\dagger}_{\lambda\mu} \\ &= \tfrac{1}{2}\sum_{\mu\lambda} |f_{\mu\lambda}|^2 \end{aligned}$$

which is simply the usual expression for the differential cross-section for scattering with an unpolarised initial beam and for which the final spins are not observed. The density matrix gives a more precise way of describing these operations. We have

$$\mathrm{Tr}\,\rho^{\mathrm{f}} = \left(\frac{\mathrm{d}\sigma}{\mathrm{d}\Omega}\right)_{\textbf{unpolarised}}$$

where the differential cross-section is that for scattering, from the initial state ρ^{i}, without observation of the final spins.

In the helicity description of spin, f is a matrix whose column index refers to the z-axis as quantisation axis, while the row index refers to the direction (θ, ϕ) of the scattered nucleon as quantisation axis. Thus if we write

$$\rho^{\mathrm{f}} = \tfrac{1}{2}(1 + \boldsymbol{P}^{\mathrm{f}}\cdot\boldsymbol{\sigma})\left(\frac{\mathrm{d}\sigma}{\mathrm{d}\Omega}\right)_{\textbf{unpolarised}}$$

then $\boldsymbol{P}^{\mathrm{f}}$ is a vector whose components refer to the axes (x', y', z') reached from the standard axes by the rotation $R(\phi, \theta, 0)$ and shown in fig. 4.1. Moreover $\boldsymbol{P}^{\mathrm{f}}$ refers to the polarisation in the rest-frame of the final spin-$\tfrac{1}{2}$ particle.

APPENDIX C

ISOSPIN AND $SU(3)$ PHASE CONVENTIONS

This appendix contains an explanation of the phase conventions used in this book for the operators and states of the internal symmetries: isospin, $SU(3)$ and charge conjugation.

C.1 Phase conventions for isospin and charge conjugation

The Condon and Shortley (CS) phase convention originally adopted for angular momentum in §3.2.2 may be taken over to isospin. It requires that the shift operators $I_\pm$ have *real* and *positive* matrix elements with respect to the standard eigenstates $|I, I_3\rangle$ of I^2 and I_3.

Thus we may make the following assignments for the nucleon isodoublet

$$|\mathrm{p}\rangle = |\mathrm{N}, \tfrac{1}{2}, +\tfrac{1}{2}\rangle, \quad |\mathrm{n}\rangle = |\mathrm{N}, \tfrac{1}{2}, -\tfrac{1}{2}\rangle \tag{C.1}$$

In the case of the pion triplet we might put $|\pi^+\rangle = |1, +1\rangle$, $|\pi^0\rangle = |1, 0\rangle$, $|\pi^-\rangle = |1, -1\rangle$. Although this is a perfectly acceptable choice, one usually finds in the literature that minus signs appear on the right of certain of these equations. Similarly for the antinucleon doublet, $|\bar{\mathrm{p}}\rangle = |\bar{\mathrm{N}}, \tfrac{1}{2}, +\tfrac{1}{2}\rangle$ and $|\bar{\mathrm{n}}\rangle = |\bar{\mathrm{N}}, \tfrac{1}{2}, -\tfrac{1}{2}\rangle$ are not the standard choice. The reason is that there is another operator, the charge conjugation operator U_C, which relates certain of these states either within the same multiplet as with π^+ and π^-, or in different multiplets as with p and $\bar{\mathrm{p}}$. It turns out that the requirement that U_C have positive matrix elements is incompatible with the CS convention for $I_\pm$. The unhealthy compromise with which we shall live is to use state symbols like those on the left of (C.1) to denote those states simply related by U_C, and those on the right for standard isospin states. Minus signs may then appear in (C.1) and its analogues.

By its definition charge conjugation changes the sign of the third component of isospin. It follows that U_C and I_3 must satisfy

$$U_C I_3 U_C^{-1} = -I_3 \tag{C.2}$$

We can find the effect of U_C on $I_\pm$ as follows. The transformed operators

$$I_{\pm}^{C} \equiv U_C I_{\pm} U_C^{-1}$$

are the isospin step operators for the charge conjugate multiplet, and we therefore require that they too obey isospin commutation relations. With this in mind we apply the operation $U_C \ldots U_C^{-1}$ to the equation

$$I_3 I_{\pm} - I_{\pm} I_3 = \pm I_{\pm}$$

to obtain

$$I_3^C I_{\pm}^C - I_{\pm}^C I_3^C = \pm I_{\pm}^C \quad \text{(C.3)}$$

Since from (C.2)

$$I_3^C = -I_3$$

(C.3) becomes

$$[I_3, I_{\pm}^C] = \mp I_{\pm}^C$$

From this equation and from the requirement that I_-^C and I_+^C are mutually Hermitian conjugate we can conclude that

$$U_C I_{\pm} U_C^{-1} = I_{\pm}^C = \alpha I_{\mp}$$

where α is $+1$ or -1.

The value α must be chosen by convention. For reasons going back to quantum field theory† it is usual to choose

$$\alpha = -1$$

so that

$$U_C I_{\pm} U_C^{-1} = -I_{\mp} \quad \text{(C.4)}$$

We note that this implies

$$U_C I_1 U_C^{-1} = -I_1 \quad \text{(C.5}a\text{)}$$

and

$$U_C I_2 U_C^{-1} = +I_2 \quad \text{(C.5}b\text{)}$$

Our basic equation is (C.4).

Consider the nucleon–antinucleon system. We set

$$|\mathrm{p}\rangle = |\mathrm{N}, \tfrac{1}{2}, +\tfrac{1}{2}\rangle, \quad |\mathrm{n}\rangle = |\mathrm{N}, \tfrac{1}{2}, -\tfrac{1}{2}\rangle \quad \text{(C.6)}$$

where, as stated above, the states on the right are standard isospin states for which the CS phase convention holds. From (C.4) it follows that

† In particular, the choice made in the text corresponds to the condition that if the field operator ϕ creates a particle x the Hermitian conjugate operator $\phi^\dagger$ creates the corresponding antiparticle $\bar{x}$ in the *same* state, i.e.

$$|x\rangle = \phi\,|\mathrm{vac}\rangle, \quad |\bar{x}\rangle = \phi^\dagger\,|\mathrm{vac}\rangle$$

where $|\mathrm{vac}\rangle$ denotes the vacuum state.

$$\begin{aligned} I_+U_C|\mathrm{p}\rangle &= -U_C I_-|\mathrm{p}\rangle \\ &= -U_C I_-|\mathrm{N}, \tfrac{1}{2}, +\tfrac{1}{2}\rangle \\ &= -U_C|\mathrm{N}, \tfrac{1}{2}, -\tfrac{1}{2}\rangle \\ &= -U_C|\mathrm{n}\rangle \end{aligned} \tag{C.7}$$

and if we denote the effect of charge conjugation by a bar over the particle label, thus

$$U_C|\mathrm{p}\rangle = |\bar{\mathrm{p}}\rangle, \quad U_C|\mathrm{n}\rangle = |\bar{\mathrm{n}}\rangle \tag{C.8}$$

then (C.7) gives

$$I_+|\bar{\mathrm{p}}\rangle = -|\bar{\mathrm{n}}\rangle \tag{C.9}$$

This shows that I_+ has a negative matrix element with respect to the states $|\bar{\mathrm{p}}\rangle$ and $|\bar{\mathrm{n}}\rangle$.

Similarly one finds

$$I_-U_C|\mathrm{n}\rangle = -U_C|\mathrm{p}\rangle$$

or

$$I_-|\bar{\mathrm{n}}\rangle = -|\bar{\mathrm{p}}\rangle$$

To define antiparticle states for which the CS convention holds we put

$$\left.\begin{aligned} |\bar{\mathrm{N}}, \tfrac{1}{2}, +\tfrac{1}{2}\rangle &= -|\bar{\mathrm{n}}\rangle \\ |\bar{\mathrm{N}}, \tfrac{1}{2}, -\tfrac{1}{2}\rangle &= +|\bar{\mathrm{p}}\rangle \end{aligned}\right\} \tag{C.10}$$

so that (C.9) becomes

$$I_+|\bar{\mathrm{N}}, \tfrac{1}{2}, -\tfrac{1}{2}\rangle = |\bar{\mathrm{N}}, \tfrac{1}{2}, +\tfrac{1}{2}\rangle$$

and the CS convention is satisfied. The position of the minus sign in (C.10) is arbitrary: it could have been attached to the $|\bar{\mathrm{p}}\rangle$ state instead.

Next we consider the pion isotriplet. In this case U_C relates particles in the same multiplet. It would be possible to analyse this case from first principles as with the nucleons and this would expose the manifold possibilities for sign conventions. Instead we shall go quickly to our chosen convention.

First we must note that

$$U_C|\pi^0\rangle = |\pi^0\rangle \tag{C.11a}$$

is not a convention, but a physical statement that the charge conjugation parity of the π^0 is $+1$.

However, we choose the positive sign in

$$U_C|\pi^+\rangle = +|\pi^-\rangle \tag{C.11b}$$

from which

$$U_C|\pi^-\rangle = +|\pi^+\rangle \tag{C.11c}$$

follows since $U_C^2 = 1$.

From (C.4) we find

$$I_+ U_C |\pi^+\rangle = -U_C I_- |\pi^+\rangle$$

and thus

$$I_+ |\pi^-\rangle = -U_C I_- |\pi^+\rangle \tag{C.12}$$

Since by (C.11) U_C has positive matrix elements with respect to the states $|\pi^+\rangle$, $|\pi^0\rangle$ and $|\pi^-\rangle$ there is a minus sign to be absorbed in (C.12).

If we set

$$|1, +1\rangle = -|\pi^+\rangle, \quad |1, 0\rangle = |\pi^0\rangle, \quad |1, -1\rangle = |\pi^-\rangle \tag{C.13}$$

then the CS convention is satisfied for the states $|I, I_3\rangle$ on the left-hand side. For example

$$I_+ U_C |\pi^+\rangle = -U_C I_- |\pi^+\rangle$$

or

$$\begin{aligned} I_+ |\pi^-\rangle &= +U_C I_- |1, +1\rangle \\ &= +2^{1/2} U_C |1, 0\rangle \\ &= +2^{1/2} |1, 0\rangle \end{aligned}$$

and so

$$I_+ |1, -1\rangle = +2^{1/2} |1, 0\rangle$$

The kaons are handled in an analogous way to the nucleons, but with hypercharge replacing baryon number. We choose

$$\left.\begin{aligned} |K^+\rangle &= |K, \tfrac{1}{2}, +\tfrac{1}{2}\rangle; \quad |K^0\rangle = |K, \tfrac{1}{2}, -\tfrac{1}{2}\rangle \\ |\overline{K^0}\rangle &= -|\overline{K}, \tfrac{1}{2}, +\tfrac{1}{2}\rangle; \quad |K^-\rangle = |\overline{K}, \tfrac{1}{2}, -\tfrac{1}{2}\rangle \end{aligned}\right\} \tag{C.14}$$

Here $|K^-\rangle$ is $|\overline{K^+}\rangle$, i.e.

$$U_C |K^+\rangle = |K^-\rangle$$

The final example is that of the Σ-hyperon

$$|\Sigma^+\rangle = -|\Sigma, 1, +1\rangle; \quad |\Sigma^0\rangle = |\Sigma, 1, 0\rangle; \quad |\Sigma^-\rangle = |\Sigma, 1, -1\rangle$$
$$|\overline{\Sigma^-}\rangle = -|\overline{\Sigma}, 1, +1\rangle; \quad |\overline{\Sigma^0}\rangle = |\overline{\Sigma}, 1, 0\rangle; \quad |\overline{\Sigma^+}\rangle = |\overline{\Sigma}, 1, -1\rangle \tag{C.15}$$

here $|\overline{\Sigma^-}\rangle$ denotes the positively charged antisigma. With these assignments (*a*) U_C sends $|x\rangle$ into $|\bar{x}\rangle$, (*b*) (C.4) is satisfied, and (*c*) $I_\pm$ have positive matrix elements with respect to $|\Sigma, 1, I_3\rangle$ and to $|\overline{\Sigma}, 1, I_3\rangle$.

The physicist reader may now be asking: if I have a positive sigma hyperon emerging from a beam line, do I describe it by $|\Sigma^+\rangle$ or by $-|\Sigma^+\rangle$ or what? The answer is that in questions involving isospin (or $SU(3)$) we are trying to relate the properties of different particles, e.g. Σ^+ with Σ^0 or $\overline{\Sigma^+}$. If $|\Sigma^+\rangle$ represents a positive sigma in a given space–spin state, then $|\Sigma^0\rangle$ represents a neutral sigma in the same space–spin state and $|\overline{\Sigma^+}\rangle$ a negative antisigma in that same state.

The only reason for introducing the states $|\Sigma, 1, I_3\rangle$ and $|\Sigma, 1, I_3\rangle$ is that tables of Clebsch–Gordan coefficients refer to states for which the CS convention holds, and so the translations from $|\Sigma^+\rangle$ etc. to $|\Sigma, 1, I_3\rangle$ must first be made.

C.2 *G*-parity

The conventional choice of α leading to (C.4) governs the definition of the *G*-parity operator. Since U_C reverses the sign of I_3 and I_1 the compensating isospin rotation must be about the second axis through π. The corresponding unitary operator is

$$P_i = e^{-i\pi I_2} \tag{C.16}$$

which has the following properties

$$P_i I_1 P_i^{-1} = -I_1 \tag{C.17a}$$

$$P_i I_2 P_i^{-1} = +I_2 \tag{C.17b}$$

$$P_i I_3 P_i^{-1} = -I_3 \tag{C.17c}$$

The correctness of these equations can be seen by using the kind of argument made in the corresponding case of angular momentum, cf. the discussion preceding (3.63).

Alternatively a formal derivation of (C.17) can be given. Since P_i plays an important role in our discussion of $SU(3)$ we shall do this. The trick is to work with

$$P_i(\beta) = e^{-i\beta I_2}$$

and set $\beta = \pi$ at the end. We consider the operator depending on β,

$$F(\beta) = e^{-i\beta I_2} I_1 e^{+i\beta I_2}$$

which can be differentiated with respect to β, giving

$$\frac{dF}{d\beta} = e^{-i\beta I_2}(-iI_2)I_1 e^{i\beta I_2} + e^{-i\beta I_2} I_1(+I_2)e^{i\beta I_2}$$

Here the order of I_2 and the exponential function of I_2 is immaterial. Thus

$$\frac{dF}{d\beta} = ie^{-i\beta I_2}[I_1, I_2]e^{i\beta I_2}$$

$$= -e^{-i\beta I_2} I_3 e^{i\beta I_2}$$

We now define

$$K(\beta) = e^{-i\beta I_2} I_3 e^{i\beta I_2}$$

So that we have

$$\frac{\mathrm{d}F}{\mathrm{d}\beta} = -K(\beta) \tag{C.18}$$

Differentiating K we find by a similar procedure

$$\frac{\mathrm{d}K}{\mathrm{d}\beta} = F(\beta) \tag{C.19}$$

(C.18) and (C.19) are a pair of simultaneous first-order differential equations to be solved subject to the initial conditions

$$F(\beta = 0) = I_1, \quad K(\beta = 0) = I_3$$

It is straightforward to obtain

$$F(\beta) = I_1 \cos\beta - I_3 \sin\beta$$

$$K(\beta) = I_1 \sin\beta + I_3 \cos\beta$$

On setting $\beta = \pi$, we arrive at the desired results (C.17*a*) and (C.17*c*).

From (C.17) we obtain the transformation law of the shift operators

$$P_{\mathrm{i}} I_{\pm} P_{\mathrm{i}}^{-1} = -I_{\mp} \tag{C.20}$$

Let us now define the G-parity operator

$$U_G = U_C \mathrm{e}^{-\mathrm{i}\pi I_2} = \mathrm{e}^{-\mathrm{i}\pi I_2} U_C$$

whose inverse is

$$U_G^{-1} = \mathrm{e}^{\mathrm{i}\pi I_2} U_C^{-1} = \mathrm{e}^{\mathrm{i}\pi I_2} U_C$$

From (C.4) and (C.20) we can compute

$$\begin{aligned} U_G I_{\pm} U_G^{-1} &= U_C \mathrm{e}^{-\mathrm{i}\pi I_2} I_{\pm} \mathrm{e}^{\mathrm{i}\pi I_2} U_C^{-1} \\ &= U_C(-I_{\mp}) U_C^{-1} \\ &= +I_{\pm} \end{aligned}$$

Similarly, using (C.2) and (C.17*b*), we find

$$U_G I_3 U_G^{-1} = +I_3$$

So U_G commutes with the algebra of isospin operators $I_{\pm}$ and I_3, and it follows that simultaneous eigenstates of G-parity and isospin exist.

We note that

$$U_G^2 = U_C \mathrm{e}^{-\mathrm{i}\pi I_2} U_C \mathrm{e}^{-\mathrm{i}\pi I_2} = U_C^2 \mathrm{e}^{-2\pi\mathrm{i} I_2}$$

Since $U_C^2 = 1$, we have

$$U_G^2 = e^{-2\pi i I_2} = (-1)^{2I} \tag{C.21}$$

since just as with ordinary angular momentum an isospin rotation by 2π leaves a state vector unchanged or multiplies it by -1 according as the total isospin I is integral or half-integral.

Returning to conventions, if we had chosen $\alpha = +1$ in (C.4) then the definition of U_G would require $e^{-i\pi I_1}$. This choice is made by Källén (1964). A mixture of the two conventions leads to opposite G-parities for the charged and neutral pions (Frazer, 1966). Although the physical consequences are unaffected, the conceptual simplicity of G-parity is thereby lost.

C.3 Phase conventions in $SU(3)$

As we have seen, many of the important consequences of $SU(3)$ invariance can be obtained by requiring both I-spin and U-spin invariance. These are both relatively easy to handle because they only require the familiar algebra of angular momentum. We choose to stress U-spin rather than V-spin because electromagnetism conserves U-spin.

From this point of view it is natural to extend the Condon and Shortley phase convention to the one already stated in § 10.3.4:

$SU(3)$ phase convention: The matrix elements of $I_\pm$ and $U_\pm$ shall be real and positive.

This also has the advantage that calculations done using I-spin and U-spin techniques and $SU(2)$ Clebsch–Gordan coefficients agree exactly with those done using the full paraphernalia of $SU(3)$ Clebsch–Gordan coefficients.

Similarly the rule (C.4) for the action of U_C on I-spin operators can be extended to U-spin. Thus

$$U_C U_\pm U_C^{-1} = -U_\mp$$

Here again the sign on the right is conventional.

For the case of the quark **3** and antiquark **3*** supermultiplets, it can be verified that our $SU(3)$ phase convention is satisfied for standard states defined as follows

$$|3, +\tfrac{1}{3}, \tfrac{1}{2}, +\tfrac{1}{2}\rangle = |u\rangle \quad |3^*, +\tfrac{2}{3}, 0, 0\rangle = |\bar{s}\rangle$$

$$|3, +\tfrac{1}{3}, \tfrac{1}{2}, -\tfrac{1}{2}\rangle = |d\rangle \quad |3^*, -\tfrac{1}{3}, \tfrac{1}{2}, +\tfrac{1}{2}\rangle = -|\bar{d}\rangle$$

$$|3, -\tfrac{2}{3}, 0, 0\rangle = |s\rangle \quad |3^*, -\tfrac{1}{3}, \tfrac{1}{2}, -\tfrac{1}{2}\rangle = |\bar{u}\rangle$$

Here we simply extended the convention for isospin multiplets, and checked the U-spin cases.

Table C.1. *The definition of the standard SU(3) states of baryon and meson octets*

$\vert 8, Y, I, I_3\rangle$	Meson	Baryon	Antibaryon
$\vert 8, +1, \frac{1}{2}, +\frac{1}{2}\rangle$	$\vert K^+\rangle$	$\vert p\rangle$	$-\vert\overline{\Xi^-}\rangle$
$\vert 8, +1, \frac{1}{2}, -\frac{1}{2}\rangle$	$\vert K^0\rangle$	$\vert n\rangle$	$\vert\overline{\Xi^0}\rangle$
$\vert 8, 0, 1, +1\rangle$	$-\vert\pi^+\rangle$	$-\vert\Sigma^+\rangle$	$-\vert\overline{\Sigma^-}\rangle$
$\vert 8, 0, 1, 0\rangle$	$\vert\pi^0\rangle$	$\vert\Sigma^0\rangle$	$\vert\overline{\Sigma^0}\rangle$
$\vert 8, 0, 1, -1\rangle$	$\vert\pi^-\rangle$	$\vert\Sigma^-\rangle$	$\vert\overline{\Sigma^+}\rangle$
$\vert 8, 0, 0, 0\rangle$	$\vert\eta^0\rangle$	$\vert\Lambda^0\rangle$	$\vert\overline{\Lambda^0}\rangle$
$\vert 8, -1, \frac{1}{2}, +\frac{1}{2}\rangle$	$-\vert\overline{K^0}\rangle$	$\vert\Xi^0\rangle$	$-\vert\bar{n}\rangle$
$\vert 8, -1, \frac{1}{2}, -\frac{1}{2}\rangle$	$\vert K^-\rangle$	$\vert\Xi^-\rangle$	$\vert\bar{p}\rangle$

Our standard states for baryon and meson octets have been chosen to agree with the conventions adopted for the corresponding isospin multiplets, and are given in table C.1.

Unfortunately the commonly-used convention of de Swart (1963) required (in our notation) that $I_\pm$ and $V_\pm$ shall have positive matrix elements. The minus signs in the commutation relations (10.14) preclude simultaneously positive matrix elements for all six operators.

CLEBSCH–GORDAN COEFFICIENTS FOR 8 × 8

We give the Clebsch–Gordan coefficients for the product **8** × **8** in the form of expressions for the two particle states $|N, Y, I, I_3\rangle$ of the definite total Y, I and I_3 belonging to the supermultiplet $N =$ **27**, **10**, **10***, $\mathbf{8}_A$, $\mathbf{8}_S$, or **1**.

The baryon–meson systems have been used as a concrete example, but of course the results apply to any two octets with appropriate translation of the symbols.

The symbol [BM] on the right-hand side denotes the state of two particles B and M coupled to a definite value of the total isospin I and I_3 as given in the state vector on the left-hand side. The coefficients of the symbols [BM] are just the isoscalar factors.

The states [BM] can be written out explicitly with the aid of the isospin C–G. coefficients given in tables 3.2 to 3.4, using (3.107) where necessary. In doing so the relations between particle states and standard isospin states given in appendix C must be used.

For example, in $|27, -1, \frac{1}{2}, +\frac{1}{2}\rangle$ we compute

$$[\Sigma\bar{K}] = \sum_{I_3 I_3'} C^{1/2 + 1/2}_{1 I_3, 1/2 I_3'} |\Sigma, 1, I_3\rangle |\bar{K}, \tfrac{1}{2}, I_3'\rangle$$

$$= -(\tfrac{1}{3})^{1/2} |\Sigma, 1, 0\rangle |\bar{K}, \tfrac{1}{2}, +\tfrac{1}{2}\rangle + (\tfrac{2}{3})^{1/2} |\Sigma, 1, +1\rangle |\bar{K}, \tfrac{1}{2}, -\tfrac{1}{2}\rangle$$

$$= +(\tfrac{1}{3})^{1/2} |\Sigma^0\rangle |\bar{K}^0\rangle - (\tfrac{2}{3})^{1/2} |\Sigma^+\rangle |K^-\rangle$$

As remarked in appendix C, our conventions differ from those of de Swart. The effect is to change the signs of some of the isoscalar factors. In the tables a dagger attached to a [BM] symbol indicates that the isoscalar factor preceding it, has the opposite sign if the de Swart convention is used as in the 'Review of particle properties'.

Baryon–meson states of the **27**

$$|27,+2,1,I_3\rangle = [\mathrm{NK}]$$

$$|27,+1,\tfrac{3}{2},I_3\rangle = (\tfrac{1}{2})^{1/2}[\mathrm{N}\pi] + (\tfrac{1}{2})^{1/2}[\Sigma\mathrm{K}]$$

$$|27,+1,\tfrac{1}{2},I_3\rangle = -(\tfrac{1}{20})^{1/2}[\mathrm{N}\pi]^\dagger + (\tfrac{1}{20})^{1/2}[\Sigma\mathrm{K}]^\dagger + (\tfrac{9}{20})^{1/2}[\Lambda\mathrm{K}] + (\tfrac{9}{20})^{1/2}[\mathrm{N}\eta]$$

$$|27,\ 0,2,I_3\rangle = [\Sigma\pi]$$

$$|27,\ 0,1,I_3\rangle = (\tfrac{1}{5})^{1/2}[\mathrm{N\bar{K}}] + (\tfrac{1}{5})^{1/2}[\Xi\mathrm{K}] + (\tfrac{3}{10})^{1/2}[\Sigma\eta] + (\tfrac{3}{10})^{1/2}[\Lambda\pi]$$

$$|27,\ 0,0,0\rangle = -(\tfrac{3}{20})^{1/2}[\mathrm{N\bar{K}}]^\dagger + (\tfrac{3}{20})^{1/2}[\Xi\mathrm{K}]^\dagger - (\tfrac{1}{40})^{1/2}[\Sigma\pi] + (\tfrac{27}{40})^{1/2}[\Lambda\eta]$$

$$|27,-1,\tfrac{3}{2},I_3\rangle = (\tfrac{1}{2})^{1/2}[\Sigma\bar{\mathrm{K}}] + (\tfrac{1}{2})^{1/2}[\Xi\pi]$$

$$|27,-1,\tfrac{1}{2},I_3\rangle = -(\tfrac{1}{20})^{1/2}[\Sigma\bar{\mathrm{K}}]^\dagger + (\tfrac{1}{20})^{1/2}[\Xi\pi]^\dagger + (\tfrac{9}{20})^{1/2}[\Lambda\bar{\mathrm{K}}] + (\tfrac{9}{20})^{1/2}[\Xi\eta]$$

$$|27,-2,1,I_3\rangle = [\Xi\bar{\mathrm{K}}]$$

BM states of the **10***

$$|10^*,+2,0,0\rangle = [\mathrm{NK}]^\dagger$$

$$|10^*,+1,\tfrac{1}{2},I_3\rangle = \tfrac{1}{2}[\mathrm{N}\pi]^\dagger + \tfrac{1}{2}[\Sigma\mathrm{K}]^\dagger + \tfrac{1}{2}[\mathrm{N}\eta] - \tfrac{1}{2}[\Lambda\mathrm{K}]$$

$$|10^*,\ 0,1,I_3\rangle = (\tfrac{1}{6})^{1/2}[\mathrm{N\bar{K}}] + (\tfrac{1}{6})^{1/2}[\Sigma\pi]^\dagger + \tfrac{1}{2}[\Sigma\eta] - \tfrac{1}{2}[\Lambda\pi] - (\tfrac{1}{6})^{1/2}[\Xi\mathrm{K}]$$

$$|10^*,-1,\tfrac{3}{2},I_3\rangle = (\tfrac{1}{2})^{1/2}[\Sigma\bar{\mathrm{K}}] - (\tfrac{1}{2})^{1/2}[\Xi\pi]$$

BM states of the **10**

$$|10,+1,\tfrac{3}{2},I_3\rangle = (\tfrac{1}{2})^{1/2}[\mathrm{N}\pi]^\dagger - (\tfrac{1}{2})^{1/2}[\Sigma\mathrm{K}]^\dagger$$

$$|10,\ 0,1,I_3\rangle = (\tfrac{1}{6})^{1/2}[\mathrm{N\bar{K}}]^\dagger + (\tfrac{1}{6})^{1/2}[\Sigma\pi] - \tfrac{1}{2}[\Sigma\eta]^\dagger + \tfrac{1}{2}[\Lambda\pi]^\dagger - (\tfrac{1}{6})^{1/2}[\Xi\mathrm{K}]$$

$$|10,-1,\tfrac{1}{2},I_3\rangle = \tfrac{1}{2}[\Sigma\bar{\mathrm{K}}] + \tfrac{1}{2}[\Xi\pi] - \tfrac{1}{2}[\Xi\eta]^\dagger + \tfrac{1}{2}[\Lambda\bar{\mathrm{K}}]^\dagger$$

$$|10,-2,0,0\rangle = [\Xi\bar{\mathrm{K}}]$$

BM states of the $\mathbf{8_S}$

$$|8_S, +1, \tfrac{1}{2}, I_3\rangle = (\tfrac{9}{20})^{1/2}[N\pi] - (\tfrac{9}{20})^{1/2}[\Sigma K] + (\tfrac{1}{20})^{1/2}[\Lambda K]^\dagger + (\tfrac{1}{20})^{1/2}[N\eta]^\dagger$$

$$|8_S, \ 0, 1, I_3\rangle = (\tfrac{6}{20})^{1/2}[N\bar{K}]^\dagger + (\tfrac{6}{20})^{1/2}[\Xi K]^\dagger - (\tfrac{1}{5})^{1/2}[\Sigma\eta]^\dagger - (\tfrac{1}{5})^{1/2}[\Lambda\pi]^\dagger$$

$$|8_S, \ 0, 0, 0\rangle = (\tfrac{1}{10})^{1/2}[N\bar{K}] - (\tfrac{1}{10})^{1/2}[\Xi K] + (\tfrac{6}{10})^{1/2}[\Sigma\pi]^\dagger + (\tfrac{2}{10})^{1/2}[\Lambda\eta]^\dagger$$

$$|8_S, -1, \tfrac{1}{2}, I_3\rangle = (\tfrac{9}{20})^{1/2}[\Sigma\bar{K}] - (\tfrac{9}{20})^{1/2}[\Xi\pi] + (\tfrac{1}{20})^{1/2}[\Lambda\bar{K}]^\dagger + (\tfrac{1}{20})^{1/2}[\Xi\eta]^\dagger$$

BM states of the $\mathbf{8_A}$

$$|8_A, +1, \tfrac{1}{2}, I_3\rangle = \tfrac{1}{2}[N\pi] - \tfrac{1}{2}[N\eta]^\dagger + \tfrac{1}{2}[\Sigma K] + \tfrac{1}{2}[\Lambda K]^\dagger$$

$$|8_A, \ 0, 1, I_3\rangle = -(\tfrac{1}{6})^{1/2}[N\bar{K}]^\dagger + (\tfrac{2}{3})^{1/2}[\Sigma\pi] + (\tfrac{1}{6})^{1/2}[\Xi K]^\dagger$$

$$|8_A, \ 0, 0, 0\rangle = (\tfrac{1}{2})^{1/2}[N\bar{K}] + (\tfrac{1}{2})^{1/2}[\Xi K]$$

$$|8_A, -1, \tfrac{1}{2}, I_3\rangle = \tfrac{1}{2}[\Sigma\bar{K}] + \tfrac{1}{2}[\Xi\pi] - \tfrac{1}{2}[\Lambda\bar{K}]^\dagger + \tfrac{1}{2}[\Xi\eta]^\dagger$$

BM state of the **1**

$$|1, 0, 0, 0\rangle = \tfrac{1}{2}[N\bar{K}] - \tfrac{1}{2}[\Xi K] - (\tfrac{3}{8})^{1/2}[\Sigma\pi]^\dagger + (\tfrac{1}{8})^{1/2}[\Lambda\eta]^\dagger$$

RECENT DEVELOPMENTS

1 Weak neutral currents and parity violation in atoms

After neutrino scattering experiments had confirmed the existence of weak neutral currents, Bouchiat and Bouchiat (1974) noted that neutral currents could lead to observable effects in atomic spectra. The unified theories of weak and electromagnetic interactions predict a parity-violating electron–nucleon interaction due to exchange of a neutral vector boson Z^0, in addition to the electromagnetic interaction due to photon exchange. Such an interaction leads to admixtures of the 'wrong parity' in atomic electron states and the consequent circular polarisation of photons emitted in transitions as discussed in § 5.2. Bouchiat and Bouchiat presented arguments to show that in heavy atoms the effect could be sufficiently enhanced to be just observable. Kriplovich (1974, and report to Tbilisi conference) noted that the parity-violating effects would lead to a rotation of the plane of polarisation (Faraday rotation) of a beam of light passing through an atomic vapour of, for example, bismuth. The expected magnitude of the effects depends on details of atomic structure calculations as well as on the details of the unified model.

Recent experimental results (Lewis *et al.*, 1977 and Baird *et al.*, 1977) find the magnitude of the Faraday rotation to be much smaller than that predicted by theory. Whether this discrepancy requires a modification of gauge theory of weak neutral currents or whether it can be attributed to our lack of knowledge of electronic structure of heavy atoms remains an open question at the time of writing.

2 Colour

A more fundamental resolution of the problem of statistics may come from the hypothesis of coloured quarks. In this case the secret variable ζ is promoted to a genuine dynamical degree of freedom. This is called *colour*, because the values 1, 2, 3 are replaced by colours, say: red, green and blue. Thus we have nine coloured quarks

$$\begin{matrix} u_r & u_g & u_b \\ d_r & d_g & d_b \\ s_r & s_g & s_b \end{matrix}$$

Han and Nambu (1965) suggested that there is an $SU(3)$ symmetry on the colour degree of freedom. Thus the three 'up' quarks u_r, u_g, u_b form a fundamental triplet of the colour group $SU(3)_c$. Similarly for the three 'down' quarks and the three 'strange' quarks. This colour $SU(3)$, denoted by $SU(3)_c$, must be distinguished from the $SU(3)$ symmetry of the u–d–s degree of freedom discussed so far, and which is now referred to as *flavour* $SU(3)$:$SU(3)_f$. The colour $SU(3)_c$ is assumed to be an exact symmetry, while we have seen that the flavour $SU(3)_f$ is only approximate.

An important argument for colour is the experimental value of the ratio

$$R = \frac{\sigma(e^+e^- \to \text{hadrons})}{\sigma(e^+e^- \to \mu^+\mu^-)}$$

The data on R can be summarised as two plateaus with resonance structure superimposed. On the first plateau from 1.5 to 3.5 GeV, $R \approx 2.5$, and on the higher plateau from 4 to 7 GeV, $R \approx 5$. At these energies the cross-section $\sigma(e^+e^- \to \text{hadrons})$ is assumed to be given by the cross-section $\sigma(e^+e^- \to q_n\bar{q}_n)$ summed over all types of quark, n. The probability for $q\bar{q} \to$ hadrons is assumed to be approximately unity. This gives for R

$$R \approx \sum_n Q_n^2$$

where Q_n is the charge of the nth type of quark (positron charge = + 1). For the simple three-quark model we find $R = \frac{2}{3}$, but for the three coloured quark model this becomes $R = 2$, which is in rough agreement with the first plateau. The second plateau may then correspond to the energy region where a fourth 'charmed' quark can be excited. (See below.)

If hadron interactions are invariant under the two independent symmetries $SU(3)_c$ and $SU(3)_f$ (neglecting $SU(3)_f$ breaking for simplicity) then hadron states should form simultaneous multiplets of both groups. The analogue in the colour picture of the requirement of total antisymmetry in χ_{secret} for baryons, is that baryon states be *singlets* of $SU(3)_c$. The equivalence of these requirements can be seen from the symmetry rules of § 11.3.1, but now applied to the colour group. We simply note that the 1_c formed from three triplets 3_c of $SU(3)_c$ is the totally antisymmetric combination.

Let us consider briefly meson states in the colour picture. There are now nine antiquarks which form a 3* of $SU(3)_f$ and also a 3* of $SU(3)_c$. Considering only the colour, we can form an $SU(3)_c$ singlet from 3_c and 3_c^*:

$$3_c \times 3_c^* = 8_c + 1_c$$

The various flavours associated with the $SU(3)_c$ singlet can be identified with the mesons as discussed before. A state previously written $|q_1 \bar{q}_2\rangle$ is now the colour-symmetric combination

$$|q_{1r}\bar{q}_{2r}\rangle + |q_{1g}\bar{q}_{2g}\rangle + |q_{1b}\bar{q}_{2b}\rangle$$

It is now a further hypothesis of the colour models that the dynamics (see below) is such that only colour singlets occur in Nature. For baryons, three coloured quarks can combine to form a singlet, cf. equation (11.16)

$$3_c \times 3_c \times 3_c = 10_c + 8_c + 8_c + 1_c$$

but two quarks cannot, since

$$3_c \times 3_c = 6_c + 3_c^*$$

nor can four quarks. Thus the hypothesis of 'colour singlets only', affords a natural rule with which to exclude the unwanted states in the $SU(3)_f$ picture.

If colour is a dynamical variable then the strong interaction must be colour dependent if we are to account for the preferred existence of colour singlets and not, for example, colour octets (at least, at energies now accessible with accelerators). One attractive theoretical possibility under active consideration is Quantum Chromodynamics (QCD). In this theory a colour octet of massless vector mesons (coloured gluons) is coupled to coloured quarks in a gauge-invariant manner. Note that the virtual emission of a vector gluon

$$q_c \rightleftharpoons q_c + V_c$$

is allowed because 3_c is contained in $8_c \times 3_c$:

$$8_c \times 3_c = 15_c + 6_c^* + 3_c$$

The gauge invariance of the theory has theoretical advantages, but it has not so far proved possible to show that the theory leads preferentially to colour singlet states.

For a recent review of colour models see Greenberg and Nelson (1977).

3 Charm

The proposals to unify weak and electromagnetic interactions by means of non-Abelian gauge theories with spontaneous symmetry breaking (Glashow, Salam, Weinberg) received an impetus when it was shown by t'Hooft and others that this type of theory is renormalisable.

Such theories predicted weak neutral currents, whose effects were subsequently observed. In such a theory strangeness-changing neutral currents, which are known not to exist to first order in G_{wk}, can be avoided in a simple way only by introducing a fourth quark (a fourth flavour) denoted by c (Glashow, Iliopoulous and Maiani, 1970). The new quark, c, is an isoscalar with the same electric charge as the u quark, and carries one unit of a new quantum number, *charm*, denoted by C. So we add to table 11.1 the following entries

	B	Y	I	I_3	Q	S	C
c	$\frac{1}{3}$	$-\frac{2}{3}$	0	0	$\frac{2}{3}$	0	+1
$\bar{c}$	$\frac{1}{3}$	$+\frac{2}{3}$	0	0	$-\frac{2}{3}$	0	−1

The charm of the u, d and s quarks is zero. The extended Gell-Mann–Nishijima formula is now

$$Q = I_3 + \tfrac{1}{2}(B + S + C)$$

and hypercharge and strangeness are now related by

$$Y = B + S - C.$$

Let us note here that with the addition of the charmed quark the ratio R of cross-sections for hadron to $\mu^+\mu^-$ production in e^+e^- collisions becomes $\frac{10}{9}$ (without colour) and $\frac{10}{3}$ (with colour). Again the colour plus flavour picture gives an R value in crude agreement with experiment above 4 GeV where a charmed quark pair can be excited.

If we consider a $c\bar{c}$ system of two heavy quarks moving non-relativistically in a potential well ('charmonium') then with some simple assumptions about the potential the spectrum of the new mesons in the range 3 to 4 GeV can be fitted very well. The $\psi(3100)$ forms the lowest 1^- level of this system and the $\psi(3685)$ and χ states may be assigned to higher levels.

All of these mesonic states have zero total charm although they are composed of charmed quarks. However, narrow $\kappa\pi$ and $\kappa\pi\pi$ resonances in e^+e^- experiments, have been interpreted as the decays of *charmed* D and F mesons, which are $(c\bar{d})$ $(c\bar{s})$ and $(c\bar{u})$ bound states. Charmed analogues of the baryons containing one charmed and two uncharmed quarks are expected and some may have been seen already. References to this work may be found in Jackson (1976) and the 'Review of particle properties' (1978).

4 *SU*(4)

How the additional flavour degree of freedom represented by the c quark is to be combined with the u, d, s flavours is as yet an unsolved problem.

A natural assumption is that the u, d, s, and c quarks participate in a broken *SU*(4) symmetry, so that hadron states form approximate *SU*(4) supermultiplets. The quarks and anti-quarks belong to the basic supermultiplets 4 and 4*. The higher supermultiplets of *SU*(4) may be obtained by an extension of the techniques we have described for *SU*(3), either by the algebra of operators, as in Chapter 10, or more easily by building particles up out of quarks, as described in Chapter 11. The weight diagrams are now three-dimensional, and have a tetrahedral symmetry. Further details may be found in Gaillard, Lee and Rosner (1975) and Jackson (1976).

More recent experimental discoveries and theoretical developments make this simple picture look less attractive. The upsilon (9500) discovered by Herb *et al.* (1977) and further mesonic resonances in the 10 GeV region point to the existence of a further quark b (called the bottom quark).

If the Glashow–Salam–Weinberg unification of electromagnetic and weak interactions based on the gauge group $SU(2) \times U(1)$ is correct then the question arises how the flavour symmetries are to be incorporated into this picture. If a unification is to be found along the lines described, for example, by Georgi and Glashow (1974) then the flavour groups *SU*(4) or *SU*(3) or even isospin are phenomenological symmetries which arise after the quarks have acquired their masses by spontaneous symmetry breaking.

Additional references

The latest editions of the 'Review of particle properties' are:

(1976) *Rev. Mod. Phys.* 48, No. 2, part 2 with a supplement on the 'New particles' in *Phys. Lett.* 68B, No. 1 (1977).

(1978) *Phys. Lett.* 75B, No. 1.

Baird, P.E.G., Brimicombe, M.W.S.M., Hunt, R.G., Roberts, G.J., Sandars, P.G.H. & Stacey, D.N. (1977). *Phys. Rev. Lett.* 39, 798.

Bouchiat, M.A. & Bouchiat, C.C. (1974). *Phys. Lett.* **48B**, 111.
Gaillard, M.K., Lee, B.W. & Rosner, J.L. (1975). *Rev. Mod. Phys.* **47**, 277.
Georgi, H. and Glashow, S.L. (1974). *Phys. Rev. Lett.* **32**, 438.
Glashow, S.L., Iliopoulos, J. & Maiani, L. (1970). *Phys. Rev.* **D2**, 1285.
Greenberg, O.W. & Nelson, C.A. (1977). *Physics Reports,* **32C**, 69.
Han, M.Y. & Nambu, Y. (1965). *Phys. Rev.* **139B**, 1006.
Herb, S.W. *et al.* (1977). *Phys. Rev. Lett.* **39**, 252.
Jackson, J.D. (1976). *Proceedings of the SLAC Summer Institute on Particle Physics,* 1976. Report SLAC-198, p. 147.
Kriplovich, I.B. (1974). *JETP Lett.* **20**, 686.
Lewis, L.L., Hollister, J.H., Soreide, D.C., Lindahl, E.G. & Fortson, E.N. (1977). *Phys. Rev. Lett.* **39**, 795.

REFERENCES

In addition to the usual literature references, we have sometimes found it appropriate, e.g. for world averages of data or for standards of notation, to refer to the compilation 'Review of particle properties' prepared by the members of the Particle Data Group and appearing each April, alternately in *Reviews of Modern Physics* and *Physics Letters*.

In particular we refer to

1973: *Rev. Mod. Phys.* **45**, no. 2, part 2, supplement.
1974: *Phys. Lett.* **50B**, no. 1.

Abashian, A. & Hafner, E.M. (1958). *Phys. Rev. Lett.* **1**, 255.

Abov, Yu. G., Krupchitskii, P.A., Bulgakov, M.I., Yermakov, O.N. & Karpikin, I.L. (1968). *Phys. Lett.* **27B**, 16.

Abov, Yu. G., Krupchitskii, P.A. & Oratovskiĭ, Yu.A. (1965). *Sov. J. Nucl. Phys.* **1**, 341.

Adair, R.K. (1955). *Phys. Rev.* **100**, 1540.

Aidzu (1953). *Proceedings of the International Conference of Theoretical Physics*, Kyoto and Tokyo, p. 200.

Ajzenberg-Selove, F. & Lauritsen, T. (1968). *Nuclear Physics*, **A114**, 1.

Alff, C., Gelfand, N., Nauenberg, U., Nussbaum, M., Schultz, J., Steinberger, J., Brugger, H., Kirsch, L., Plano, R., Berley, D. & Prodell, A. (1965). *Phys. Rev.* **137B**, 1105.

Alikanov, A.I., Galaktionov, Yu.V., Gorodkov, Yu.V., Eliseev, G.P. & Lyubimov, V.A. (1960). *Sov. Phys. J.E.T.P.* **11**, 1380

Ayres, D.S., Cormack, A.M., Greenberg, A.J., Kenney, R.W., Caldwell, D.O., Elings, V.B., Hesse, W.P. & Morrison, R.J. (1968). *Phys. Rev. Lett.* **21**, 261.

Backenstoss, G., Hyams, B.D., Knop, G., Marin, P.C. & Stierlin, U. (1961). *Phys. Rev. Lett.* **6**, 415.

Bailey, J., Bartl, W., Bochmann, G. von, Brown, R.C.A., Farley, F.J.M., Jostlein, H., Picasso, E. & Williams, R.W. (1968). *Phys. Lett.* **B28**, 287.

Baird, J.K., Miller, P.D., Dress, W.B. & Ramsey, N.F. (1969). *Phys. Rev.* **179**, 1285.

Baltay, C., Barash, N., Franzini, P., Gelfand, N., Kresch, L., Lütjens, G., Severiens, J.C., Steinberger, J., Tycko, D. & Zanello, D. (1965). *Phys. Rev. Lett.* **15**, 591.

Baltay, C., Franzini, P., Kim, J., Kirsch, L., Zanello, D., Lee-Franzini, J., Loveless, R., McFadyen, J. & Yarger, H. (1966). *Phys. Rev. Lett.* **16**, 1224.

Bardon, M., Franzini, P. & Lee, J. (1961). *Phys. Rev. Lett.* **7**, 23.

Bardon, M., Norton, P., Peoples, J., Sachs, A.M. & Lee-Franzini, J. (1965). *Phys. Rev. Lett.* **14**, 449.

Barnes, V.E., Connolly, P.L., Crennell, D.J., Culwick, B.D., Delaney, W.C., Fowler, W.B., Hagerty, P.E., Hart, E.L., Horwitz, N., Hough, P.V.C., Jenson, J.E., Koop, J.K., Lai, K.W., Leitner, J., Lloyd, J.L., London, G.W., Morris, T.W., Oren, Y., Palmer, R.B., Prodell, A.G., Radojičić, D., Rahm, D.C., Richardson, C.R., Samios, N.P., Sanford, J.R., Shutt, R.P., Smith, J.R., Stonehill, D.L., Strand, R.C., Thorndike, A.M., Webster, M.S., Willis, W.J. & Yamamoto, S.S. (1964). *Phys. Rev. Lett.* **12**, 204.

Beg, M.A.B., Lee, B.W. & Pais, A. (1964). *Phys. Rev. Lett.* **13**, 514.
Behrends, F. (1968). In E. Loebl (editor), *Group Theory*, vol. 2. New York: Academic Press.
Behrends, R.E., Dreitlein, J., Fronsdal, C. & Lee, W. (1962). *Rev. Mod. Phys.* **34**, 1.
Bell, J.S. & Mandl, F. (1958*a*). *Proc. Phys. Soc.* **A71**, 273.
Bell, J.S. & Mandl, F. (1958*b*). *Proc. Phys. Soc.* **A71**, 867.
Bell, J.S. & Steinberger, J. (1966). *Oxford International Conference on Elementary Particles, Proceedings.* Chilton: Rutherford High Energy Laboratory.
Berman, S. (1965). In A. Zichichi (editor), *Symmetries in Particle Physics.* New York: Academic Press.
Bernstein, J., Feinberg, G. & Lee, T.D. (1965). *Phys. Rev.* **139B**, 1650.
Bilenky, S.M. (1958). *Nuovo Cim.* **10**, 1049.
Bilenky, S.M. & Ryndin, R.M. (1965). *Phys. Lett.* **18**, 346.
Blatt, J. & Weisskopf, V.F. (1952). *Theoretical Nuclear Physics.* New York: John Wiley & Sons.
Block, M.M., Lendinari, L. & Monari, L. (1962). *Proceedings of the Annual International High Energy Physics Conference*, p. 371. Geneva: CERN.
Bock, P. & Schopper, H. (1965). *Phys. Lett.* **16**, 284.
Bodansky, D., Braithwaite, W.J., Shreve, D.C., Storm, D.W. & Weitkamp, W.G. (1966). *Phys. Rev. Lett.* **17**, 589.
Bodansky, D., Eccles, S.F., Farwell, G.W., Rickey, M.E. & Robison, P.C. (1959). *Phys. Rev. Lett.* **2**, 101.
Boehm, F. & Kankeleit, E. (1965). *Phys. Rev. Lett.* **14**, 312.
Bott-Bodenhausen, M., Cladwell, D.O., Fabjan, C.W., Gruhn, C.R., Peak, L.S., Rochester, L.S., Sauli, F., Steirlin, U., Tirler, R., Winstein, B. & Zahniser, D. (1972). *Phys. Lett.* **40B**, 693.
Burgy, M.T., Krohn, V.E., Novey, T.B., Ringo, G.R. & Telegdi, V.L. (1958). *Phys. Rev. Lett.* **1**, 324.
Byers, N. & Burkhardt, H. (1961). *Phys. Rev.* **121**, 281.
Byers, N. & Fenster, S. (1963). *Phys. Rev. Lett.* **11**, 52.
Calaprice, F.P., Commins, E.D., Gibbs, H.M., Wick, G.L. & Dobson, D.A. (1967). *Phys. Rev. Lett.* **18**, 918.
Camerini, U., Hautman, R.L., March, R.L., March, R.H., Murphree, D., Gidal, G., Kalmus, G.E., Powell, W.M., Pu, R.T., Sandler, C.L., Natali, S. & Villani, M. (1965). *Phys. Rev. Lett.* **14**, 989.
Carruthers, P. (1966). *Introduction to Unitary Symmetry*. New York: Interscience.
Cartwright, W.F., Richman, C., Whitehead, N.M. & Wilcox, H.A. (1953). *Phys. Rev.* **91**, 677.
Chamberlain, O., Segré, E., Tripp, R., Wiegand, C. & Ypsilantis, T. (1954). *Phys. Rev.* **93**, 1430.
Chen, J.R., Sanderson, J., Appel, J.A., Gladding, G., Goitein, M., Hanson, K., Imrie, D.C., Kirk, T., Madaras, R., Pound, R.V., Price, L., Wilson, R. & Zajde, C. (1968). *Phys. Rev. Lett.* **21**, 1279.
Cheston, W.B. (1951). *Phys. Rev.* **83**, 1118.
Christ, N. & Lee, T.D. (1966*a*). *Phys. Rev.* **143**, 1310.
Christ, N. & Lee, T.D. (1966*b*). *Phys. Rev.* **148**, 1520.
Christensen, J.H., Cronin, J.W., Fitch, V.L. & Turley, R. (1964). *Phys. Rev. Lett.* **13**, 138.
Cnops, A.M., Finocchiaro, G., Lassalle, J.C., Mittner, P., Zanella, P., Dufey, J.P., Gobbi, B., Pouchon, M.A. & Muller, A. (1966). *Phys. Lett.* **22**, 546.
Coleman, S. (1965). In *Trieste Seminar on High Energy Physics and Elementary Particles.* Vienna: IAEA.

Coleman, S. (1966). In A. Zichichi (editor), *Strong and Weak Interactions, Present Problems*. New York: Academic Press.
Courant, H., Filthuth, H., Franzini, P., Glasser, R.G., Minguzzi-Ranzi, A., Segar, A., Willis, W., Bernstein, R.A., Day, T.B., Kehoe, B., Herz, A.J., Sakitt, M., Sechi-Zorn, B., Seeman, N. & Snow, G.A. (1963). *Phys. Rev. Lett.* **10**, 409.
Crawford, Jr, F.S., Cresti, M., Good, M.L., Gottstein, K., Lyman, E.M., Solmitz, F.T., Stevenson, M.L. & Ticho, H.K. (1957). *Phys. Rev.* **108**, 1102.
Cronin, J.W. & Overseth, O.E. (1963). *Phys. Rev.* **129**, 1795.
Culligan, G., Frank, S.G.F., Holt, J.R., Kluyver, J.C. & Massam, T. (1957). *Nature, Lond.* **180**, 751.
Dalitz, R.H. (1952). *Proc. Phys. Soc.* **A65**, 175.
Dalitz, R.H. (1962). *Strange Particles and Strong Interactions.* Oxford University Press for the Tata Institute, Bombay.
Dalitz, R.H. & Horgan, R. (1973). *Nucl. Phys.* **B66**, 135. See also: Horgan, R. (1974). *Nucl. Phys.* **B71**, 514.
de Swart, J.J. (1963). *Rev. Mod. Phys.* **35**, 916.
Dieterle, B.D., Arens, J.F., Chamberlain, O., Gramais, P.D., Hansroul, M.J., Holloway, L.E., Johnson, C.H., Schultz, C., Steiner, H., Shapiro, G. & Weldon, D. (1968). *Phys. Rev.* **167**, 1191.
Dobrzynski, L., Xuong, N.-H., Montanet, L., Tomas, M., Duboc, J. & Donald, R.A. (1966). *Phys. Lett.* **22**, 105.
Dress, W.B., Baird, J.K., Miller, P.D. & Ramsey, N.F. (1968). *Phys. Rev.* **170**, 1200.
Dress, W.B., Miller, P.D. & Ramsey, N.F. (1973). *Phys. Rev.* **D7**, 3147.
Durbin, R., Loar, H. & Steinberger, J. (1951). *Phys. Rev.* **81**, 894.
Edmonds, A.R. (1957). *Angular Momentum in Quantum Mechanics*. Princeton University Press.
Eisler, F., Plano, R., Prodell, A., Samios, N., Schwartz, M., Steinberger, J., Bassi, P., Borelli, V., Puppi, G., Tanako, H., Waloschek, P., Zoboli, V., Conversi, M., Franzini, P., Manelli, I., Santangelo, R., Silvestrini, V., Brown, G.L., Glaser, D.A. & Graves, C. (1958). *Nuovo. Cim.* **7**, 222.
Faiman, D. & Hendry, A.W. (1968). *Phys. Rev.* **173**, 1720.
Faiman, D. & Hendry, A.W. (1969). *Phys. Rev.* **180**, 1572.
Federman, P., Rubinstein, H.R. & Talmi, I. (1966). *Phys. Lett.* **22**, 208.
Feinberg, G. (1958). *Phys. Rev.* **109**, 1019.
Feinberg, G. (1960). *Phys. Rev.* **120**, 640.
Feinberg, G. (1965). *Phys. Rev.* **140B**, 1403.
Feldman, G. & Fulton, T. (1958). *Nucl. Phys.* **8**, 106.
Feldman, G. & Matthews, P.T. (1965). *Ann. Phys.* **31**, 469.
Fermi, E. (1955). *Nuovo. Cim. Suppl.* **2**, 17.
Fermi, E. & Yang, C.N. (1949). *Phys. Rev.* **76**, 1739.
Feynman, R.P. & Gell-Mann, M. (1958). *Phys. Rev.* **109**, 193.
Feynman, R.P., Kislinger, M. & Ravndal, F. (1971). *Phys. Rev.* **D3**, 2706.
Feynman, R.P., Leighton, R.B. & Sands, M. (1963). *The Feynman Lectures on Physics*, vol. 1, ch. 11. Reading, Mass: Addison-Wesley.
Frauenfelder, H., Bobone, R., von Goeler, E., Levine, N., Lewis, H.R., Peacock, R., Rossi, A. & de Pasquali, G. (1957). *Phys. Rev.* **106**, 386.
Frazer, W.R. (1966). *Elementary Particles*, pp. 39–40. Englewood Cliffs, N.J: Prentice-Hall.
Friedman, J.I. & Telegdi, V. (1957). *Phys. Rev.* **105**, 1681.
Garwin, R.L., Lederman, L. & Weinrich, M. (1957). *Phys. Rev.* **105**, 1415.
Gasiorowicz, S. (1966). *Elementary Particle Physics*. New York: John Wiley & Sons.

Gatto, R. (1957). *Phys. Rev.* **109**, 610.
Gell-Mann, M. (1961). *California Institute of Technology Synchrotron Lab. Report CTSL-20*. (Reprinted in Gell-Mann and Ne'eman (1964).)
Gell-Mann, M. (1962). *Phys. Rev.* **125**, 1067.
Gell-Mann, M. (1964). *Phys. Lett.* **8**, 214.
Gell-Mann, M. & Ne'eman, Y. (1964). *The Eightfold Way*. New York: Benjamin.
Glasser, R.G., Kehoe, B., Engelmann, P., Schneider, H. & Kirsch, L.E. (1966). *Phys. Rev. Lett.* **17**, 603.
Goldhaber, M. (1958). *Annual International Conference on High Energy Physics*, p. 233. Geneva: CERN.
Goldhaber, M., Grodzins, L. & Sunyar, A.W. (1958). *Phys. Rev.* **109**, 1015.
Golub, R. & Pendlebury, J.M. (1972). *Contemp. Phys.* **13**, 519.
Gourdin, M. (1967). *Unitary Symmetries and Their Application to High Energy Physics*. Amsterdam: North-Holland.
Grawert, G., Luders, G. & Rollnik, H. (1959). *Fortsch. Physik.* **7**, 291.
Greenberg, O.W. (1964). *Phys. Rev. Lett.* **13**, 598.
Grodzins, L. & Genovese, F. (1961). *Phys. Rev.* **121**, 228.
Gursey, F. & Radicati, L.A. (1964). *Phys. Rev. Lett.* **13**, 173.
Harari, H. (1968). In J. Prentki and J. Steinberger (editors), *14th International Conference on High-Energy Physics*. Geneva: CERN.
Hamermesh, M. (1962). *Group Theory and its Application to Physical Problems*. Reading, Mass.: Addison-Wesley.
Hamilton, W.D. (1968). *Prog. Nucl. Phys.* **10**, 1.
Heisenberg, W. (1932). *Z. Physik*, **77**, 1.
Henley, E.M. (1969). *Ann. Rev. Nucl. Sci.* **19**, 367.
Henley, E.M. & Jacobsohn, B.A. (1959). *Phys. Rev.* **113**, 225, 234.
Hillman, P., Johansson, A. & Tibell, G. (1958). *Phys. Rev.* **110**, 1218.
Hwang, C.F., Ophel, T.R., Thorndike, E.H. & Wilson, R. (1960). *Phys. Rev.* **119**, 352.
Ikeda, M., Ogawa, S. & Ohnuki, Y. (1959). *Progr. Theor. Phys.* **22**, 715.
Ikeda, M., Ogawa, S. & Ohnuki, Y. (1961). *Progr. Theor. Phys. Suppl.* **19**, 44.
Inglis, D.R. (1953). *Rev. Mod. Phys.* **25**, 390.
Jacob, M. & Wick, G.C. (1959). *Ann. Phys.* **7**, 404.
Jones, D.P., Murphy, P.G. & O'Neill, P.L. (1958). *Proc. Phys. Soc.* **A72**, 429.
Kabir, P. (1968). *The CP Puzzle: strange decays of the neutral kaon*. London: Academic Press.
Källén, G. (1964). *Elementary Particle Physics*. Reading, Mass.: Addison-Wesley.
Koch, W. (1964). *CERN Yellow Report 64–13*, vol. 2.
Kokkedee, J.J.J. (1969). *The Quark Model*. New York: Benjamin.
Landau, L.D. (1957). *Nucl. Phys.* **3**, 127.
Lander, R.L., Powell, W.M. & White, H.S. (1959). *Phys. Rev. Lett.* **3**, 236.
Layter, J.G., Appel, J.A., Kotlewski, A., Lee, W., Stein, S. & Thaler, J.J. (1973). *Phys. Rev. Lett.* **29**, 316.
Lee, T.D., Oehme, R. & Yang, C.N. (1957). *Phys. Rev.* **106**, 340.
Lee, T.D. & Wolfenstein, L. (1965). *Phys. Rev.* **138B**, 1490.
Lee, T.D. & Wu, C.S. (1966). *Ann. Rev. Nucl. Sci.* **16**, 471.
Lee, T.D. & Yang, C.N. (1956*a*). *Phys. Rev.* **104**, 254.
Lee, T.D. & Yang, C.N. (1956*b*). *Nuovo. Cim.* **3**, 749.
Lee, T.D. & Yang, C.N. (1957). *Phys. Rev.* **108**, 1645.
Lee, T.D. & Yang, C.N. (1958). *Phys. Rev.* **109**, 1755.
Leitner, J., Nordin, Jr, P., Rosenfeld, A.H., Solmitz, F.T. & Tripp, R.D. (1959). *Phys. Rev. Lett.* **3**, 238.
Levi-Setti, R. (1969). In G. von Dardel (editor), *Proceedings of the Lund International Conference on Elementary Particles*. Lund: Institute of Physics.

Lipkin, H. (1973). *Physics Reports*, **8C**, 173.
Lobashov, V.M., Nazarenko, V.L., Saenko, L.F., Smotritskii, L.M. & Kharkevich, G.I. (1966). *Sov. Phys. J.E.T.P. Lett.* **3**, 173.
Lobkowicz, F., Melissinos, A.C., Nagashima, Y., Tewksbury, S., von Briesen, Jr, H. & Fox, J.D. (1966). *Phys. Rev. Lett.* **17**, 548.
Low, F.E. (1966). In *Particle Symmetries and Axiomatic Field Theory (1965 Brandeis Summer Institute Lectures)*, vol. 2. New York: Gordon & Breach.
Luders, G. (1954). *Kgl. Dan. Mat. fys. Medd.* **28**, no. 5.
Luders, G. & Zumino, B. (1957). *Phys. Rev.* **106**, 385.
Macq, P.C., Crowe, K.M. & Haddock, R.P. (1957). *Phys. Rev.* **112**, 2061.
Maglić, B.C., Kalbfleisch, G.R. & Stevenson, M.L. (1961). *Phys. Rev. Lett.* **7**, 137.
Mandl, F. (1957). *Quantum Mechanics.* London: Butterworths.
Marshak, R.E. (1951). *Phys. Rev.*, **82**, 313.
Matthews, P.T. (1967). In E.H.S. Burhop (editor), *High Energy Physics,* vol. 1. New York: Academic Press.
Meyer, S.L., Anderson, E.W., Bleser, E., Lederman, L.M., Rosen, J.L., Rothberg, J. & Wang, I.-T. (1963). *Phys. Rev.* **132**, 2693.
Michel, L. (1953). *Nuovo. Cim.* **10**, 319.
Miller, P.D., Dress, W.B., Baird, J.K. & Ramsey, N.F. (1967). *Phys. Rev. Lett.* **19**, 381.
Morpurgo, G. (1965). *Physics*, **2**, 95.
Mott, N.F. (1929). *Proc. Roy. Soc.* **A124**, 425, appendix.
Ne'eman, Y. (1962). *Nucl. Phys.* **26**, 222.
Okubo, S. (1962). *Prog. Theor. Phys.* **27**, 949.
Pais, A. (1959). *Phys. Rev. Lett.* **3**, 242.
Pais, A. (1964). *Phys. Rev. Lett.* **13**, 175.
Pais, A. (1966). *Rev. Mod. Phys.* **38**, 215.
Pauli, W. (1955). In W. Pauli (editor), *Niels Bohr and the Development of Physics.* London: Pergamon.
Player, M.A. & Sandars, P.G.H. (1970). *J. Phys.* **C3**, 1620.
Racah, G. (1965). *Ergebn. exakt Naturw.* **37**, 38. (Reprint of 1951 Princeton Lectures).
Rich, A. & Crane, H.R. (1966). *Phys. Rev. Lett.* **17**, 271.
Roper, L.D. & Wright, R.M. (1965). *Phys. Rev.* **138B**, 921.
Rose, M.E. (1957). *Elementary Theory of Angular Momentum.* New York: John Wiley & Sons.
Rosen, L. & Brolley, Jr, J.E. (1959). *Phys. Rev. Lett.* **2**, 98.
Rowe, E.G.P. & Squires, E.J. (1969). *Rep. Prog. Phys.* **32**, 273.
Sakata, S. (1956). *Prog. Theor. Phys.* **16**, 686.
Sakita, B. (1964). *Phys. Rev.* **136**, 1756.
Samios, N.P., Goldberg, M. & Meadows, B.T. (1974). *Rev. Mod. Phys.* **46**, 49.
Sandars, P.G.H. (1968). *Proc. Phys. Soc.* **B1**, 499.
Schiff, L.I. (1968). *Quantum Mechanics*, 3rd edition. New York: McGraw-Hill.
Schubert, K.R., Wolff, B., Chollet, J.C., Gaillard, J.-M., Jane, M.R., Ratcliffe, T.J. & Repellin, J.-P. (1970). *Phys. Lett.* **31B**, 662.
Shull, C.G. & Nathans, R. (1967). *Phys. Rev. Lett.* **19**, 384.
Smith, J.H., Purcell, E.M. & Ramsey, N.F. (1957). *Phys. Rev.* **108**, 120.
Smorodinskii, Yu. A. (1965). *Sov. Phys. Uspekhi*, **7**, 63.
Stein, T.S., Carrico, J.P., Lipworth, E. & Weisskopf, M.C. (1969). *Phys. Rev.* **186**, 39.
Steinberger, J. (1970). *CERN Yellow Report 70–1.*

Streater, R.F. & Wightman, A.S. (1964). *PCT, Spin and Statistics and All That.* New York: Benjamin.
Tanner, N. (1957). *Phys. Rev.* **107**, 1203.
Thomas, L.H. (1926). *Nature, Lond.* **117**, 514.
Thomas, L.H. (1927). *Phil. Mag.* (7), **3**, 1.
Thornton, S.T., Jones, C.M., Bair, J.K., Maucusi, M.D. & Willard, H.B. (1968). *Phys. Rev. Lett.* **21**, 447.
Thrall, R.M. (1941). *Duke Math. J.* **8**, 611.
Treiman, S.B. (1959). *Phys. Rev.* **113**, 355.
Tripp, R.D. (1965). *Ann. Rev. Nucl. Sci.* **15**, 325.
Ueda, Y. & Okubo, S. (1963). *Nucl. Phys.* **49**, 345.
van Royen, R. & Weisskopf, V.F. (1967). *Nuovo Cim.* **50A**, 617.
von Witsch, W., Richter, A. & Brentano, P. (1967). *Phys. Rev. Lett.* **19**, 524.
Watson, K.M. (1954). *Phys. Rev.* **95**, 228.
Weitkamp, W.G., Storm, D.W., Shreve, D.C., Braithwaite, W.J. & Bodansky, D. (1968). *Phys. Rev.* **165**, 1233.
Weyers, J. (1973). *Lectures at the Louvain Summer School. CERN Report TH. 1743.*
Wigner, E.P. (1927). *Z. Physik.* **43**, 624; erratum: **45**, 601.
Wigner, E.P. (1932). *Göttingen Nachr.* **1932**, 546.
Wigner, E.P. (1937). *Phys. Rev.* **51**, 106.
Wigner, E.P. (1939). *Ann. Math.* **40**, 149.
Wilkinson, D.H. (1958). *Phys. Rev.* **109**, 1603.
Wolfenstein, L. (1964). *Phys. Rev. Lett.* **13**, 562.
Wolfenstein, L. (1966). *Nuovo Cim.* **42**, 17.
Wolfenstein, L. & Ashkin, J. (1952). *Phys. Rev.* **85**, 947.
Wu, C.S., Ambler, E., Hayward, R.W., Hoppes, D.D. & Hudson, R.P. (1957). *Phys. Rev.* **105**, 1413.
Wu, T.T. & Yang, C.N. (1964). *Phys. Rev. Lett.* **13**, 380.
Xuong, N.-H., Lynch, G.R. & Hinrichs, C.K. (1961). *Phys. Rev.* **124**, 575.
Zulkarneyev, R. Ya., Kiselev, V.S., Nadezhdin, V.S. & Satarov, V.I., as reported by Kazarinov, Yu. M. (1967). *Rev. Mod. Phys.* **39**, 509.
Zweig, G. (1964). *CERN Reports TH. 401* and *TH.412.*
Zweig, G. (1965). In A. Zichichi (editor), *Symmetries in Elementary Particle Physics.* New York: Academic Press.

INDEX